The 3rd International Conference on Advances in Mechanical Engineering

The 3rd International Conference on Advances in Mechanical Engineering

Editors

Muhammad Mahabat Khan
Mohammad Javed Hyder
Muhammad Irfan
Manzar Masud

MDPI

Basel • Beijing • Wuhan • Barcelona • Belgrade • Novi Sad • Cluj • Manchester

Editors
Muhammad Mahabat Khan
Capital University of Science and Technology
Islamabad, Pakistan

Mohammad Javed Hyder
Capital University of Science and Technology
Islamabad, Pakistan

Muhammad Irfan
Capital University of Science and Technology
Islamabad, Pakistan

Manzar Masud
Capital University of Science and Technology
Islamabad, Pakistan

Editorial Office
MDPI
St. Alban-Anlage 66
4052 Basel, Switzerland

This is a reprint of articles from the Proceedings published online in the open access journal *Engineering Proceedings* (ISSN 2673-4591) (available at: https://www.mdpi.com/2673-4591/45/1).

For citation purposes, cite each article independently as indicated on the article page online and as indicated below:

Lastname, A.A.; Lastname, B.B. Article Title. *Journal Name* **Year**, *Volume Number*, Page Range.

ISBN 978-3-0365-9246-6 (Hbk)
ISBN 978-3-0365-9247-3 (PDF)
doi.org/10.3390/books978-3-0365-9247-3

Contents

About the Editors

Muhammad Mahabat Khan

Muhammad Mahabat Khan (Dr.) is an Associate Professor and Head of the Mechanical Engineering Department at Capital University of Science and Technology, Islamabad, Pakistan. He obtained his Ph.D. in Computation Fluid Dynamics from Ecole Centrale de Lyon, France. He authored numerous WOS-indexed Journal articles and is currently the reviewer for several international journals and conferences. His research interests include Heat Transfer, Turbulent Flows, Thermal Energy Storage, and Multiphase Flows.

Mohammad Javed Hyder

Mohammad Javed Hyder (Dr.) is a Professor at the Capital University of Science and Technology (CUST), Islamabad, Pakistan. His research focuses on Renewable Energy Stirling Engine Design and Development, Mechanical System Design, Turbomachinery, Mechanical Behavior of Materials, and Computational Engineering. He received the Best University Teachers Award from the Higher Education Commission, Pakistan, and the Performance Gold Medal from the PIEAS governing body. He has teaching experience of 40 years. He taught more than 50 Mechanical Engineering and Engineering Management subjects to graduate and undergraduate students.

Muhammad Irfan

Muhammad Irfan (Dr.) is an Associate Professor of Mechanical Engineering at the Capital University of Science and Technology, Islamabad. He obtained his Ph.D. from Koc University, Turkey, where he worked on developing a front-tracking multiphase phase change solver for the evaporation and combustion of fuel droplets. His research interests include phase change material-based energy storage devices, performance enhancements of photovoltaic panels, multiphase flows, and ejector refrigeration systems.

Manzar Masud

Manzar Masud is a lecturer at the Department of Mechanical Engineering, Capital University of Science and Technology, Pakistan. He received his BS in aerospace engineering from the Institute of Space Technology, Pakistan, and an MS in Mechanical Engineering from HITEC University, Pakistan. Currently, he is a PhD candidate in the Department of Mechanical Engineering at SMME, NUST. His research interests include computational mechanics, material characterization of composite materials, mechanical behavior of bio-hybrid composite materials, and solid mechanics.

engineering proceedings

MDPI

Editorial

Statement of Peer Review †

Muhammad Mahabat Khan, Mohammad Javed Hyder *, Muhammad Irfan and Manzar Masud

Department of Mechanical Engineering, Capital University of Science and Technology (CUST), Islamabad 44000, Pakistan; drmahabat@cust.edu.pk (M.M.K.); m.irfan@cust.edu.pk (M.I.); manzar.masud@cust.edu.pk (M.M.)

* Correspondence: javed.hyder@cust.edu.pk; Tel.: +92-51-111555666 (ext. 282)

† All the papers published in the volume are presented at the 3rd International Conference on Advances in Mechanical Engineering (ICAME-23), Islamabad, Pakistan, 24 August 2023.

In submitting conference proceedings to *Engineering Proceedings*, the volume editors of the proceedings certify to the publisher that all papers published in this volume have been subjected to peer review administered by the volume editors. Reviews were conducted by expert referees to the professional and scientific standards expected of a proceedings journal.

- Type of peer review: single blind;
- Conference submission management system: through email;
- Number of submissions sent for review: 81;
- Number of submissions accepted: 53;
- Acceptance rate: 65%;
- Average number of reviews per paper: 1;
- Total number of reviewers involved: 20.

Conflicts of Interest: The authors declare no conflict of interest.

Disclaimer/Publisher's Note: The statements, opinions and data contained in all publications are solely those of the individual author(s) and contributor(s) and not of MDPI and/or the editor(s). MDPI and/or the editor(s) disclaim responsibility for any injury to people or property resulting from any ideas, methods, instructions or products referred to in the content.

Citation: Khan, M.M.; Hyder, M.J.; Irfan, M.; Masud, M. Statement of Peer Review. *Eng. Proc.* **2023**, *45*, 9. https://doi.org/10.3390/engproc2023045009

Published: 8 September 2023

engineering proceedings MDPI

Editorial

Preface: The Third International Conference on Advances in Mechanical Engineering 2023 (ICAME-23) †

Muhammad Mahabat Khan, Mohammad Javed Hyder *, Muhammad Irfan and Manzar Masud

Department of Mechanical Engineering, Capital University of Science and Technology (CUST), Islamabad 44000, Pakistan; drmahabat@cust.edu.pk (M.M.K.); m.irfan@cust.edu.pk (M.I.); manzar.masud@cust.edu.pk (M.M.)

* Correspondence: javed.hyder@cust.edu.pk; Tel.: +92-51-111555666 (ext. 282)

† All the papers published in the volume are presented at the 3rd International Conference on Advances in Mechanical Engineering (ICAME-23), Islamabad, Pakistan, 24 August 2023.

Citation: Khan, M.M.; Hyder, M.J.; Irfan, M.; Masud, M. Preface: The Third International Conference on Advances in Mechanical Engineering 2023 (ICAME-23). *Eng. Proc.* **2023**, *45*, 10. https://doi.org/10.3390/engproc2023045010

Published: 8 September 2023

The third International Conference on Advances in Mechanical Engineering 2023 (ICAME-23) was held on 24 August 2023 by the Department of Mechanical Engineering, Capital University of Science and Technology (CUST). In ICAME-23, forty-three (43) national presenters presented their research papers physically in the Mechanical Engineering Department, Capital University of Science and Technology (CUST), whereas ten (10) international presenters along with four (4) keynote speakers did so online. ICAME is held annually and welcomes high-quality theoretical and empirical original research papers, case studies, and review papers from researchers, academicians, professionals, practitioners, and students from all over the world. All the articles have been single-blind reviewed.

ICAME-23 accepts research papers in the disciplines of:

- Experimental and computational fluid dynamics;
- Thermodynamics and heat transfer analysis;
- Machines and mechanisms;
- Design and solid mechanics;
- Manufacturing and production and industrial engineering;
- Engineering and technology management;
- Renewable energy and environmental engineering;
- Bioengineering;
- Materials and failure analysis;
- Other related fields.

The ICAME-23 conference provided a platform to the national and international speakers/participants to present their state-of-the-art research to diverse audiences from academia and industry. The international speakers were also offered an opportunity to participate virtually. For the complete details of the keynote speakers and conference program, please visit the official webpage of the conference, https://icame.cust.edu.pk.

The ICAME-23 advisory committee consists of

- Mr. Mian Amer Mahmood, Patron (Chancellor);
- Dr. M. Mansoor Ahmed, Co-Patron (Vice Chancellor);
- Dr. Imtiaz Ahmed Taj, General Advisor (Dean Faculty of Engineering);
- Dr. M. Mahabat Khan, Principal Advisor (Head, Department of Mechanical Engineering);
- Dr. M. Javed Hyder, Chair/Conference Secretary (Director Research, Department of Mechanical Engineering).

The ICAME-23 organizing committee consists of

- Dr. Salman S. Warsi;
- Dr. M. Irfan;

- Dr. Ghulam Asghar;
- Mr. Saif Ullah;
- Mr. Manzar Masud;
- Mr. Syed Hassan Shah;
- Mr. Tauseef Ahmed;
- Ms. Shummaila Rasheed;
- Mr. Mohammad Zulfiqar;
- Mr. Rizwan Siddiqui;
- Mr. Muhammad Haroon;
- Mr. Saad Abdul Raheem;
- Mr. Sarmad Ali;
- Mr. Rizwan Khalil;
- Mr. Muhammad Ahmed;
- Mr. Huzaifa Waseem Malik;
- Mr. Raza Muhammad;
- Mr. Zahoor Ahmed;
- Mr. Qaiser Mehmood;
- Mr. Zafar Iqbal;
- Mr. Zunair Nasir Kayani.

The conference proceedings are published in the Engineering Proceeding Journal of MDPI (Multidisciplinary Digital Publishing Institute), Switzerland. All the articles are individually indexed and are citable via Digital Object Identification (DOI). An electronic book of the conference proceedings with ISBN is also digitally available.

Four keynote speakers enlightened the participants and presenters of ICAME-23. ***Keynote Speaker 1: Prof. Tim Wentz*** delivered the talk titled ***"Back to the Future: Our Industry in 2030"***. Prof. Tim Wentz is an Emeritus Professor of Construction Management at the University of Nebraska—Lincoln, where he is still actively involved in various faculty endeavors, such as the Chancellor's Sustainability and Resilience Commission. Tim has served the Society in many different capacities. He started his ASHRAE career in 1976 after graduating from the University of Nebraska and worked in a grassroots capacity for many years. Tim has also been very active in the Mechanical Contractors Association of America (MCAA), where he has served as a trustee for the Mechanical Contractors Education and Research Fund (MCERF) and is currently a member of the National Education Initiative (NEI) faculty, and also of the Institute for Project Management (IPM) faculty.

Keynote Speaker 2: Dr. Muhammad Wakil Shahzad delivered the talk titled ***"An advance water treatment system operated with solar energy"***. Dr. Muhammad Wakil Shahzad is an Associate Professor and Head of Subject in the Mechanical and Construction Engineering Department at Northumbria University (NU), Newcastle Upon Tyne, United Kingdom. He is an expert in renewable energy storage and its applications for water treatment, hybrid desalination processes, heating/cooling, solar to alternative fuels, and life cycle cost analysis. He has won many international awards for his innovative desalination cycle, including the Global Innovation Award 2020, and is listed among the top 2% researchers in energy 2021/2022.

Keynote Speaker 3: Dr. Senol Piskin delivered the talk titled ***"Current trends in Biomedical Engineering"***. Dr. Senol Piskin is currently serving at Istinye University, Istanbul, Turkey, and he is holder of a Marie Sklodowska-Curie Fellowship. During his Ph.D., he modeled the cardiovascular system in several physical aspects and his thesis has been selected as one of the best Academic Thesis Applied in Industry by the Technology Development Foundation of Turkey (TTGV). He has modeled palliative surgeries in collaboration with cardiovascular surgeons, cardiologists, and radiologists. For this project, he was awarded the grand prize in the "New Ideas, New Businesses" innovation and entrepreneurship competition organized by the Middle East Technical University, Turkey.

Keynote Speaker 4: Muhammad Saleem Sumbal delivered the talk titled ***"Digitalization, Knowledge Workers and Future of Work: A post Covid-19 perspective"***.

Dr. Muhammad Saleem Sumbal is currently working as Research Assistant Professor at The Hong Kong Polytechnic University, Hong Kong (QS ranking: 65). Previously, he worked at the NUST Business School, National University of Sciences and Technology (NUST), Pakistan, and at the Royal Bank of Scotland, UK. He completed his Ph.D. on prestigious the Hong Kong Fellowship Scheme from The Hong Kong Polytechnic University and MSc in Computer Vision and Robotics on a prestigious Erasmus Mundus Fellowship from Heriot Watt University, UK; University of Girona, Spain; and University of Burgundy, France.

Conflicts of Interest: The authors declare no conflict of interest.

engineering proceedings

MDPI

Proceeding Paper

Numerical Study of Gas-Sand Two-Phase Flow Erosion in a Standard 90° Elbow †

Nauman Khan, Muhammad Rehan Khan *, Sati Ullah, Tariq Talha, Muhammad Ali Khan and Zubair Sajid

Department of Mechanical Engineering, College of Electrical and Mechanical Engineering, National University of Sciences and Technology, Islamabad 44000, Pakistan; dnaumankhan890@gmail.com (N.K.); satiullahmaikash507@gmail.com (S.U.); tariq.talha@ceme.nust.edu.pk (T.T.); mak.ceme@ceme.nust.edu.pk (M.A.K.); zubair.sajid@ceme.nust.edu.pk (Z.S.)

* Correspondence: mrehan.khan@ceme.nust.edu.pk

† Presented at the Third International Conference on Advances in Mechanical Engineering 2023 (ICAME-23), Islamabad, Pakistan, 24 August 2023.

Abstract: Erosive wear is a common problem in hydrocarbon industry pipelines transporting a multi-phase flow. Erosion is more critical in flow-changing devices, like tees, elbows, U- bends, valves, and chokes. By performing numerical simulations, this research investigation focuses on pre-eminent parameters that influence elbow wear, including erodent size and flow velocity. In this analysis, particle diameters of 150 and 300 μm and velocities of 11, 15, and 23 m/s were used. At a velocity of 11 m/sec, the erosion rate induced by a particle size of 300 μm was 1.62 times higher than with a particular size of 150 μm. Similarly, when the velocity was increased from 15 m/s to 23 m/s for the same particle size of 300 μm, the erosion rate was increased by 3.23 times.

Keywords: elbow erosion; two-phase flow; numerical analysis

Citation: Khan, N.; Khan, M.R.; Ullah, S.; Talha, T.; Khan, M.A.; Sajid, Z. Numerical Study of Gas-Sand Two-Phase Flow Erosion in a Standard 90° Elbow. *Eng. Proc.* **2023**, *45*, 28. https://doi.org/10.3390/engproc2023045028

Academic Editors: Mohammad Javed Hyder, Muhammad Mahabat Khan, Muhammad Irfan and Manzar Masud

Published: 12 September 2023

1. Introduction

In the hydrocarbon industries, erosion is a common problem that leads to pipeline leakage, component damage, and even overall pipeline system failure. Erosive wear takes place when a fluid-carrying erodent consistently impacts the target surface [1]. This consistent impact force applied by the erodent causes gradual material loss from the target material [2]. The angle of impaction between carrier fluid and target surface plays a crucial role in defining the erosion rate [3]. Therefore, flow-changing devices, like elbows, valves, chokes, tees, U-bends, and offset elbows are comparatively more prone to erosive wear and tear [4].

The standard 90° elbow is widely used in the hydrocarbon industry as a flow-changing device [5]. Erosive wear in the 90° elbows has been the central focus of various experimental and numerical investigations [6]. Erosion in elbows is quite large compared to straight pipes [7]. Though erosion is inevitable and cannot be avoided fully, engineers and scientists have utilized a diverse range of erosion mitigation techniques to minimize devastating damages to elbows due to erosive wear and tear [8].

Design modification is a popular technique for mitigating erosive wear and tear in elbows. A vertex chamber was examined in comparison to a standard 90° elbow [9]. The basic geometric parameters, such as pipe diameter, domain size, and curvature radius, were not modified. A semi-sphere was constructed at the upstream section. The research work concluded that the erosion rate in the vertex chamber was considerably reduced, and in the worst-case scenario it was half of the 90° elbow.

This research study presents a numerical approach to predict the erosion rate and mechanism in a standard 90° elbow. The key focus of this research work aims at analyzing the influence of two-phase, air-solid flow on the erosion of the elbow at different locations.

To decouple the effect of two-phase, air-sand flow on erosion in a standard 90° elbow, it is subjected to a range of flow velocities and mass flow rates.

This research work numerically validates the experimental and computational claims made by Viera et al. [10].

2. Methods

The 3D model displayed in Figure 1 was created in SOLIDWORKS. To compare and analyze the results, the geometry of Viera et al. [10] was utilized for the simulations. Flow enters the pipe through a straight section that is 1000 mm in length. This section contains a 90° elbow with r/D equal to 1.5 followed by another straight segment that is a 600 mm flow outlet. The geometry and other details are shown in Figure 1. The inside diameter (D) of the pipe is equal to 76.2 mm. In this study, a grid with 2.7 million cells was used for all the simulation cases due to its optimal computational cost and accuracy.

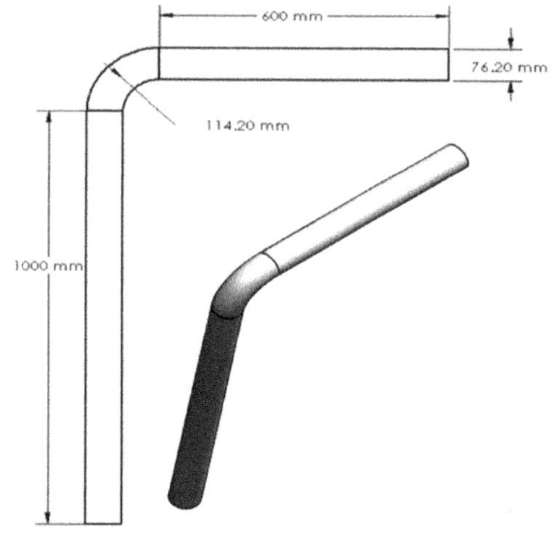

Figure 1. Geometry dimensions.

The experimental results of Viera et al. [10] were validated using CFD simulations. In Viera et al.'s experiment, various tests were performed at different air velocities and different sand flow rates. The experimental results were then validated by CFD simulations. For the vertical horizontal (V-H) configuration tests, sand sizes of 150 and 300 μm were used.

3. Results and Discussion

This research work presents a numerical approach to verify the experimental findings of Viera et al. [10]. Six cases with different velocities and particle sizes were carried out. The operating conditions were consistent with those used in the experiments. The numerical results, which are displayed in Table 1, were in close agreement with the experimental results.

Table 1. Comparison of numerical and experimental results.

Tests	Air Velocity (m/s)	Particle Size (μm)	Sand Rate (kg/day)	Erosion Rate (mm/year) Experimental Results	Erosion Rate (mm/year) Numerical Results	%Error
1	11	150	254	6.5	7.31	12.46
2	11	300	288	16.9	19.2	16.36
3	15	150	237	13.2	15.03	13.86
4	15	300	192	19.3	22.83	18.29
5	23	150	257	36.2	42.33	16.93
6	23	300	227	80.3	96.70	20.42

A literature study reveals that elliptical and V-shape erosion scars appear on the largest inner surface bend (extrados) of the elbow due to erosive wear. In this research work, both elliptical and V-shape scars were developed, as shown in Figure 2a. The erosion rate is negligible from 0° to 20° and increases gradually from 20° to 55° of the 90° elbow, with severe erosion observed at 45° to 55°. Figure 2b shows the erosion rate at different curvature angles of the standard 90° elbow.

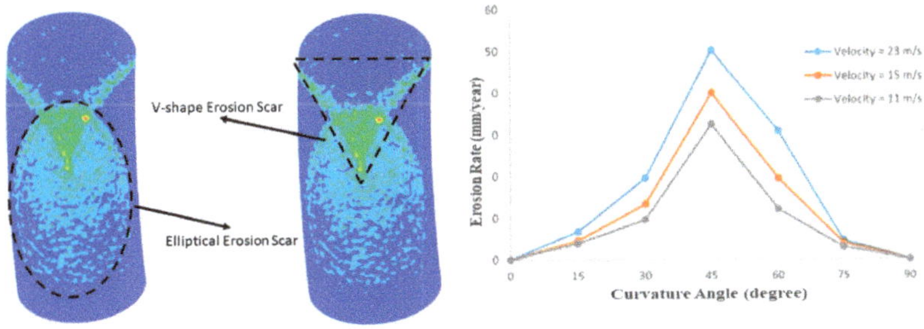

(**a**) (**b**)

Figure 2. (**a**) Elliptical and V-shaped erosion scar; (**b**) erosion at various curvature angles.

It was observed that the erosive wear rate increased with erodent size diameter, as shown in Figure 3. Erodents of 150 and 300 μm size were used in this case study, and carrier fluid velocity was kept at 23 m/s for both particle sizes.

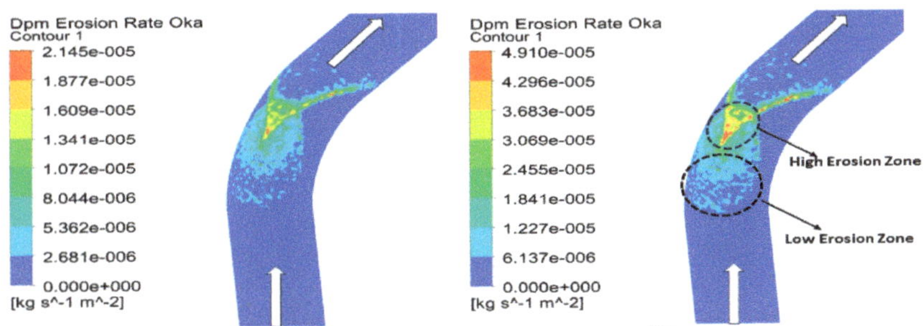

(**a**) (**b**)

Figure 3. Erosion rate for different particle sizes: (**a**) 150 μm (**b**) 300 μm.

The influence of fluid velocity on erosion rate was analyzed. Two flow velocities, 11 and 23 m/s were considered in this case. The particle size of 300 μm was used for both velocities. It was concluded that the erosion rate increases with the fluid velocity.

It is evident from the numerical results that the highest pressure is observed at the outer surface, and the lowest pressure is achieved at the inner surface of the elbow. Gradual pressure drop is observed from outer surface towards inner surface. The pressure distribution across the entire elbow is presented in Figure 4.

Particle trajectory shows that particles enter through the inlet and follow a straight path and then move in an irregular pattern after hitting the elbow. The majority of the erodent impacts the outer wall, which produces maximum wear. The inner wall of the pipe is a no-erodent zone with minimum erosion, as shown in Figure 5.

Figure 4. Pressure variation.

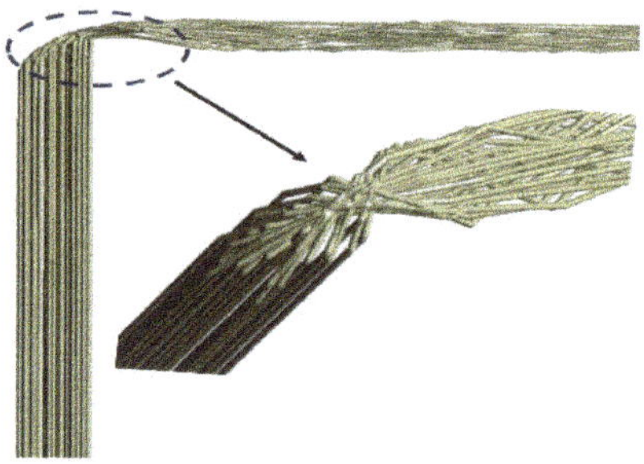

Figure 5. Particle track.

4. Conclusions

A CFD-based numerical approach was utilized to validate previous experimental and numerical results of erosion in a 90° elbow bend. The influence of different parameters on erosive wear rate and future recommendations are extensively discussed below.

1. A detailed study of the impact of particle size on erosive wear was conducted. It was found that erosive wear increased with the particle diameter size when the flow velocity and mass flow rate were kept constant.
2. An increase in erosive wear rate with flow velocity was observed. Maximum erosion was detected at 45° to 55° of the largest inner surface bend (extrados) of the elbow. V-shape and elliptical erosion scars were produced at the extrados of the elbow.
3. Maximum pressure was detected at the extrados of the elbow due to maximum particle impact. A gradual decrease in pressure was observed from extrados towards intrados.
4. Maximum flow velocity was observed at the intrados of the elbow due to low pressure. A gradual decrease in velocity was observed from intrados towards extrados.

The influence of non-spherical sand particles could be studied in future for a better understanding of sand erosion, as sand particles are non-spherical in nature.

Author Contributions: Conceptualization, writing—original draft, formal analysis, software, and validation: N.K.; supervision, investigation, and methodology: M.R.K., S.U. and T.T.; writing—review and editing: M.A.K. and Z.S. All authors have read and agreed to the published version of the manuscript.

Funding: This research work did not receive any funding from external sources.

Institutional Review Board Statement: Not applicable.

Informed Consent Statement: Not applicable.

Data Availability Statement: The data presented in this study are available on request from the corresponding authors.

Acknowledgments: The authors acknowledge the College of Electrical and Mechanical Engineering (CEME), NUST, for providing the high-performance computing system.

Conflicts of Interest: The authors declare no conflict of interest.

References

1. Solnordal, C.B.; Wong, C.Y.; Boulanger, J. An experimental and numerical analysis of erosion caused by sand pneumatically conveyed through a standard pipe elbow. *Wear* **2015**, *336–337*, 43–57. [CrossRef]
2. Rahman, S.; Khan, R.; Niazi, U.M.; Legutko, S.; Khan, M.A.; Ahmed, B.A.; Petrů, J.; Hajnyš, J.; Irfan, M. Performance Prediction of Erosive Wear of Steel for Two-Phase Flow in an Inverse U-Bend. *Materials* **2022**, *15*, 16. [CrossRef]
3. Zhao, X.; Cao, X.; Zhang, J.; Cao, H.; Xie, Z.; Xiong, N. Numerical investigation and dimensionless erosion laws of solid particle erosion in plugged tees. *Powder Technol.* **2022**, *402*, 117342. [CrossRef]
4. Mazumder, Q.H.; Shirazi, S.A.; McLaury, B.S.; Shadley, J.R.; Rybicki, E.F. Development and validation of a mechanistic model to predict solid particle erosion in multiphase flow. *Wear* **2005**, *259*, 203–207. [CrossRef]
5. Kesana, N.R.; Grubb, S.A.; McLaury, B.S.; Shirazi, S.A. Ultrasonic Measurement of Multiphase Flow Erosion Patterns in a Standard Elbow. *J. Energy Resour. Technol. Trans. ASME* **2013**, *135*, 3. [CrossRef]
6. Zhou, H.; Zhang, Y.; Bai, Y.; Zhao, H.; Lei, Y.; Zhu, K.; Ding, X. Study on reducing elbow erosion with swirling flow. *Colloids Surf. A Physicochem. Eng. Asp.* **2021**, *630*, 127537. [CrossRef]
7. Li, A.; Wang, Z.; Zhu, L.; Wang, Z.; Shi, J.; Yang, W. Design optimization of guide vane for mitigating elbow erosion using computational fluid dynamics and response surface methodology. *Particuology* **2022**, *63*, 83–94. [CrossRef]
8. Zhu, H.; Li, S. Numerical analysis of mitigating elbow erosion with a rib. *Powder Technol.* **2018**, *330*, 445–460. [CrossRef]
9. Duarte, C.A.R.; de Souza, F.J.; Santos, V.F.D. Mitigating elbow erosion with a vortex chamber. *Powder Technol.* **2016**, *288*, 6–25. [CrossRef]
10. Vieira, R.E.; Mansouri, A.; McLaury, B.S.; Shirazi, S.A. Experimental and computational study of erosion in elbows due to sand particles in air flow. *Powder Technol.* **2016**, *288*, 339–353. [CrossRef]

engineering proceedings

MDPI

Proceeding Paper

Investigation of the Simultaneous Cooling and Heating Using a Thermoelectric Peltier †

Abdur Rehman Mazhar *, Ali Ubaid, Syed Muhammad Hamza Shah, Suhaib Masood and Muneeb Zafar Alvi

Department of Mechanical Engineering, College of Electrical & Mechanical Engineering CEME, National University of Sciences & Technology NUST, Islamabad 47301, Pakistan; aubaid.me41ceme@student.nust.edu.pk (A.U.); smhamza.me41ceme@student.nust.edu.pk (S.M.H.S.); suhaibmasood888@gmail.com (S.M.); mmueeb.me41ceme@student.nust.edu.pk (M.Z.A.)

* Correspondence: arehman.mazhar@ceme.nust.edu.pk

† Presented at the Third International Conference on Advances in Mechanical Engineering 2023 (ICAME-23), Islamabad, Pakistan, 24 August 2023.

Abstract: Substitution of vapor compression refrigeration systems (VCRS) by using Thermoelectric Peltier offers numerous advantages. The Peltier effect is the inverse of the Seebeck effect: as a current flows between two dissimilar materials, it causes one side to grow hotter, making the other side cooler. This effect was experimentally exploited to simultaneously cool and heat water. In 60 min, 1 L of water was cooled from 40 °C to 16 °C and another liter heated from 25 °C to 45 °C. Power consumption was about 0.0828 kWh with a COP of 1.33, which can be enhanced using improved methods of heat dissipation on the Peltier device.

Keywords: Thermoelectric Peltier; HVAC; water dispenser

Citation: Mazhar, A.R.; Ubaid, A.; Shah, S.M.H.; Masood, S.; Alvi, M.Z. Investigation of the Simultaneous Cooling and Heating Using a Thermoelectric Peltier. *Eng. Proc.* **2023**, *45*, 13. https://doi.org/10.3390/engproc2023045013

Academic Editors: Mohammad Javed Hyder, Muhammad Mahabat Khan, Muhammad Irfan and Manzar Masud

Published: 8 September 2023

1. Introduction

The demand for efficient and eco-friendly cooling systems has led to the exploration of new refrigeration technologies. One such technology is the use of Thermoelectric Peltier's, which operates on the principle of the Seebeck effect. Thermoelectric Peltier cooling systems do not require harmful refrigerants, making them more environmentally friendly without any mechanically moving parts in comparison to the status quo Vapor Compression Refrigeration System (VCRS). They can also potentially reduce energy consumption and improve efficiency by directly cooling or heating water without the need for a compressor or heat exchanger.

The primary objective of this project was to develop a prototype of a water refrigeration system using Thermoelectric Peltier's. The prototype comprises an AC supply, Thermoelectric Peltiers, aluminum and copper heat sinks, thermal paste, a DC pump, pipes, and containers for water. The system is designed to operate on the principle of the Reverse Seebeck Effect, where a current is passed through the Thermoelectric Peltiers to create a temperature difference across its junctions [1]. The fins of the copper heat sink are submerged in one container containing hot water, while cold water is circulated through an aluminum heat sink block via pipes.

The use of a thermoelectric cooling system in this project is an important step towards developing sustainable and eco-friendly cooling technologies for drinking water. Traditional refrigeration systems rely on the use of harmful refrigerants, such as chlorofluorocarbons (CFCs) and mechanically moving parts, which have negative environmental impacts. The use of thermoelectric cooling systems reduces the reliance on harmful refrigerants and provides a more sustainable solution for cooling applications.

The thermoelectric cooling system has several potential applications in various fields, such as the food industry, medical settings, and research laboratories. It is also compact and with low noise, making it suitable for a variety of applications, including domestic, industrial, and scientific settings [2]. The capital and operational cost of such a system is lower than a VCRS, with lower risks of refrigerant leakage and contamination, especially of drinkable water in this application involving close proximity with humans.

2. Methodology

A comprehensive literature review is primarily conducted for Thermoelectric Peltier applications for domestic applications [3].

Based on this literature review, it is clear that there is a major research gap in using Thermoelectric Peltiers for applications involving drinkable water. Additionally, there are limited studies investigating the effects of simultaneous heating and cooling with modified heat sinks attached to the device. This study aims to develop and test a Thermoelectric Peltier-based water dispenser for domestic use. This would be a competing technology for conventional VCRS-based water dispensers. Additionally, this paper and further linked papers of this study would serve as a guide to novice researchers interested in developing water refrigeration and purification systems using active carbon filtration and Thermoelectric Peltier technology.

An overview of the experimental setup is illustrated in Figure 1. Two compartments having a volume of 1 L are filled with water. The Thermoelectric Peltier device attached to a heat sink is placed in the left water compartment, as illustrated in Figure 2. The right compartment has a DC motor connected to a pump with pipes to circulate water to one side of the Peltier device placed in the left compartment. After operation, the system simultaneously heats the water in the right compartment whilst cooling the water in the left compartment using the Peltier effect. A Thermoelectric Peltier operates on the reverse Seebeck effect in which a temperature difference develops across the surface of the module when a DC current is applied to it. At a microscopic level, charge carriers in the material diffuse from the hot to the cold side of the module, creating a temperature gradient.

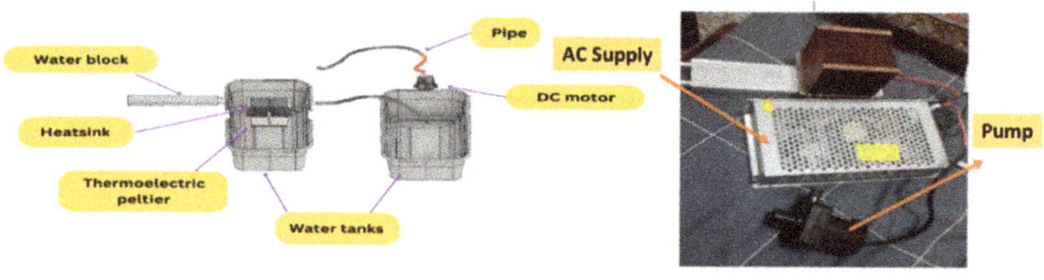

Figure 1. Schematic of the assembly of the experimental setup with the AC-DC conversion kit, pump, and the heatsink attached to the Thermoelectric Peltier.

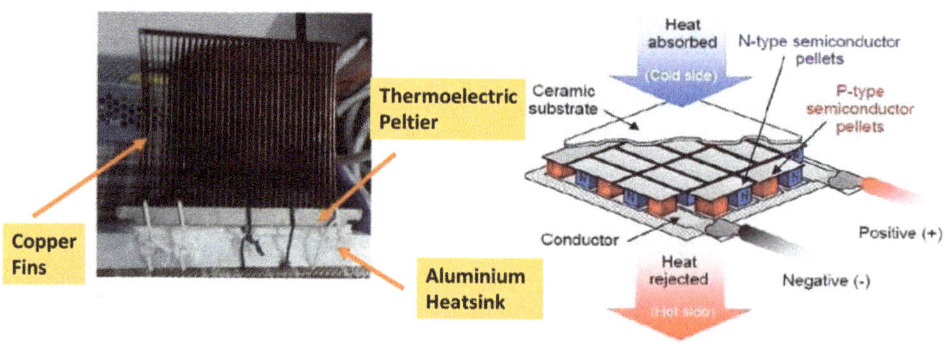

Figure 2. Assembly of the heatsink and TEC1-12706 Thermoelectric Peltier.

The system is plugged into a normal domestic 220V AC switch with a conversion kit. A layout of a uniquely designed heat sink for this specific application and the structure of the Thermoelectric Peltier module TEC1-12706 is presented in Figure 2.

Additionally, details and specifications of the equipment are presented in Table 1.

Table 1. Components and their specifications.

Components	Specifications
DC pump	12 V, 8 W, 10 L/m
Thermoelectric peltiers (TEC1-12706)	12 V, 4.4 A
Power supply	12 V, 30 A
Water block	Aluminum
Thermoelectric peltiers (TEC1-12706)	Bismuth Telluride (Bi2Te3)
Heatsink	Copper
Pitch of the fins in the heatsink	2 mm
Number of fins	30
Water tanks	PETG (3D-printed)
Pipes	Polyethylene

3. Results and Analysis

During the cooling process, the initial temperature of the water is set to 40 °C. After cooling, the temperature successfully decreases to 16 °C in just a span of 60 minutes of continuous operation. For the heating process, the initial temperature is set to 25 °C, and through heating, the temperature is increased to 45 °C. Throughout the entire operation, the device consumed approximately 0.0828 kWh of energy. This corresponds to a change in temperature of approximately 0.3 °C/minute for both heating and cooling, which is a little below the value recorded in the literature primarily because of the fact that this study investigates simultaneous heating and cooling contrary to only one form of heat transfer found in the literature [3].

The Coefficient of Performance (COP) of the device is calculated to be around 1.3296, indicating that the device is efficient in converting energy simultaneously to produce a heating and cooling effect. This value of the COP is in line with results from the literature [3]. Had the calculation of the COP taken both effects into consideration, it would have been about 2.66. To further improve the performance of the device, the process could be refined with enhanced methods of heat dissipation. In the event of standalone heating and cooling, the COP improved further.

Figure 3 presents the variation of the cold and hot water temperature with time over the duration of the experiment. The variation is a linear process, as evident from Figure 3, with minor fluctuations due to reading variations of the thermocouples. Had the device operated for further time or with a heatsink having a higher fin pitch ratio, the plot would have asymptotically converged to the temperature of the Thermoelectric Peltier.

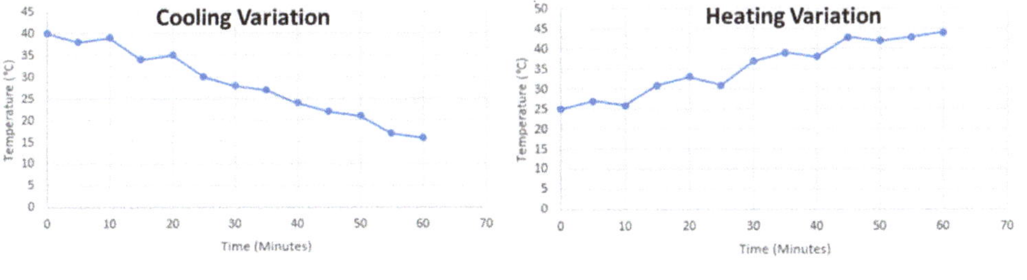

Figure 3. Temperature variation in the cold and hot waters for the duration of the experiment.

A clear shortcoming of the Peltier not illustrated in these plots of Figure 3 is the fact that it heats up after 60 min of continuous operation, as observed in the literature as well [3].

This is similar to the heating of a compressor in a VCRS operating continuously. For this, three approaches are to be investigated in further studies of this project:

1. Usage of a more efficient and effective heat sink coupled with added passive heat dissipation technologies, probably with a higher fin pitch ratio.
2. Usage of cascaded Thermoelectric Peltier to minimize the load on a single device to provide a better-combined effect.
3. Usage of an automated time-limited operation of the Peltier using a PLC controller coupled with a variable input current supply. This approach is similar to the usage of variable speed compressors in modern VCRS.

A summary of the results of this experimental investigation is presented in Table 2:

Table 2. Control parameters in the heating and cooling compartments.

Control Parameters	Cooling Compartment	Heating Compartment
Volume of water stored	1 L	1 L
Initial Temperature	40 °C	25 °C
Final Temperature	16 °C	45 °C
Time taken	60 min	60 min
Flow rate through cooling block	10 L per minute	10 L per minute

4. Conclusions and Future Recommendations

In conclusion, the development of an innovative water refrigeration system employing Thermoelectric Peltiers represents a notable stride towards realizing a more sustainable and ecologically sound water dispenser as a competitor to VCRS-based dispensers. Furthermore, the system's compact form factor and low acoustic emissions, along with a reasonable COP, contribute to its inherent appeal in domestic applications. Further work regarding the development of heatsinks, cascaded devices, and variable speed and time-based operation of the Peltier device is recommended.

Author Contributions: Conceptualization, A.R.M.; Methodology, A.U. and S.M.H.S.; Formal analysis, S.M. and M.Z.A.; Investigation, A.U. and S.M.H.S.; Resources, A.R.M.; Writing—original draft preparation, A.U., S.M.H.S., S.M. and M.Z.A.; Writing—review and editing, A.R.M.; Supervision, A.R.M.; Project administration, A.R.M. All authors have read and agreed to the published version of the manuscript.

Funding: This research received no external funding.

Institutional Review Board Statement: Not Applicable.

Informed Consent Statement: Not Applicable.

Data Availability Statement: Not Applicable.

Conflicts of Interest: The authors declare no conflict of interest.

References

1. Lin, L.; Meng, X.; Miao, Z. Analysis and verification of Seebeck effect and Peltier effect co-existence on a thermoelectric device. *SSRN Electron. J.* **2022**. [CrossRef]
2. Witek, K. Technological aspects of Semiconductor Thermogenerator (TEG) assembly. *Int. J. Model. Optim.* **2013**, *3*, 390. [CrossRef]
3. Aabid, A. A Systematic Review of Thermoelectric Peltier Devices. *Fluid Dyn. Mater. Process.* **2022**, *19*, 187–206. [CrossRef]

engineering proceedings

Proceeding Paper

Inclined Rayleigh–Benard Convection: Role of Critical Aspect Ratio in Vertical Cavities †

Wajeeha Siddiqui

Department of Mechanical Engineering, CEME, National University of Sciences and Technology, Islamabad 44000, Pakistan; wajeeha.siddiqui16@me.ceme.edu.pk

† Presented at the Third International Conference on Advances in Mechanical Engineering 2023 (ICAME-23), Islamabad, Pakistan, 24 August 2023.

Abstract: Inclined Rayleigh–Benard Convection (RBC) is numerically investigated in a two-dimensional vertical cavity; the critical aspect ratio and the critical Rayleigh number are discussed. It is established that beyond the 6500 Rayleigh number, secondary cell formation starts in the cavity. But this phenomenon is not visible at lower aspect ratios. The presence of secondary cells is directly related to heat transfer across the cavity. In recent times, insulated glazing units (IGUs) have been considered for better thermal performance, typically in energy-efficient buildings. The study gains significance by gauging the performance and optimization of IGUs.

Keywords: Rayleigh–Benard convection; aspect ratio; Rayleigh number

Citation: Siddiqui, W. Inclined Rayleigh–Benard Convection: Role of Critical Aspect Ratio in Vertical Cavities. *Eng. Proc.* **2023**, *45*, 24. https://doi.org/10.3390/engproc2023045024

Academic Editors: Mohammad Javed Hyder, Muhammad Mahabat Khan, Muhammad Irfan and Manzar Masud

Published: 11 September 2023

1. Introduction

Convection is a common and natural mode of heat transfer between moving fluids as well as solids. It sets in the fluid when there is any kind of thermal inhomogeneity [1]. The study of the convection phenomenon is vigorously important in the disciplines of basic sciences, engineering, and technology [2]. The existence of convection phenomena inside earth layers is depicted in Figure 1.

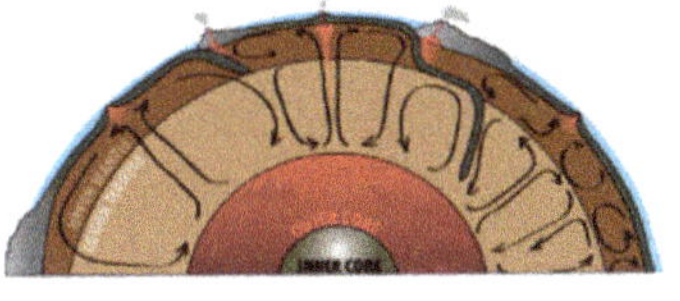

Figure 1. Convection currents inside the Earth's layers.

Convection can occur in a fluid due to the horizontal or vertical temperature gradient. The typical Rayleigh–Benard configuration (RBC) consists of a lower heated wall and an upper cold wall [3]. This RBC problem is the simplest and one of the earliest ones to be investigated. The two important parameters here are the Rayleigh number (Ra) and the Prandtl number (Pr). The Rayleigh number is a dimensionless parameter and is defined as the product of the Grashof number, which compares the buoyancy and viscosity effects within fluid, and the Prandtl number, which compares momentum diffusivity and thermal diffusivity. The formed patterns are cells in 2-dimension and rolls in 3-dimension. A very critical part is carried out by the lateral boundaries and the large extent of the third dimension. Vertical cavities with a varying aspect ratio were studied with the same phenomenological approach in this research. It is deduced that, in addition to the critical Rayleigh number, the critical aspect ratio also plays a significant role in heat transfer and heat flux.

The inclined RBC (RBC with a horizontal temperature gradient) has several practical uses; for instance, in developed countries, almost 40% of total energy is consumed only in buildings. Fenestration products, such as windows, perform as thermal holes and are among the main causes of energy loss in a building. Therefore, window thermal performance analysis and optimization have become important concerns to predict and control energy loss and make buildings more energy efficient. This arrangement is also used in solar collectors and the cooling of electronic chips and equipment.

Lartigue et al. [4] studied the phenomenon in AR = 40 for the presence, rotatory motion, and global displacement of secondary cells. In another study, the melting front of purely incompressible material amongst liquid and solid phases is studied as a case of the RBC problem [5]. D'Orazio et al. [6] studied small aspect ratios for heat transfer evolution, bifurcations, and hysteresis phenomena. Nakhoul et al. [7] reported high AR nanoparticles using near-field, surface-coupled, sophistically controlled polarized pulses with the aim of preventing bacterial proliferation. The phenomenon is extensively studied and reported by Bergman et al. [8], Yang et al. [9], and Poel [10].

2. Problem Statement and Numerical Methodology

A vertical cavity filled with air (Pr = 0.71) at atmospheric pressure is considered for $1 \leq \text{AR} \leq 40$ and $\text{Ra} \leq 10{,}000$. The assumptions and numerical methodology deployed here are the same as those in [11]. The schematic is shown in Figure 2.

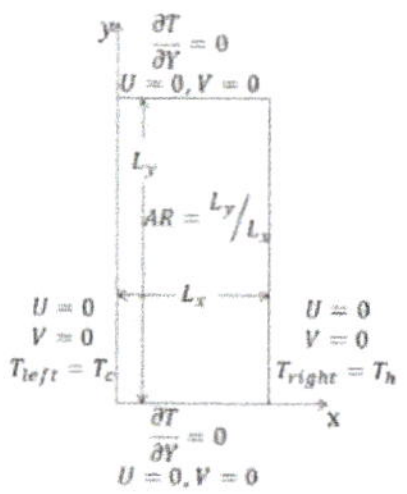

Figure 2. Schematic of the vertical cavity under study [11].

Dimensionless forms of Navier–Stokes equations are used along with the artificial compressibility method [12] to couple velocity and pressure, and a forward in time central in space (FTCS) explicit scheme is employed to discretize the governing equations.

3. Results and Discussions

3.1. *The Critical Rayleigh Number*

Numerical simulations are performed, and different output parameters are compared. It is established from [11] that at AR greater than AR_{cr}, the RB instability sets in the flow at 6500 Ra; beyond this, secondary cells are formed, as shown in Figure 3.

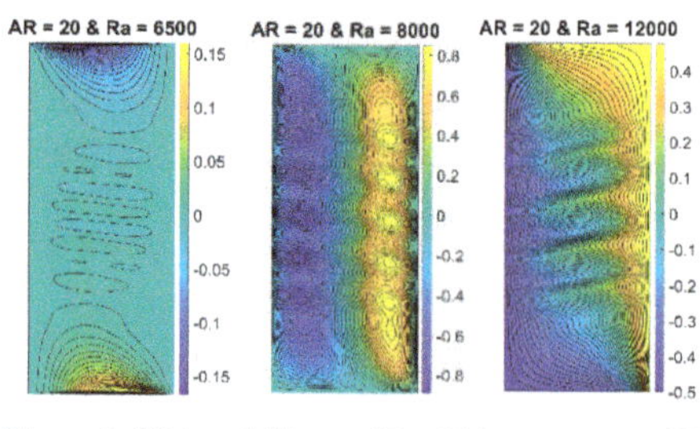

Figure 3. RB instability at AR = 20 in contours of U−velocity, V−velocity, and temperature (L to R).

The first critical Ra (Ra_{cr}), at which convection starts in the cavity in the case of a horizontal temperature gradient, is found to be as low as 370. The convection currents and thus the resultant heat transfer is initially weak and present only in the vicinity of vertical walls. The second critical-Ra ($Ra_{cr,2}$), at which Rayleigh–Benard instability sets in the flow, is found to be constant for all ARs greater than critical-AR i.e., 6500.

The same instability patterns are visible in other parameters as well (Figure 3). The higher the heat transfer rate required, the higher Ra is desirable and vice versa.

3.2. The Critical Aspect Ratio

The threshold value of aspect ratio; the critical AR (AR_{cr}), decides the boundary for the existence of RB instability in the flow. It is implied through the numerical study that RB instability will not occur in the flow for AR < AR_{cr}. Other instabilities might be present at a higher Ra in this case. The difference is displayed in Figure 4.

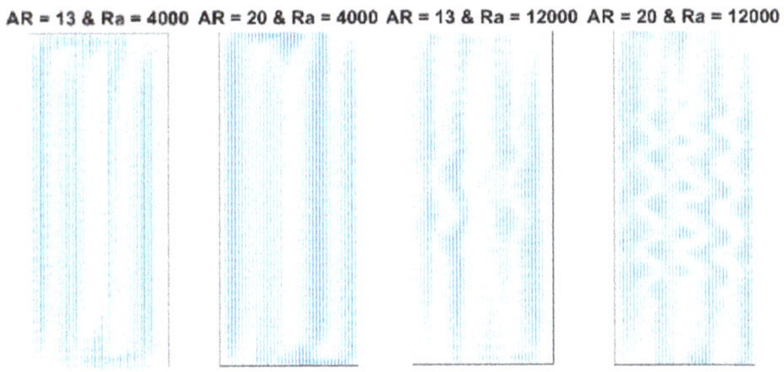

Figure 4. Velocity vector field comparison at Ra = 4000 and 12,000; AR = 13 and 20.

The reason for instability leading to secondary cell formation lies in the flow mechanics of Rayleigh–Benard (RB) instability; RB instability arises when the convection intensity continues to increase until the counterclockwise rotating single cell approaches and fills the core zone. At this stage, the streamlines of the flow begin to fluctuate due to the opposing walls' interference, as depicted by the schematic in Figure 5. Three mesh sizes are examined and tested to determine the AR-critical value to be AR_{cr} = 14.

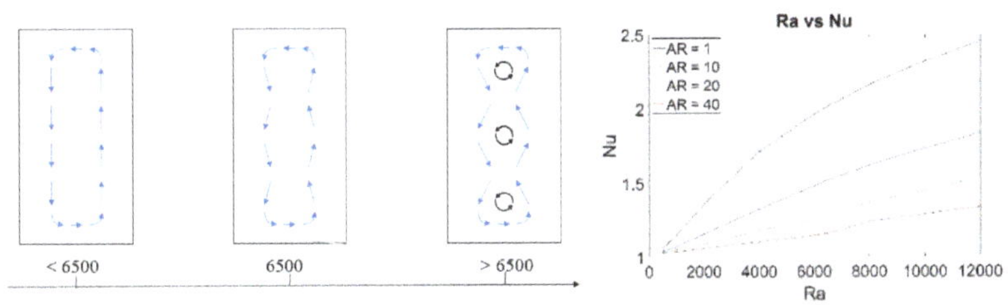

Figure 5. Formation procedure of secondary cells in RBC; Nu versus Ra plot.

From a practical application point of view, the heat transfer rate jumps high after critical aspect ratio and after critical Ra; therefore, these should be kept towards the lower side if lower heat transfer is anticipated.

3.3. Nusselt Number Versus Rayleigh Number

Generally, average Nu increases as Ra increases for all AR, as shown in Figure 5. There are slight bumps in slopes for Ra > 6500 for ARs larger than the critical value, indicating a significant rise in the heat transfer rate. It is noted that other instabilities present at lower Ra do not contribute to the added heat transfer rate as much as the RB instability does. A simplified dimensional analysis for IGU is done for Ra = 8000 for case 1: AR = 10 and case 2:

AR = 40 to further substantiate the argument. In the usual summer season in Islamabad, the average maximum temperature remains around 45 °C, and the ideal room temperature is taken as 26 °C. The real focus is to compare the design conditions for the internal air-filled cavity of a double-glazed window. It is perfectly visible from Table 1 that the heat flux in the case of a higher AR is significantly smaller than that of a lower AR.

Table 1. Comparison of two ARs for Ra = 8000.

Parameters	Case 1: AR = 10	Case 2: AR = 40
Nu	1.6172	1.2397
L_x	0.0171 m	0.0171 m
L_y	0.171 m	0.684 m
$h = (Nu \cdot k)/L$	2.482 $Wm^{-2}K^{-1}$	1.903 $Wm^{-2}K^{-1}$
Heat Flux $= q/A = h(T_h - T_c)$	47.158 Wm^{-2}	36.157 Wm^{-2}

4. Conclusions

For a wide range of Ra, natural air flows and RB instability are numerically studied for a vertical cavity from a square cavity up to AR = 40 with horizontal temperature gradients. It is found that a single counter-rotating cell is formed for AR < AR_{cr}. As Ra increases, this single cell deforms but does not break down into smaller cells. For AR $\geq$ AR_{cr}, the secondary flow sets in the cavity after 6500 Ra; the RB instability originates at this specific Ra, and secondary cells are formed beyond it. Two different cases have been compared to establish that this theatrical study provides the desired effect.

Funding: This research was funded by the National University of Sciences and Technology, Islamabad under Grant 00000116261.

Institutional Review Board Statement: Not applicable.

Informed Consent Statement: Not applicable.

Data Availability Statement: Not applicable.

Acknowledgments: The author is grateful to Imran Akhtar and Naveed Mazhar for their assistance.

Conflicts of Interest: The author declares no conflict of interest.

References

1. Ahlers, G.; Grossmann, S.; Lohse, D. Heat transfer and large scale dynamics in turbulent Rayleigh-Bénard convection. *Rev. Mod. Phys.* **2009**, *81*, 503. [CrossRef]
2. Zoubir, A.; El Otmani, R.; Bouferra, R.; Essaleh, L. Effects of thermal stratification on natural convection in a symmetrically heated channel: Comparison of characteristic quantities between air and water. *Case Stud. Therm. Eng.* **2023**, *45*, 102950. [CrossRef]
3. Koneti, L.; Venkatasubbaiah, K. A comparative heat transfer study of water and liquid gallium in a square enclosure under natural convection. *Int. J. Fluid Mech. Res.* **2023**, *50*, 39–44. [CrossRef]
4. Lartigue, B.; Lorente, S.; Bourret, B. Multicellular natural convection in a high aspect ratio cavity: Experimental and numerical results. *Int. J. Heat Mass Transf.* **2000**, *43*, 3157–3170. [CrossRef]
5. Favier, B.; Purseed, J.; Duchemin, L. Rayleigh–Bénard convection with a melting boundary. *J. Fluid Mech.* **2019**, *858*, 437–473. [CrossRef]
6. D'Orazio, M.C.; Cianfrini, C.; Corcione, M. Rayleigh–Bénard convection in tall rectangular enclosures. *Int. J. Therm. Sci.* **2004**, *43*, 135–144. [CrossRef]
7. Nakhoul, A.; Rudenko, A.; Maurice, C.; Reynaud, S.; Garrelie, F.; Pigeon, F.; Colombier, J.P. Boosted Spontaneous Formation of High-Aspect Ratio Nanopeaks on Ultrafast Laser-Irradiated Ni Surface. *Adv. Sci.* **2022**, *9*, 2200761. [CrossRef] [PubMed]
8. Bergman, T.L.; Lavine, A.S.; Incropera, F.P.; Dewitt, D.P. *Fundamentals of Heat and Mass Transfer*; John Wiley & Sons: Hoboken, NJ, USA, 2011.
9. Yang, Y.-H.; Zhu, X.; Wang, B.-F.; Liu, Y.-L.; Zhou, Q. Experimental investigation of turbulent Rayleigh-Bénard convection of water in a cylindrical cell: The Prandtl number effects for Pr > 1. *Phys. Fluids* **2020**, *32*, 015101. [CrossRef]
10. Van Der Poel, E.P.; Stevens, R.J.; Lohse, D. Comparison between two-and three-dimensional Rayleigh–Bénard convection. *J. Fluid Mech.* **2013**, *736*, 177–194. [CrossRef]

11. Siddiqui, W.; Abbas, Z.; Akhtar, I.; Khalid, M.S.U. Nusselt Number Dependence on Aspect Ratio and Rayleigh Number: A Numerical Study of Rayleigh-Benard Instability. In Proceedings of the Fluids Engineering Division Summer Meeting, Toronto, ON, Canada, 2 August 2022; p. V002T005A029.
12. Chorin, A.J. A numerical method for solving incompressible viscous flow problems. *J. Comput. Phys.* **1997**, *135*, 118–125. [CrossRef]

engineering proceedings

MDPI

Proceeding Paper

Design and Fabrication of Digital Microfluidics Device for Lab-on-a-Chip Applications †

Hammas Ullah, Faraz Shoaib, Syed Danyal Zahid, Mohammad Iymaan Mahmood, Mubashar Ali, Moazzam Ali, Hamza Abbas and Ali Turab Jafry *

Faculty of Mechanical Engineering, Ghulam Ishaq Khan Institute of Engineering Sciences and Technology, Topi 23460, Pakistan; hammas.ullah@giki.edu.pk (H.U.); farazshoaib42@gmail.com (F.S.); sdanyal.zahid@gmail.com (S.D.Z.); muhammadiymaanmahmood@gmail.com (M.I.M.); gme2187@giki.edu.pk (M.A.); gme2330@giki.edu.pk (M.A.); gme2223@giki.edu.pk (H.A.)

* Correspondence: ali.turab@giki.edu.pk

† Presented at the Third International Conference on Advances in Mechanical Engineering 2023 (ICAME-23), Islamabad, Pakistan, 24 August 2023.

Abstract: In digital microfluidics, discrete liquid droplets move on a dielectric surface with the help of AC or DC voltage. Digital microfluidics system has extended the scope of microfluidics by providing automated, precise flow control for point-of-care devices capable of medical diagnostics, environmental, and chemical sensing platforms. In this paper, we introduce a digital microfluidics platform based on electrowetting on a dielectric using PCB substrate designed on Easy EDA software. The platform's performance was analyzed at various DC voltages between 200 V and 400 V for droplet actuation. PCB fabrication, the utilization of easily accessible and cheap materials such as cooking oil and grafting tape as a dielectric layer, allowed the fabrication to be affordable and simple. The results indicate that droplet actuation takes place at 220 V. The droplet velocity obtained was 2.6 mm/s and 1.51 mm/s using silicon oil and cooking oil, respectively, at 400 V. Our proposed digital microfluidics platform will play a vital role in droplet actuation and can be used for diagnostic applications such as DNA analysis and cell culture due to its simple fabrication and cost-effective technique.

Keywords: digital microfluidics; electrowetting; contact angle; lab-on-a-chip

Citation: Ullah, H.; Shoaib, F.; Zahid, S.D.; Mahmood, M.I.; Ali, M.; Ali, M.; Abbas, H.; Jafry, A.T. Design and Fabrication of Digital Microfluidics Device for Lab-on-a-Chip Applications. *Eng. Proc.* **2023**, *45*, 7. https://doi.org/10.3390/engproc2023045007

Academic Editors: Mohammad Javed Hyder, Muhammad Mahabat Khan, Muhammad Irfan and Manzar Masud

Published: 8 September 2023

1. Introduction

Microfluidics refers to the behavior and control of liquids constrained to volumes near to the microlevel range. The behavior of liquids in the micro domain differs greatly from macroscopic fluids in terms of surface tension, diffusion, and fast thermal relaxation. Electrowetting-on-dielectric (EWOD) is a technique in which we manipulate droplets on a dielectric surface with the help of an electric field developed via the potential difference between electrodes. EWOD applications are suitable for a variety of lab-on-a-chip applications. This capacity makes it possible to manage complex protocol needed for laboratory applications. It is also referred to as digital microfluidics (DMF), which has two types of configurations, i.e., open configuration and closed configuration [1]. The objectives of this platform are to actuate microliter droplets, automate the fluid transportation, and detect protocols for biological and chemical processes. For a variety of droplet activities, closed DMF devices are usually preferred due to their higher functionality and control compared to the open platform. Overall, DMF has the advantages of less reagent usage, quick heat transfer, and simple integration with other analytical methods [2].

There are different configurations used for DMF devices. For instance, a multilayer PCB configuration has been previously implemented using a replaceable membrane of a dielectric layer. The authors used an electrode array, with an ITO glass plate on top, and a replaceable hydrophobic Teflon membrane, which played a critical role in enhancing the reliability of the device [3]. A decade ago, microfluidics devices were not commercially available and were only used for research purposes. However, over the last few years, many researchers have asked the question whether a true lab-on-a-chip device is possible or not. At present, these devices are entering the commercial sector and are being used for a number of applications, most of which are in the field of diagnostics and cell culture [4]. Some researchers have recently considered using high-voltage metal-oxide-semiconductor field-effect transistors (MOSFETS) to overcome its limitations, like reducing power dissipation and increasing switching speed between ON and OFF voltages for electrodes [5].

In this study, we have used a grafting tape dielectric layer with silicon oil and cooking oil as hydrophobic layers, respectively, for the smooth actuation of a droplet on printed circuit board. The oil layer was used to slide the droplet on the PCB substrate with minimal thickness for the smooth movement of the droplet. Using two-dimensional digital microfluidics platforms can provide dynamic configurability and simplicity in automation. The key advantages of DMF devices are the use of a minimum sample volume to limit reagent usage and processing time. Moreover, it enables full automation using Arduino without requiring any manual input.

2. Theoretical Background

External electric field counteracts the surface tension forces, which affect the contact angle of a liquid droplet on a solid surface. The Young–Lippmann equation provides this change in contact angle, θ_V:

$$\cos(\theta_V) = \cos(\theta_0) + \frac{\epsilon_r \epsilon_0 V^2}{2\gamma d}$$

In this equation, the contact angle, following the application of voltage, is denoted as θ; ϵ_0 and ϵ_r are the permittivity of vacuum and the relative permittivity of the material, respectively; d is the thickness of the dielectric layer; γ represents the surface tension of liquid; and V stands for the applied DC voltage to an electrode. The variable θ_0 is the initial contact angle of the droplet before it is placed under the influence of the electric field.

3. Materials and Methods

Distilled water was used in all experiments. Printed circuit board (PCB) was purchased from Smart PCBs (Rawalpindi, Pakistan). The size of the electrodes was 2 × 2 mm and the gap between two successive electrodes was 0.2 mm. Grafting tape acting as dielectric layer was used with a thickness of 0.03 mm. Power supply having DC 1000 V and 0.5% ripple maximum current of 1 milliampere was supplied by Qosain Scientific (Lahore, Pakistan).

First, we prepared our sample by pouring the cooking oil on the surface of the electrodes. Afterward, grafting tape was placed on top of the electrodes. The oil layer helped remove any air between the grafting tape and electrodes. For smooth droplet movement, sliding layers of either silicon oil or cooking oil (purchased from the local market) were applied. By applying DC voltage between two electrodes, the droplet actuation was observed with the help of a mobile phone camera. We found contact angles and the velocity of the droplet by performing image analysis using ImageJ software.

4. Results and Discussion

Figure 1 shows the results and validation of droplet movement by applying DC voltage. We found that the droplet movement starts at 220 V. The experimental results further prove that the droplet actuation speed increases with an increase in voltage (Figure 2). To observe other factors affecting speed, we changed the hydrophobic coating from silicon oil to cooking oil and repeated the same experimental procedure. From the scatter-line plot of the two oils, we can observe that for voltages up to 250 volts, the droplet speed on both oil surfaces is roughly the same; however, from 250 volts onward, there is a clear gap between them. The cooking oil yields better results for droplet speed reaching 2.6 mm/s at 400 V compared to 1.51 mm/s when the droplet is on silicon oil. Similar experiments were conducted while changing the electrode size and ultimately, the droplet volume. It is observed that a droplet with lower volume has better actuation results with higher voltages. Moreover, at higher voltages, the difference in droplet speed increases when comparing the three droplet volumes.

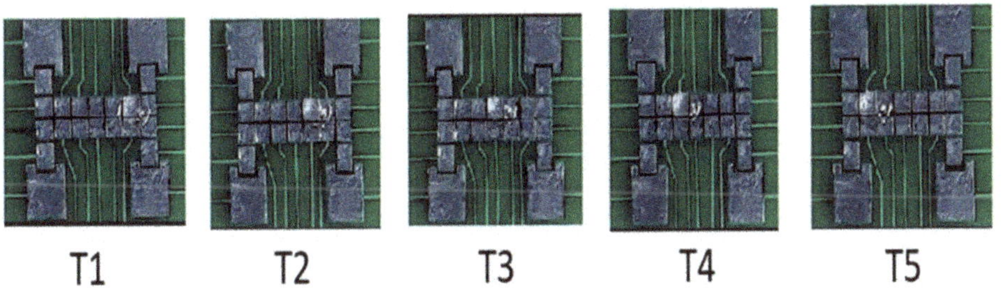

Figure 1. Droplet actuation in 5 frames.

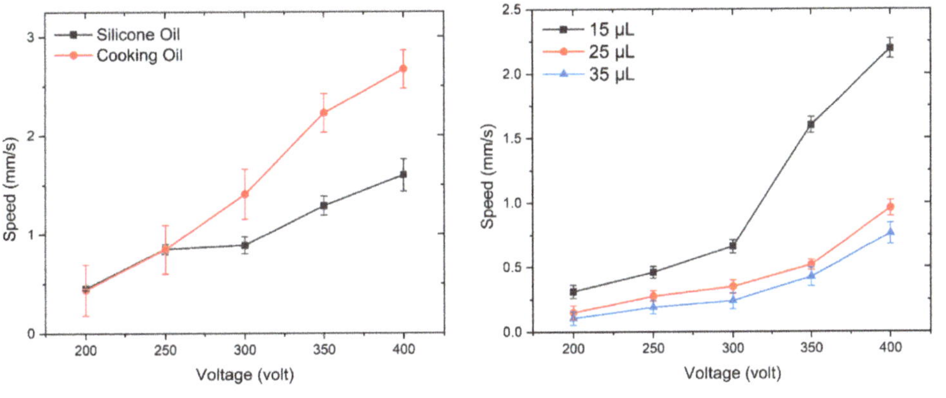

(**a**) (**b**)

Figure 2. (**a**) Droplet speed against applied voltage for silicon and cooking oil; (**b**) droplet speed with changing droplet volume.

We attribute this change in velocity to the viscosity of the oils used. The viscosity of silicon oil was comparably higher (around 3 times) compared to cooking oil. Since cooking oil is less viscous, it is therefore easy for the water droplet to slide across the electrodes as compared to silicon oil.

5. Conclusions

In this paper, we fabricated a digital microfluidics platform that uses electrowetting as a mechanism for the precise manipulation of microliter volumes of liquid samples. We conducted experiments to analyze the performance of the DMF platform at different voltage levels, ranging from 200 V to 400 V. The fabrication involves an open configuration design that allows for flexibility and versatility. Additionally, we utilized easily accessible materials such as cooking oil and grafting tape, which makes the platform cost-effective and practical for implementation at an industrial scale. Furthermore, we successfully achieved our targets, including developing a cost-efficient platform, achieving linear actuation for a precise liquid manipulation, and merging various components to create a functional system. The droplet velocity ranges from 0.5 mm/s to 2.6 mm/s under various conditions. The platform shows promising results and potential for future development for applications related to medical diagnostics, environmental monitoring, and chemical sensing applications.

Author Contributions: Conceptualization, H.U., F.S. and S.D.Z.; methodology, F.S., S.D.Z. and M.I.M.; software, S.D.Z.; validation, H.U., M.A. (Mubashar Ali) and M.A. (Moazzam Ali); investigation, H.U. and F.S.; resources, H.A.; writing—original draft preparation, H.U., F.S., S.D.Z. and M.I.M.; writing—review and editing, H.U. and A.T.J.; supervision, A.T.J.; project administration, A.T.J.; funding acquisition, A.T.J. All authors have read and agreed to the published version of the manuscript.

Funding: This work was funded by National Research Program for Universities (NRPU-14816) by Higher Education Commission (HEC) of Pakistan.

Institutional Review Board Statement: Not applicable.

Informed Consent Statement: Not applicable.

Data Availability Statement: The data is available on request from the corresponding author due to confidentiality.

Conflicts of Interest: The authors declare no conflict of interest.

References

1. Cho, S.K.; Moon, H.; Kim, C.-J. Creating, transporting, cutting, and merging liquid droplets by electrowetting-based actuation for digital microfluidic circuits. *J. Microelectromechanical Syst.* **2003**, *12*, 70–80.
2. Choi, K.; Ng, A.H.C.; Fobel, R.; Wheeler, A.R. Digital microfluidics. *Annu. Rev. Anal. Chem.* **2012**, *5*, 413–440. [CrossRef] [PubMed]
3. Xing, Y.; Liu, Y.; Chen, R.; Li, Y.; Zhang, C.; Jiang, Y.; Lu, Y.; Lin, B.; Chen, P.; Tian, R.; et al. A robust and scalable active-matrix driven digital microfluidic platform based on printed-circuit board technology. *Lab Chip* **2021**, *21*, 1886–1896. [CrossRef] [PubMed]
4. Li, J. Current commercialization status of electrowetting-on-dielectric (EWOD) digital microfluidics. *Lab Chip* **2020**, *20*, 1705–1712. [CrossRef] [PubMed]
5. Alistar, M.; Gaudenz, U. OpenDrop: An integrated do-it-yourself platform for personal use of biochips. *Bioengineering* **2017**, *4*, 45. [CrossRef] [PubMed]

engineering proceedings

Proceeding Paper

Experimental Investigation of Impact of Cool Roof Coating on Bifacial and Monofacial Photovoltaic Modules †

Khalid Mahmood *, Abid Hussain, Muhammad Arslan and Bilal Tariq

Department of Mechanical Engineering, University of Engineering and Technology Taxila, Taxila 47050, Pakistan; abid.hussain@uettaxila.edu.pk (A.H.); m.arslan5330@gmail.com (M.A.); thebilaltariq@gmail.com (B.T.)

* Correspondence: 96khalidghumman@gmail.com

† Presented at the Third International Conference on Advances in Mechanical Engineering 2023 (ICAME-23), Islamabad, Pakistan, 24 August 2023.

Abstract: Cool roof coatings are being applied on rooftops to reflect solar irradiance back into the atmosphere. Bifacial photovoltaic modules can receive solar irradiance from the front and rear side. The purpose of this study is to investigate the impact of reflected irradiance caused by cool roof coating (CRC) on the performance of bifacial photovoltaic (bPV) modules and monofacial photovoltaic (mPV) modules. According to the experiments, cool roof coating is an efficient way to increase the output and efficiency of solar modules. Experimental results show that for cool roof coating based on TiO_2 and FC resin with an average albedo of 0.63, the peak power output of the bifacial photovoltaic module is increased by 3.29% and its highest peak power efficiency is 18.1%. A peak power bifacial gain of 15.6% is seen. Due to a 1.3 °C decrease in temperature, the monofacial photovoltaic module power output also increased by 0.24%.

Keywords: bifacial photovoltaic module (bPV); monofacial photovoltaic module (mPV); cool roof coating; albedo; bifacial gain; efficiency of photovoltaic module

Citation: Mahmood, K.; Hussain, A.; Arslan, M.; Tariq, B. Experimental Investigation of Impact of Cool Roof Coating on Bifacial and Monofacial Photovoltaic Modules. *Eng. Proc.* **2023**, *45*, 38. https://doi.org/10.3390/engproc2023045038

Academic Editors: Mohammad Javed Hyder, Muhammad Mahabat Khan, Muhammad Irfan and Manzar Masud

Published: 13 September 2023

1. Introduction

Solar energy is considered as the most significant and highly developing source of renewable energy due to its global availability. Bifacial photovoltaic (bPV) modules are quickly gaining a market share due to their increased power production [1]. The amount of light absorbed on the bifacial solar cell's rear side determines the rise in bifacial photovoltaic module production [2]. The solar irradiance reflected from the surrounding surfaces enhances the bifacial photovoltaic module efficiency, as well as the power output [3]. Researchers ran numerous trials with various configurations to increase the bifacial photovoltaic module's yield. A 26% increase in bifacial gain is observed when flat mirrors are used behind the bifacial photovoltaic module [3]. The results of the experiment, which involved a bifacial solar module installed on a green roof with plant leaves, showed that for silver leaves, the increase in the photovoltaic module's bifacial gain is up to 13%. In contrast, white sand's irradiance reflection (albedo) of 0.47 results in an output gain of 4% [4]. A PV system with an aluminum sheet reflector was built, increasing the system's annual output by 15% [5]. The albedo of fresh snow typically ranges from 0.4 to 0.8, and due to snow's high reflectivity of up to 80%, a bifacial photovoltaic system's total energy output increased by up to 20% [6]. Cool roof coatings, with high solar reflectance values, primarily aim to reflect solar radiation back into the atmosphere, reducing the energy requirements for cooling [7]. By adding reflectors, the bifacial gain was raised to an optimum value. According to the simulation results, using a standard ground cover with soil and vegetation, an albedo of 0.25 provides a bifacial gain of less than 10%. In addition, bifacial gain was increased to 20% when a white concrete surface with an albedo of 0.5 was used as the test condition [8].

2. Methodology

2.1. Preparation of Cool Roof Coating

Titanium dioxide (TiO_2) was chosen as a covering pigment for a cool roof coating due to its high reflection in the near infrared region (NIR), ranging from 750 to 2500 nm. TiO_2 powder was purchased from Vinkimya Private Limited, based in Lahore. The coating was prepared by utilizing a low-cost and straightforward procedure. Water was used as the solvent, with small amounts of antifoaming, dispersing, and wetting agents added to it. To achieve even mixing, the solution was swirled for 10 min. Then, while still swirling, FC resin was added as a filler and adhesive to prevent the coating from peeling after application. TiO_2 was added after a period of stirring at an amount of 50% by weight (after drying), and the stirring speed was raised. To get rid of the bubbles created by stirring, the mixture was allowed to stand for a while [9]. After that, the coating was applied on a glass substrate as a test specimen and on the installation site of photovoltaic modules with a paint spryer machine up to a thickness of 0.5 mm (dried) and was allowed to air dry.

2.2. Experimental Setup

Figure 1 shows the experimental setup with CRC (a) and without CRC (b). The experimental setup consisted of a TiO_2 cool roof coating covering the ground, mPV, bPV, and an adjustable stand. Photovoltaic modules were installed on the roof of MED UET Taxila, Pakistan (33.7° N, 72.8° E latitude angle). The tilt angle of the PV module was kept at almost 34°. Solar irradiance, temperature, and photovoltaic module output were measured by using a pyrometer, laser thermometer and digital multimeter, respectively.

(**a**) (**b**)

Figure 1. Bifacial and monofacial photovoltaic modules with CRC (**a**) and bifacial and monofacial photovoltaic modules without CRC (**b**).

3. Results and Discussion

The reflectance spectrum of a single-layer TiO_2 resin-based cool roof coating with 0.5 mm thickness was obtained by using a spectrophotometer (DR 6000) and a glass substrate test specimen, and the results are shown Figure 2a.

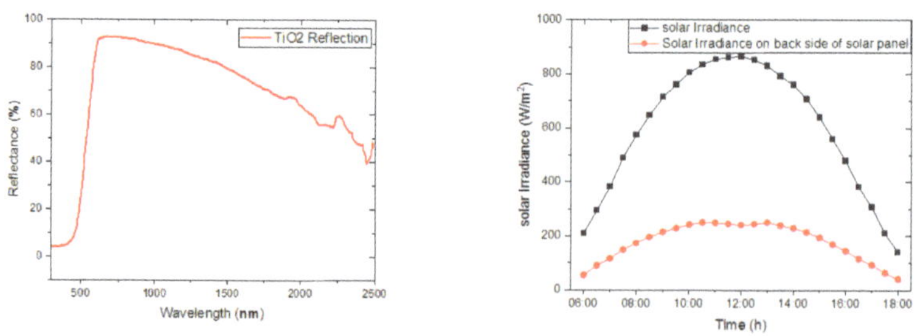

(**a**) (**b**)

Figure 2. Spectral reflection of the TiO_2 coating (**a**) and solar irradiance on the front and back surface of the PV module (**b**).

The TiO_2 cool roof coating exhibits 0.61 spectrum reflectance for ultraviolet region, 0.64 in the visible spectrum, and 0.723 for the near infrared region, and the total reflection observed for the cool roof coating is 0.63. Figure 2b illustrates the amount of solar irradiance incident on the front side and irradiance on the rear side of the bifacial photovoltaic module due to reflection. The irradiance on the front side varied from 141.3 W/m^2 to 863.6 W/m^2. The reflected irradiance on the rear side of the bifacial photovoltaic module ranged from 40.2 W/m^2 to 248 W/m^2 due to the cool roof coating.

The temperature of photovoltaic modules with and without the cool roof coating is illustrated in Figure 3a. It is observed that the temperatures of the photovoltaic modules with the cool roof coating are lower those of the photovoltaic modules without the cool roof coating. The maximum temperature for the monofacial photovoltaic module without CRC is 66.5 °C at 13:30. The monofacial photovoltaic module with CRC, the bifacial photovoltaic module without CRC, and the bifacial photovoltaic module have temperatures of 65.2 °C, 63.9 °C, and 62.7 °C, respectively. The temperature of bifacial photovoltaic module is lower than that of the monofacial photovoltaic module, due to the transparency of bifacial photovoltaic module. The highest peak power is produced by the bifacial photovoltaic module with the cool roof coating, with an efficiency of 18.09%, as shown in Figure 3b.

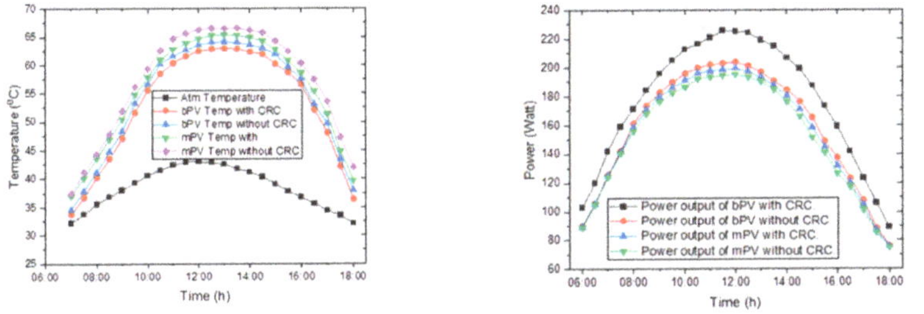

(a) (b)

Figure 3. Comparison of PV module temperature (**a**) and power output (**b**).

The peak power of the bifacial photovoltaic module is enhanced by 3.29% with the usage of the cool roof coating, as shown in Figure 4a. The bifacial photovoltaic module also outperforms the monofacial photovoltaic module with and without the cool roof coating, by 5.28% and 5.52%, respectively. The efficiency of the bifacial photovoltaic module is more than that of the monofacial photovoltaic module because of the rear irradiance and the lower module temperature due to the transparency of the bifacial photovoltaic module. The bifacial gain of the bifacial photovoltaic module is shown in Figure 4b. The bifacial gain measured at peak power is 15.6%. The highest bifacial increase was 21.68% in the evening.

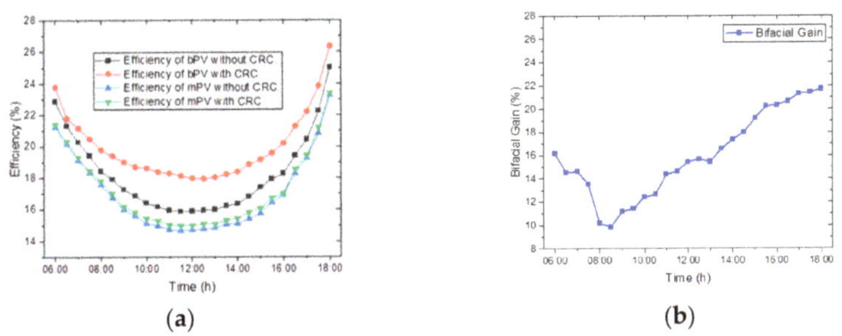

(a) (b)

Figure 4. Efficiency of PV modules (**a**) and the bifacial gain of the bifacial photovoltaic module (**b**).

4. Conclusions

The purpose of this study was to investigate the impact of a cool roof coating on the performance of bPV and mPV modules which was developed by using TiO_2 and FC resin.

A considerable increase in power and efficiency has been observed for the bifacial PV module. According to the experimental findings, cool roof coatings provide enough reflection (albedo) to improve the efficiency of bifacial PV modules. It is observed that the peak power efficiency of the bifacial PV module increased by 3.29% and ranges up to 18.1%, with the average reflection (albedo) of the cool roof coating being 0.63. A peak power bifacial gain of up to 15.6% is observed, while the highest bifacial gain is 21.68%. The performance of the monofacial PV module is also increased by 0.24% due to the 1.3 °C decrease in the temperature caused by using the cool roof coating.

Author Contributions: K.M.: Conceptualization, methodology, formal analysis, investigation, resources, data curation, B.T. and M.A.: data curation, writing—original draft preparation, A.H.: writing—review and editing, supervision. All authors have read and agreed to the published version of the manuscript.

Funding: This research received no external funding.

Institutional Review Board Statement: Not applicable.

Informed Consent Statement: Not applicable.

Data Availability Statement: Not applicable.

Conflicts of Interest: The authors declare no conflict of interest.

References

1. Shoukry, I.; Libal, J.; Kopecek, R.; Wefringhaus, E.; Werner, J. Modelling of Bifacial Gain for Stand-alone and in-field Installed Bifacial PV Modules. *Energy Procedia* **2016**, *92*, 600–608. [CrossRef]
2. Cuevas, A.; Luque, A.; Eguren, J.; del Alamo, J. 50 Per cent more output power from an albedo-collecting flat panel using bifacial solar cells. *Sol. Energy* **1982**, *29*, 419–420. [CrossRef]
3. Lo, C.K.; Lim, Y.S.; Rahman, F.A. New integrated simulation tool for the optimum design of bifacial solar panel with reflectors on a specific site. *Renew. Energy* **2015**, *81*, 293–307. [CrossRef]
4. Rodriguez-Pastor, D.A.; Ildefonso-Sanchez, A.F.; Soltero, V.M.; Peralta, M.E.; Chacartegui, R. A new predictive model for the design and evaluation of bifacial photovoltaic plants under the influence of vegetation soils. *J. Clean. Prod.* **2023**, *385*, 135701. [CrossRef]
5. Ganesan, K.; Winston, D.P.; Sugumar, S.; Jegan, S. Performance analysis of n-type PERT bifacial solar PV module under diverse albedo conditions. *Sol. Energy* **2023**, *252*, 81–90. [CrossRef]
6. Taomoto, Y.; Hosokawa, K.; Yagami, M.; Hanzawa, H.; Ohkawa, T.; Iwamoto, K. Bifacial tracking system in snowy region. In Proceedings of the BiFiPV 2016 Workshop, Miyazaki, Japan, 29–30 September 2016.
7. Santamouris, M.; Synnefa, A.; Karlessi, T. Using advanced cool materials in the urban built environment to mitigate heat islands and improve thermal comfort conditions. *Sol. Energy* **2011**, *85*, 3085–3102. [CrossRef]
8. Sun, X.; Khan, M.R.; Deline, C.; Alam, M.A. Optimization and performance of bifacial solar modules: A global perspective. *Appl. Energy* **2018**, *212*, 1601–1610. [CrossRef]
9. He, Y.; Xia, Z.; Wang, R.; Yan, Y.; Hu, K.; Sun, H.; Liu, X. An easily prepared and long-term effective cooling coating that can be cooled to sub-ambient temperature without polyethylene film. *Sol. Energy* **2022**, *246*, 1–13. [CrossRef]

engineering proceedings

MDPI

Proceeding Paper

Effect of Storage Reservoir on Fluid Velocity in Lateral Flow Paper Device †

Hamza Abbas, Mubashar Ali, Nauman Naeem, Hammas Ullah, Moazzam Ali and Ali Turab Jafry *

Faculty of Mechanical Engineering, Ghulam Ishaq Khan Institute of Engineering Sciences and Technology, Topi 23460, Pakistan; gme2223@giki.edu.pk (H.A.); gme2187@giki.edu.pk (M.A.); gme2331@giki.edu.pk (N.N.); hammas.ullah@giki.edu.pk (H.U.); gme2330@giki.edu.pk (M.A.)

* Correspondence: ali.turab@giki.edu.pk

† Presented at the Third International Conference on Advances in Mechanical Engineering 2023 (ICAME-23), Islamabad, Pakistan, 24 August 2023.

Abstract: Paper-based microfluidics provides a versatile platform for numerous applications, including diagnostics at the point of care, environmental detection, and food quality control. Utilizing paper as a substrate offers numerous benefits, including flexibility, biodegradability, low cost, and capillary-driven flow. Despite significant advancements in paper-based microfluidic devices, regulating fluid flow effectively within the paper channel remains a persistent challenge. In order to resolve this problem, our research proposes a method for manipulating fluid flow in the paper channel by altering the surface contact between the storage medium and the paper channel. According to the results, increasing the contact area between the reservoir and the paper surface decreases the flow velocity and increases the fluidic time delay. This is due to the increased diffusion area between the storage medium and the paper channel, which creates resistance to liquid flow. Therefore, the liquid takes longer to travel through the paper channel, resulting in a decrease in velocity relative to contact areas of 50% or 70%. We are confident that this method of flow control can enable lateral flow sensing devices with enhanced flow variation options for use in medical, environmental, or food quality applications.

Keywords: microfluidics; paper-based microfluidics; contact area; flow control; reservoirs

Citation: Abbas, H.; Ali, M.; Naeem, N.; Ullah, H.; Ali, M.; Jafry, A.T. Effect of Storage Reservoir on Fluid Velocity in Lateral Flow Paper Device. *Eng. Proc.* **2023**, *45*, 5. https://doi.org/10.3390/engproc2023045005

Academic Editors: Mohammad Javed Hyder, Muhammad Mahabat Khan, Muhammad Irfan and Manzar Masud

Published: 8 September 2023

1. Introduction

Microfluidics is an interdisciplinary field concerned with the precise manipulation and control of microscopic fluid volumes [1]. It involves the study and development of devices and systems capable of managing microliter or nanoliter volumes of fluids [2]. Microfluidics enables a variety of applications, including chemical analysis, biomedical diagnostics, drug delivery, and DNA sequencing, by capitalizing on the unique properties of fluids at these scales. The ability of this technology to miniaturize and automate laboratory processes has revolutionized industries, providing benefits such as decreased sample and reagent consumption, quicker analysis times, and increased portability [3]. Microfluidics has enormous potential for advancing research, diagnostics, and technological advancements in a variety of fields. The field of microfluidics includes a subfield known as "paper-based microfluidics", which employs paper or porous membranes to facilitate fluid transport via capillary action.

First, paper-based microfluidics can operate without pumps or additional apparatus, making them portable and self-contained. This feature facilitates preparation and simplifies the experimental or diagnostic procedure. In addition, these devices are less expensive than conventional microfluidic devices, making them accessible to a broader audience. Furthermore, the disposal of paper-based microfluidic devices is easier and more eco-friendly [4]. Another notable advantage of paper is the simplicity of production. The manufacturing process is relatively simple and can be accomplished using standard printing techniques or straightforward cutting and folding techniques. This accessibility permits rapid prototype development and scalability [5]. These point-of-care devices offer expeditious and dependable outcomes for a wide range of medical conditions, enabling timely interventions and ultimately improving patient outcomes [6].

The manipulation and regulation of fluid flow within paper substrates have been the subject of extensive research, employing diverse methodologies. For instance, in 2010, Fu et al. demonstrated a successful approach by modifying the dimensions of the filter paper, thereby influencing the fluid flow velocity through the material [7]. Similarly, Gao et al. explored the control of flowrate through the adoption of various paper cutting techniques [8], whereas Mendez et al. employed two-dimensional patterning on the paper surface to achieve flow control [9]. Another notable method involved wax treatment to alter surface wettability in paper, effectively retarding the flow of liquids [10]. These recent advancements in the realm of flow manipulation within paper-based microfluidic systems have found practical applications in diverse fields, including food safety.

However, despite the significant advancements that have been made in flow control techniques for paper microfluidics, there has been a discernible lack of investigation into the impact the reservoirs in paper fluidics. Hence, in this research, we present a new way to alter the flowrate in porous membranes (filter paper). We investigate fluid flow through a porous medium, especially concentrating on how modifying the contact between the storage reservoir and paper strip may affect the flowrate in the paper channel. This is carried out so that we can better understand how a fluid can flow through a porous medium. We hope this will contribute to the current body of knowledge and improve the adaptability of lateral flow strips for different biomedical and food quality applications.

2. Theoretical Background

The correlation between the travel distance (L) and time (t) for one-dimensional flow in paper, as outlined by the Lucas–Washburn equation, is illustrated by Equation (1) presented hereinafter. This equation posits that the distance traversed L is directly proportional to $\sqrt{t}$ [11].

$$L = \sqrt{\frac{\gamma r \cos(\theta)}{2\mu}} \cdot \sqrt{t}$$

This equation incorporates several variables such as contact angle (θ), time (t), the liquid's surface tension (γ), dynamic viscosity (μ), and pore size (r). Another famous model for porous media for the fully wetted condition is given by Darcy's law [12].

$$Q = -\frac{kA\Delta P}{\mu l_0}$$

where Q is the volumetric flowrate, A is the cross-sectional area across flow direction, k is the permeability, and ΔP represents the pressure differential across the system.

3. Materials and Methods

The materials utilized in the experiment consisted of Whatman filter paper No. 1 and an absorbing pad. The working fluid employed was octanoic acid. Figure 1 illustrates the experimental setup, featuring an infinite reservoir volume and adjustable contact between the absorbing pad (serving as the storage reservoir) and the paper strip. A paper channel was cut and connected to this reservoir, wherein the length of fluid travel needs to be measured. The length, width, and thickness of the paper strip were 50 mm, 17 mm, and 0.18 mm, respectively, and those of the absorbent pad were 10 mm, 17 mm, and 1 mm, respectively. The absorbent pad and the paper strip were cut with a CO_2 laser. ImageJ was utilized for measuring the length of travel using image analysis.

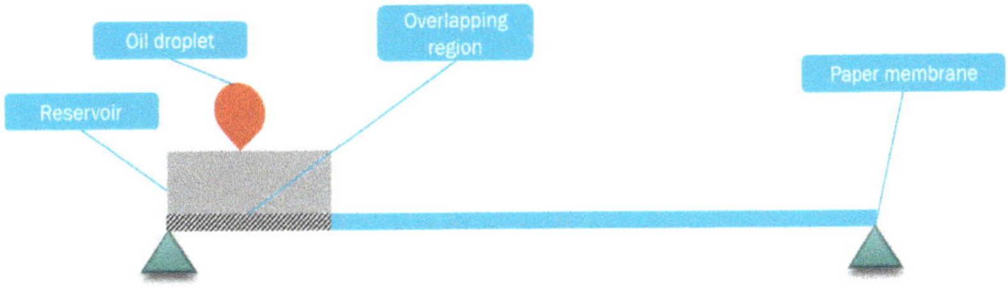

Figure 1. Experimental setup with reservoir and paper membrane connected for fluid flow with varying contact area.

4. Results and Discussion

In Figure 2, the length–time curve of the experiment is depicted, focusing on the utilization of an infinite reservoir with paper strips and varying the contact area between the reservoir and the paper strip. This study explores the impact of the contact area while maintaining an infinite fluid volume within the storage reservoir. We observed that, initially, the gradient of length is steep, indicating fast flow. The contact areas of 20% and 50% are nearly identical and differ as the flowrate decreases with time. The minimum flowrate achieved is at 100% contact and the maximum flowrate is achieved at 50% contact between the storage reservoir and the paper strip. The reason is that the fluid must travel from the absorbing pad into the paper strip and then in the whole length of the paper, because as the contact is increased, more fluid must be diffused into the paper strip to start the fluid flow; hence, the flowrate decreases. At 20% contact, the storage has an ample amount of fluid however, because of the lower contact area, the flowrate is lower than the maximum flowrate at 50% contact area. Hence, we can use this factor to increase or decrease the flowrate in the paper strip.

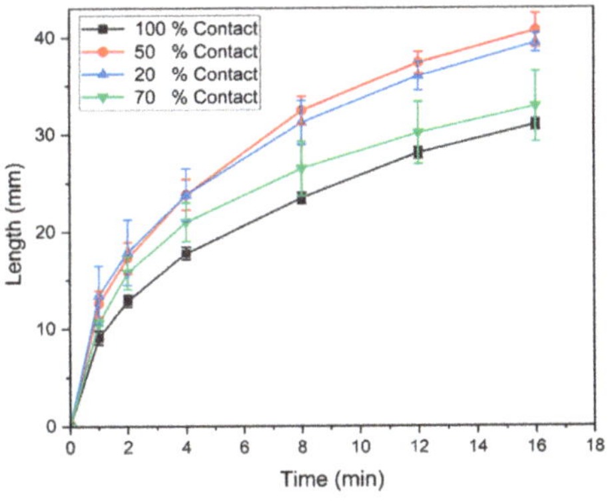

Figure 2. Flowrate at different contact areas between the reservoir and paper channel.

5. Conclusions

In summary, our study introduces a new method for flow variation by manipulating the contact area between the storage reservoir and the paper strip. We deduce that in addition to the length of the paper channel, the contact area among the reservoir and main channel, as well as the fluid available in the reservoir, affects the dynamics of imbibition in porous membrane. We have successfully demonstrated that the lowest and highest flowrates within the paper strip correspond to 100% and 50% contact. These findings enable the addition of lateral flow paper strips to benefit from flow control. By precisely adjusting the contact area, we can optimize flowrates and enhance the accuracy of environmental data collection. Overall, our research offers new insight into flow control and its potential applications in various domains.

Author Contributions: Conceptualization, H.A., M.A. (Mubashar Ali) and A.T.J.; methodology, H.A. and M.A. (Mubashar Ali); validation, N.N. and H.U.; formal analysis, H.A.; resources, M.A. (Moazzam Ali); writing—original draft preparation, H.A. and M.A. (Mubashar Ali); writing—review and editing, H.A. and A.T.J.; supervision, A.T.J.; project administration, A.T.J.; funding acquisition, A.T.J. All authors have read and agreed to the published version of the manuscript.

Funding: We are extremely grateful for the financial support by the National Research Programs for Universities (NRPU-14816) and the Technology Transfer Support Fund (TTSF-74) kindly provided by the Higher Education Commission (HEC) of Pakistan.

Institutional Review Board Statement: Not applicable.

Informed Consent Statement: Not applicable.

Data Availability Statement: The data presented in this study are available on request from the corresponding author. The data is not publicly available due to confidentiality.

Conflicts of Interest: The authors declare no conflict of interest.

References

1. Tian, W.-C.; Finehout, E. Introduction to Microfluidics. In *Microfluidics for Biological Applications*; Springer: Berlin/Heidelberg, Germany, 2008; pp. 1–34.
2. Cottet, J.; Renaud, P. Introduction to Microfluidics. In *Drug Delivery Devices and Therapeutic Systems*; Elsevier: Amsterdam, The Netherlands, 2021; pp. 3–17.
3. Dietzel, A. A brief introduction to microfluidics. In *Microsystems for Pharmatechnology: Manipulation of Fluids, Particles, Droplets, and Cells*; Springer Nature: Berlin, Germany, 2016; pp. 1–21.
4. Li, X.; Tian, J.; Garnier, G.; Shen, W. Fabrication of paper-based microfluidic sensors by printing. *Colloids Surf. B Biointerfaces* **2010**, *76*, 564–570. [CrossRef] [PubMed]
5. Gong, M.M.; Sinton, D. Turning the page: Advancing paper-based microfluidics for broad diagnostic application. *Chem. Rev.* **2017**, *117*, 8447–8480. [CrossRef] [PubMed]
6. Soum, V.; Park, S.; Brilian, A.I.; Kwon, O.S.; Shin, K. Programmable paper-based microfluidic devices for biomarker detections. *Micromachines* **2019**, *10*, 516. [CrossRef] [PubMed]
7. Fu, E.; Lutz, B.; Kauffman, P.; Yager, P. Controlled reagent transport in disposable 2D paper networks. *Lab Chip* **2010**, *10*, 918–920. [CrossRef] [PubMed]
8. Gao, F.; Mei, D.; Wang, Y.; Szanyi, J.; Peden, C.H. Selective catalytic reduction over Cu/SSZ-13: Linking homo-and heterogeneous catalysis. *J. Am. Chem. Soc.* **2017**, *139*, 4935–4942. [CrossRef] [PubMed]
9. Elías, A.L.; Botello-Méndez, A.R.; Meneses-Rodríguez, D.; Jehová González, V.; Ramírez-González, D.; Ci, L.; Muñoz-Sandoval, E.; Ajayan, P.M.; Terrones, H.; Terrones, M. Longitudinal cutting of pure and doped carbon nanotubes to form graphitic nanoribbons using metal clusters as nanoscalpels. *Nano Lett.* **2010**, *10*, 366–372. [CrossRef] [PubMed]
10. Noh, H.; Phillips, S.T. Fluidic timers for time-dependent, point-of-care assays on paper. *Anal. Chem.* **2010**, *82*, 8071–8078. [CrossRef] [PubMed]
11. Cai, J.; Jin, T.; Kou, J.; Zou, S.; Xiao, J.; Meng, Q. Lucas–Washburn equation-based modeling of capillary-driven flow in porous systems. *Langmuir* **2021**, *37*, 1623–1636. [CrossRef] [PubMed]
12. Gao, Y.; Lin, Q.; Bijeljic, B.; Blunt, M.J. Pore-scale dynamics and the multiphase Darcy law. *Phys. Rev. Fluids* **2020**, *5*, 013801. [CrossRef]

engineering proceedings

Proceeding Paper

Comparative Investigation of the Thermal Conductivity of Water-Based Nanofluids with and without the Combination of Alumina and Carbon Nanotubes †

Muhammad Talha Bin Shoaib, Hafiz Muhammad Rizwan, Muhammad Naveed Gull, Muhammad Taha Khan, Muhammad Tahir Qureshi, Muhammad Shayan Ali Malik and Taqi Ahmad Cheema *

Faculty of Mechanical Engineering, Ghulam Ishaq Khan Institute of Engineering Sciences and Technology, Topi 23460, Pakistan; u2020368@giki.edu.pk (M.T.B.S.); hafiz.rizwan@giki.edu.pk (H.M.R.); gme2220@giki.edu.pk (M.N.G.); u2020363@giki.edu.pk (M.T.K.); u2020364@giki.edu.pk (M.T.Q.); u2020357@giki.edu.pk (M.S.A.M.)

* Correspondence: tacheema@giki.edu.pk

† Presented at the Third International Conference on Advances in Mechanical Engineering 2023 (ICAME-23), Islamabad, Pakistan, 24 August 2023.

Abstract: In the present study, the thermal conductivity of two distinct water-based nanofluids of single-walled CNTs and Al_2O_3, as well as their hybrid solution, was investigated experimentally. The Al_2O_3 and CNTs nanoparticle concentrations are taken to be $2\%v/v$, while the hybrid solution contained $2\%v/v$ of both Al_2O_3 and CNTs nanoparticles. A PSS (Polly styrene sulphonic acid) solution was used as a surfactant to increase the suspension time of the nanoparticles to avoid sedimentation. The dispersion and breaking of the particles of CNT and Al_2O_3 into nano size were performed employing a probe sonicator and bath sonicator. Moreover, a hot plate magnetic stirrer was used to obtain a consistent liquid mixture. The experiments are performed on the Computer Controlled Thermal Conductivity of Liquid and Gases (TCLGC) unit available in the heat transfer lab at GIKI. The results concluded that the thermal conductivity of water-based single-walled CNT nanofluids was higher compared to Al_2O_3 and their hybrid solution. Therefore, Al_2O_3 and a hybrid solution are less desirable for thermal conduction compared to CNTs.

Keywords: thermal conductivity; CNTs; Al_2O_3; hybrid solution; nanofluids

Citation: Shoaib, M.T.B.; Rizwan, H.M.; Gull, M.N.; Khan, M.T.; Qureshi, M.T.; Malik, M.S.A.; Cheema, T.A. Comparative Investigation of the Thermal Conductivity of Water-Based Nanofluids with and without the Combination of Alumina and Carbon Nanotubes. *Eng. Proc.* **2023**, *45*, 21. https://doi.org/10.3390/engproc2023045021

Academic Editors: Mohammad Javed Hyder, Muhammad Mahabat Khan, Muhammad Irfan and Manzar Masud

Published: 11 September 2023

1. Introduction

Enhancing thermophysical characteristics has been the primary concern of contemporary nanofluid technology development. This is because many industrial applications, such as those in power production, air conditioning, vehicles, and solar collectors, undoubtedly make use of heat transfer processes. The heat transfer rate in thermal systems is constrained by the low thermal conductivity of heat transfer fluids like water. Therefore, to enhance the thermal conductivity of water, a suitable volume fraction of nanoparticles can be used to improve the heat transfer process [1]. The thermal conductivity of nanofluids depend on many factors, such as the size of nanoparticles, the concentration of nanoparticles, and the shape of nanoparticles [2]. The thermal conductivity of hybrid solutions mostly shows that the hybrid solution increases the thermal conductivity of the solution. For different nanofluids, different volume ratios have been observed for maximum heat transfer, and at different temperatures, the thermal conductivity of hybrid solutions differs [3].

Al_2O_3 and water have been used to enhance the thermal conductivity of the solution. Several other tests have been conducted on Al_2O_3 to determine the optimal size and shape for maximum thermal conductivity [4,5]. The carbon nanotubes were classified based on the number of layers: single-walled and multi-walled. Their thermal conductivity was mostly dependent on their molecular network and density. The criteria for thermal conductivity in carbon nanotubes changed with the change in the number of layers [6]. In the present study, the TCLGC was used to measure the thermal conductivity of water-based nanofluids with Al_2O_3 and single-walled CNTs and their hybrid solution to determine which solution was the best possible conductor.

2. Materials and Methods

2.1. Characterization of Nanofluids

The experiments were set up to characterize three different samples of nanofluids. A single-walled CNT $2\%v/v$ solution was the first sample of the nanofluid, as shown in Figure 1a, and the second was a $2\%v/v$ solution of Al_2O_3 in Figure 1b. The third one was the $(2 + 2)\%v/v$ hybrid solution of Alumina and CNTs, as shown in Figure 1c. A PSS (poly-styrene sulfonic acid) solution, as shown in Figure 2a, was used as a surfactant to avoid the occurrence of sedimentation. The PSS formed a negative charge layer, which produced electrostatic repulsion in the solution, preventing the suspension from settling out. The probe sonicator, in Figure 2b was used to break down particles into nanosize and evenly disperse them into the solution. The bath sonicator, Figure 2c, and hot plate magnetic stirrer in Figure 2d were used to mix the solution. The bath sonicator generated high-frequency waves, which produced intense mechanical forces that agitated the liquid, creating cavitation bubbles, which mixed the nanoparticles with the liquid. The magnetic stirrer continuously stirred the solution to obtain a homogenous liquid mixture. The solution was stirred at 1000 rpm for half an hour in all three samples. The settling time of the nanoparticles was found to be 72 h for Alumina and 36 h for CNTs.

2.2. Experimental Analysis

The apparatus used for the investigation of thermal conductivity in different nanofluids is shown in Figure 3. The Computer Controlled Thermal Conductivity of Liquid and Gases Unit "TCLGC" shown in Figure 3 consists of an aluminum cylinder, a brass jacket, a heating power element, and temperature and flow sensors. The aluminum cylinder serves as the core body of the apparatus. There are three temperature sensors, ST-1, ST-2, and ST-3, which are installed at different distances of the radius. There is also a variable heating power element (AR-1) to adjust the heat generation. A brass jacket is placed on the outer radius of the cylinder, which is equipped with a cooling water system. The governing equations for heat conduction and thermal conductivity are given as follows: The heat conducted is calculated employing Equation (1)

$$\dot{Q}_{conducted} = \dot{Q}_{generated} - \dot{Q}_{lost} \tag{1}$$

The thermal conductivity is calculated using Equation (2)

$$k = \frac{\dot{Q}_{conducted}}{2\pi L(T_i - T_e)} \ln(\frac{r_e}{r_i}) \tag{2}$$

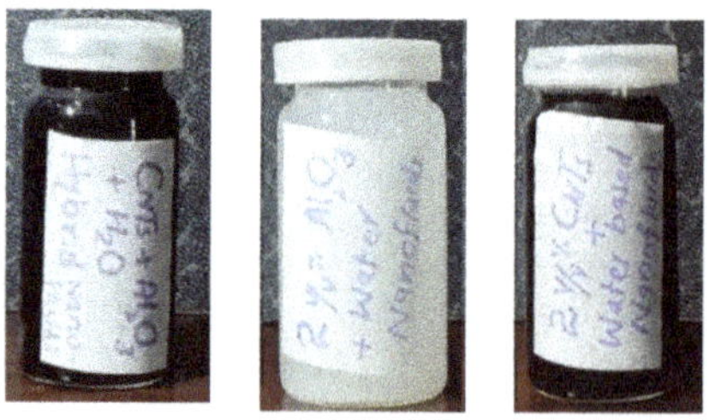

(a) (b) (c)

Figure 1. Samples of water-based nanofluids: (**a**) Hybrid solution (2% + 2%) v/v of Al_2O_3 and CNT; (**b**) $2\%v/v$ solution of Al_2O_3 solution; (**c**) $2\%v/v$ solution of single-walled CNT.

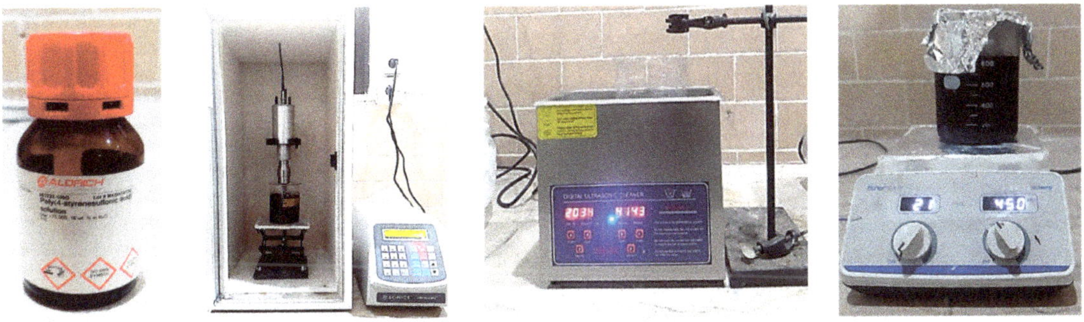

(**a**) (**b**) (**c**) (**d**)

Figure 2. Synthesis of nanofluids: (**a**) PSS; (**b**) Probe sonicator; (**c**) Bath sonicator; (**d**) Hot plate. magnetic stirrer.

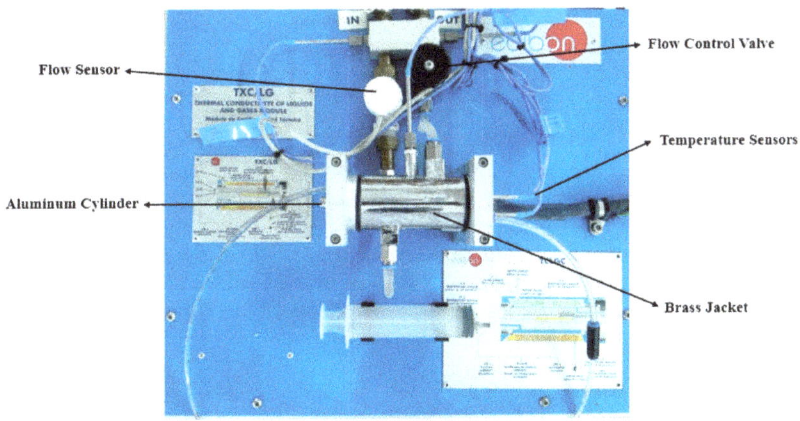

Figure 3. The Computer Controlled Thermal Conductivity of Liquid and Gases Unit.

3. Results and Discussion

Figure 4 represents the results of the experiments performed for the thermal conductivity of the different nanofluids. The relationship between temperature difference and the thermal conductivity of the nanofluids is shown in Figure 4a, whereas the relationship between the heat transferred to the working fluid and the thermal conductivity of the different samples of nanofluids is shown in Figure 4b. It can be seen from Figure 4a that at all values of temperature difference, it was evident that water-based Al_2O_3 nanofluids had a lower thermal conductivity than all the other samples because, at the nanoscale, the alumina particles experience a phonon-scattering phenomenon in fluids that reduce the thermal energy transfer capacity of water-based alumina nanofluids. For all the samples, the difference in the value of thermal conductivity in the nanofluids was greater at lower temperature differences. However, when the temperature difference increased, the variation in thermal conductivity tended to become smaller for all the samples. This is because larger heat fluxes are provided to heat transfer fluids in their flow domain and improve the rate of conduction in heat transfer. In Figure 4b, the effect of heat conduction on the thermal conductivity of different nanofluids is presented. The results show similar trends as seen in Figure 4a, as at low values of heat conduction, the value of thermal conductivity in all the samples was higher; however, as the value of heat conduction increased, the variation in the value of thermal conductivity became smaller. It can be seen from the results that the nanofluid solution of CNTs had the highest value of thermal conductivity compared to the hybrid solution and the Alumina. Therefore, it can be concluded that water-based CNT nanofluids are suitable for thermal conduction compared to water-based Al_2O_3 nanofluids and their hybrid solution.

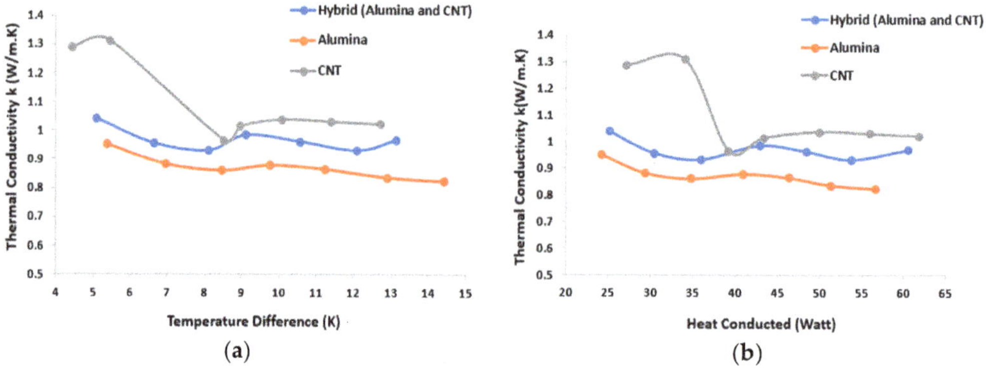

Figure 4. Effect of heat transfer parameters on the thermal conductivity of water-based nanofluids of CNTs and Al_2O_3 and their hybrid solution: (**a**) Temperature difference; (**b**) Heat conducted.

4. Conclusions

The present study successfully examines the water-based nanofluids of CNTs and Al_2O_3 and their hybrid solution. The CNTs of water-based nanofluids showed the greatest improvement in thermal conductivity compared to the hybrid solution consisting of both CNTs and Al_2O_3 nanoparticles. However, the solution of Al_2O_3 nanoparticles showed the least improvement in thermal conductivity. Overall, the findings from this investigation highlight the potential of using nanoparticles, particularly CNTs, to enhance the thermal conductivity of water.

Author Contributions: Conceptualization, T.A.C. and H.M.R.; methodology, M.T.B.S.; validation, M.N.G., M.T.K. and M.T.Q.; formal analysis, M.S.A.M.; investigation, M.T.B.S.; resources, M.T.Q.; writing—original draft preparation, M.T.B.S.; writing—review and editing, H.M.R.; supervision, T.A.C. All authors have read and agreed to the published version of the manuscript.

Funding: This research received no external funding.

Institutional Review Board Statement: Not applicable.

Informed Consent Statement: Not applicable.

Data Availability Statement: The data presented in this study will be available on request from the corresponding author.

Acknowledgments: The authors acknowledge the support of Interdisciplinary Engineering Research Group (IERG) of GIK Institute, Topi, 23460, Pakistan.

Conflicts of Interest: The authors declare no conflict of interest.

References

1. Eastman, J.A.; Choi, U.S.; Li, S.; Thompson, L.J.; Lee, S. Enhanced Thermal Conductivity through the Development of Nanofluids. *MRS Online Proc. Libr. OPL* **1996**, *457*, 3. [CrossRef]
2. Jang, S.P.; Choi, S.U.S. Effects of Various Parameters on Nanofluid Thermal Conductivity. *J. Heat Transf.* **2007**, *129*, 617–623. [CrossRef]
3. Sajid, M.U.; Ali, H.M. Thermal conductivity of hybrid nanofluids: A critical review. *Int. J. Heat Mass Transf.* **2018**, *126*, 211–234. [CrossRef]
4. Beck, M.P.; Yuan, Y.; Warrier, P.; Teja, A.S. The effect of particle size on the thermal conductivity of alumina nanofluids. *J. Nanopart. Res.* **2009**, *11*, 1129–1136. [CrossRef]

5. Timofeeva, E.V.; Routbort, J.L.; Singh, D. Particle shape effects on thermophysical properties of alumina nanofluids. *J. Appl. Phys.* **2009**, *106*, 014304. [CrossRef]
6. Kumanek, B.; Janas, D. Thermal conductivity of carbon nanotube networks: A review. *J. Mater. Sci.* **2019**, *54*, 7397–7427. [CrossRef]

engineering proceedings

MDPI

Proceeding Paper

Dynamic Contact Angle Variation with Applied Voltage and Droplet Volume in Digital Microfluidics †

Moazzam Ali, Hammas Ullah, Mubashar Ali, Nauman Naeem, Hamza Abbas and Ali Turab Jafry *

Faculty of Mechanical Engineering, Ghulam Ishaq Khan Institute of Engineering Sciences and Technology, Topi 23460, Pakistan; gme2330@giki.edu.pk (M.A.); hammas.ullah@giki.edu.pk (H.U.); gme2187@giki.edu.pk (M.A.); gme2331@giki.edu.pk (N.N.); gme2223@giki.edu.pk (H.A.)

* Correspondence: ali.turab@giki.edu.pk

† Presented at the Third International Conference on Advances in Mechanical Engineering 2023 (ICAME-23), Islamabad, Pakistan, 24 August 2023.

Abstract: Digital microfluidics allows for controlled droplet movement by applying AC or DC voltages. In this research, we investigated the dynamic contact angle variation of droplets at different voltages and droplet volumes in a digital microfluidics platform. Volumes of 10 μL, 14 μL, and 18 μL were investigated with voltages ranging from 250 V to 400 V. The goal was to investigate how variations in voltage and droplet volume affected the contact angle, specifically by tracking variations in the advancing and receding contact angles. The findings showed that as voltage rises, the contact angle decreases more noticeably in terms of both advancing and receding angles. This shows that higher voltages boost the electrowetting effect and lead to better droplet dispersal and substrate wetting. Furthermore, it was found that across the studied voltage range, the effect of volume on the contact angle was largely consistent. The relation between voltage, volume, and contact angle in electrowetting is better understood from our analysis.

Keywords: digital microfluidics; electrowetting; contact angle; voltage; droplet volume

Citation: Ali, M.; Ullah, H.; Ali, M.; Naeem, N.; Abbas, H.; Jafry, A.T. Dynamic Contact Angle Variation with Applied Voltage and Droplet Volume in Digital Microfluidics. *Eng. Proc.* **2023**, *45*, 31. https://doi.org/10.3390/engproc2023045031

Academic Editors: Mohammad Javed Hyder, Muhammad Mahabat Khan, Muhammad Irfan and Manzar Masud

Published: 12 September 2023

1. Introduction

Digital microfluidic platforms have emerged as powerful tools for the precise manipulation and control of minute liquid droplets in diverse scientific and technical domains [1]. Using different electrodes, these platforms generate electric fields that govern the behavior and movement of droplets on surfaces. The contact angle, a fundamental parameter denoting the angle at the liquid–solid interface, plays a crucial role in influencing the behavior, stability, and wetting properties of these droplets [2]. Digital microfluidic (DMF) devices offer several advantages, such as accelerated heat transfer, reduced reagent consumption, and the potential for seamless integration [3].

The contact angle and droplet behavior are also greatly affected by droplet volume. The size, shape, and surface tension of the droplet are all influenced by its volume, which also influences the contact angle and forces driving droplet motion. It is essential to comprehend how voltage and volume together affect the contact angle to optimize the functionality and dependability of digital microfluidics. While previous studies have investigated the effect of either voltage or volume on the contact angle, there is a gap in our understanding of the combined influence of these parameters. Therefore, this research paper aims to systematically investigate and quantify the impact of voltage and volume on the contact angles of droplets using electrowetting. By varying the applied voltage within the range of 250 V to 400 V and utilizing droplets of different volumes, we seek to analyze the resulting changes in the advancing as well as receding contact angles.

The outcomes of this study will provide valuable insights into the fundamental physics governing droplet behavior on digital microfluidic platforms.

2. Theoretical Background

The theoretical background for electrowetting on a dielectric is governed by the Young–Lippmann equation, given below [4].

$$\cos(\theta_V) = \cos(\theta_0) + \frac{\epsilon_r \epsilon_0 V^2}{2\gamma d}$$

Herein, ϵ_r and ϵ_0 are the relative and vacuum permittivity. The thickness of the dielectric layer is represented by d, γ represents the surface tension, and V is the applied voltage. The initial and final contact angles are θ_0 and θ_V.

Young's equation is used to equate forces at the droplet interface as presented in Figure 1 [5].

$$\gamma_{lg} \cos\theta + \gamma_{sl} = \gamma_{sg}$$

$$\cos\theta = \frac{\gamma_{sg} - \gamma_{sl}}{\gamma_{lg}}$$

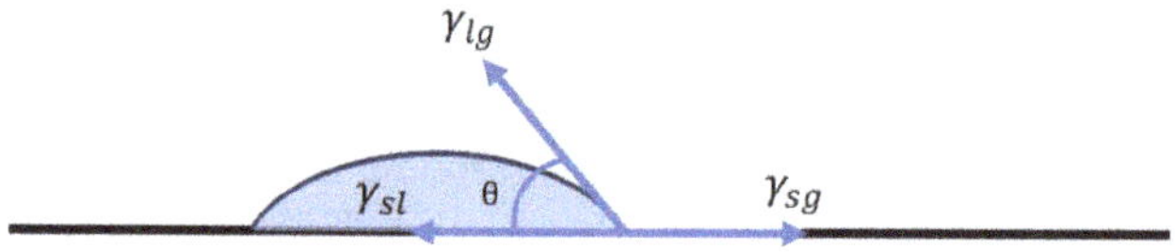

Figure 1. Schematic of surface tension of a liquid droplet acting at the solid–liquid-gas interface.

3. Materials and Methods

Distilled water was used in all experiments. The digital microfluidics platform was fabricated using a printed circuit board (PCB) from Smart PCBs (Rawalpindi, Pakistan). This was coated with grafting tape as the dielectric layer. To ease droplet movement, cooking oil was used to slide the water droplet. A 1000 V DC power supply with 0.5% ripple and maximum current of 1 milliampere was supplied by Qosain Scientific (Lahore, Pakistan).

Figure 2 represents the schematic illustration of our experiment. Once the potential difference was applied, the contact angles and velocity of the droplet were captured by a mobile camera. Quantitative analysis was performed via image analysis using ImageJ software. The size of the electrodes was 2 mm by 2 mm, and the gap between two successive electrodes was 200 µm. Experimentation was performed on water droplets with volumes of 10 µL, 14 µL, and 18 µL. We repeated each experiment 5 times to obtain accurate results.

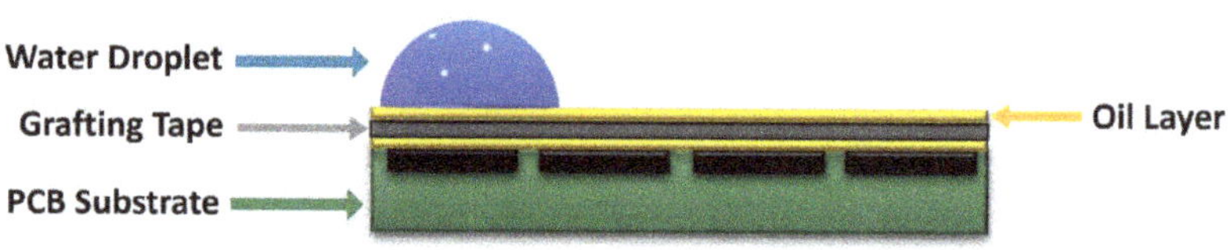

Figure 2. Schematic illustration of the digital microfluidics device.

4. Results and Discussion

The obtained results were analyzed to determine the relationship between voltage, volume, and contact angle. We examined the influence of voltage on the electrowetting effect, which modifies the contact angle (advancing and receding), by applying an electric field, as illustrated in Figures 3 and 4.

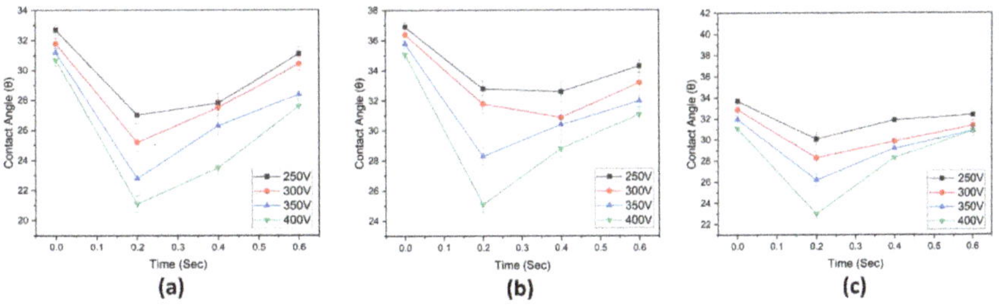

Figure 3. Advancing contact angle of the droplet at different volumes: (**a**) 10 µL, (**b**) 14 µL, (**c**) 18 µL.

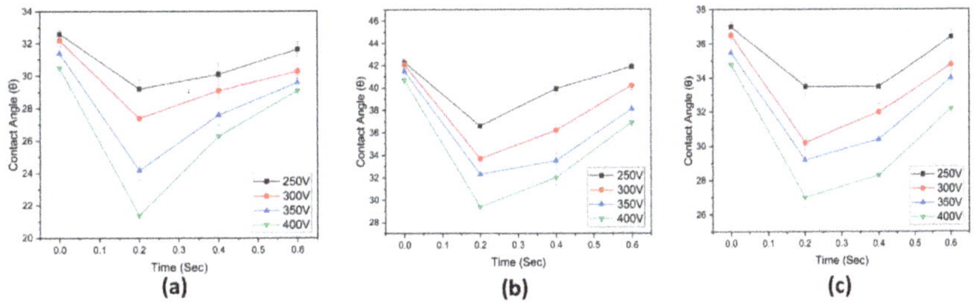

Figure 4. Receding contact angle of the droplet at different volumes: (**a**) 10 µL, (**b**) 14 µL, (**c**) 18 µL.

The results demonstrate that increasing voltage leads to a greater reduction in the advancing as well as the receding contact angles.

As the applied voltage increases, the electric field intensifies at the droplet–substrate interface, leading to enhanced polarization of the liquid molecules in that region. Consequently, the interfacial tension is reduced, prompting the liquid droplet to exhibit increased spreading on the solid surface, resulting in a decrease in the contact angle. The magnitude of this reduction in contact angle becomes more pronounced with higher voltages as the electrowetting effect gains prominence.

When the volume of the droplet is increased (Figure 5), the change in contact angle decreases due to the geometry of the droplet. As the volume of the droplet increases, there is a corresponding widening of its base, leading to an elongation of the three-phase contact line where the droplet interfaces with the solid surface and the surrounding medium. This results in an increase in the total interfacial energy at the contact line. To restore equilibrium, the excess energy arising from the elongated contact line necessitates compensation. The liquid droplet accomplishes this by adopting a larger contact angle. Consequently, with increasing droplet volume, the contact angle expands, leading to a decrease in the change of contact angle with volume.

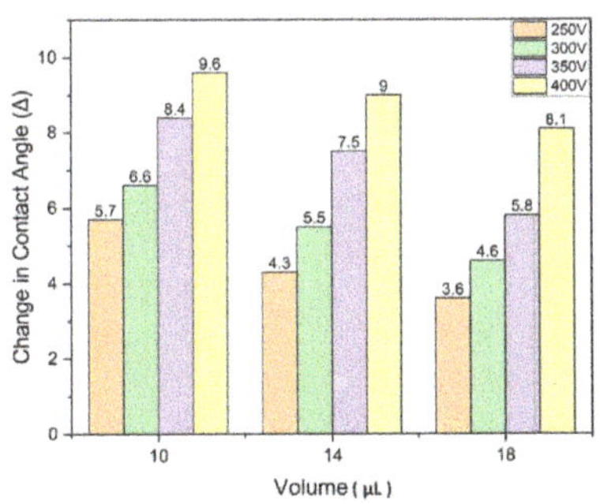

Figure 5. Comparison of change in contact angle with the applied voltage and volume of the water droplet.

5. Conclusions

In conclusion, the findings from this research demonstrate that increasing voltage on a digital microfluidics platform leads to a corresponding increase in the reduction in the contact angle, in terms of both the advancing and receding angles. This indicates that higher voltages enhance the electrowetting effect, resulting in improved droplet spreading and wetting on the substrate. On the other hand, while voltage has a significant influence on the contact angle, changes in droplet volume at a constant voltage show a consistent effect on the contact angle. Specifically, as the droplet volume increases at a constant voltage, the change in the contact angle reduces. These findings contribute to a better understanding of the relationship between voltage, droplet volume, and contact angle in digital microfluidic platforms, enabling researchers to optimize droplet manipulation techniques for improved performance and reliability in number of applications, such as lab-on-a-chip systems, bioassays, and chemical synthesis.

Author Contributions: Conceptualization, M.A. (Moazzam Ali) and H.U.; methodology, M.A. (Moazzam Ali), H.U. and M.A. (Mubashar Ali); software, N.N.; validation, M.A. (Moazzam Ali) and H.U.; investigation, H.A.; resources, H.A.; writing—original draft preparation, M.A. (Moazzam Ali) and H.U.; writing—review and editing, M.A. (Moazzam Ali) and A.T.J.; supervision, A.T.J.; project administration, A.T.J.; funding acquisition, A.T.J. All authors have read and agreed to the published version of the manuscript.

Funding: This research work was funded by the National Research Program for Universities (NRPU-14816) by the Higher Education Commission (HEC) of Pakistan.

Institutional Review Board Statement: Not applicable.

Informed Consent Statement: Not applicable.

Data Availability Statement: The data is available on request from the corresponding author due to confidentiality.

Conflicts of Interest: The authors declare no conflict of interest.

References

1. Jafry, A.T.; Lee, H.; Tenggara, A.P.; Lim, H.; Moon, Y.; Kim, S.-H.; Lee, Y.; Kim, S.-M.; Park, S.; Byun, D.; et al. Double-sided electrohydrodynamic jet printing of two-dimensional electrode array in paper-based digital microfluidics. *Sens. Actuators B Chem.* **2019**, *282*, 831–837. [CrossRef]
2. Sung Kwon, C.; Hyejin, M.; Chang-Jin, K. Creating, transporting, cutting, and merging liquid droplets by electrowetting-based actuation for digital microfluidic circuits. *J. Microelectromechanical Syst.* **2003**, *12*, 70–80. [CrossRef]
3. Choi, K.; Ng, A.H.C.; Fobel, R.; Wheeler, A.R. Digital Microfluidics. *Annu. Rev. Anal. Chem.* **2012**, *5*, 413–440. [CrossRef] [PubMed]
4. Li, J.; Kim, C.-J.C. Current commercialization status of electrowetting-on-dielectric (EWOD) digital microfluidics. *Lab Chip* **2020**, *20*, 1705–1712. [CrossRef] [PubMed]
5. Alistar, M.; Gaudenz, U. OpenDrop: An Integrated Do-It-Yourself Platform for Personal Use of Biochips. *Bioengineering* **2017**, *4*, 45. [CrossRef]

engineering proceedings

MDPI

Proceeding Paper

Uncovering the Cooling Potential by Water Circulation on the Hot Side of a Peltier Module †

Fahad Iqbal, Muhammad Naveed Gull, Hafiz Muhammad Rizwan, Daniyal Amir, Arsalan Saleem, Sami Ul Haq and Taqi Ahmad Cheema *

Faculty of Mechanical Engineering, Ghulam Ishaq Khan Institute of Engineering Sciences and Technology, Topi 23460, Pakistan; u2020114@giki.edu.pk (F.I.); ngull3004@gmail.com (M.N.G.); hafiz.rizwan@giki.edu.pk (H.M.R.); u2020105@giki.edu.pk (D.A.); u2020350@giki.edu.pk (A.S.); u2020087@giki.edu.pk (S.U.H.)

* Correspondence: tacheema@giki.edu.pk

† Presented at the Third International Conference on Advances in Mechanical Engineering 2023 (ICAME-23), Islamabad, Pakistan, 24 August 2023.

Abstract: Thermolectric cooling offers several advantages over conventional refrigeration systems due to its light weight, environmental friendliness, silent operation, and no moving parts. In this work, a thermoelectric water cooling system is created in which water flowing at various flowrates removes heat produced on the hot side of the Peltier module. The cooling effect and coefficient of performance (COP) of the cooling system are experimentally determined at various flowrates of water on the hot side of the module. The cooling effect produced in water increases with the increase in flowrate. A similar trend is noticed for the COP of the system. The maximum cooling effect produced in water is 1363 W at 43 mL/s. The maximum COP of the system is 3.99.

Keywords: thermoelectric cooling; Peltier module; cooling effect; coefficient of performance

Citation: Iqbal, F.; Gull, M.N.; Rizwan, H.M.; Amir, D.; Saleem, A.; Haq, S.U.; Cheema, T.A. Uncovering the Cooling Potential by Water Circulation on the Hot Side of a Peltier Module. *Eng. Proc.* **2023**, *45*, 34. https://doi.org/10.3390/engproc2023045034

Academic Editors: Mohammad Javed Hyder, Muhammad Mahabat Khan, Muhammad Irfan and Manzar Masud

Published: 13 September 2023

1. Introduction

The Peltier effect governs how thermoelectric coolers (TEC) function. According to the Peltier effect, a temperature differential results from applying an electric potential across the junction. The electric current passes across the junctions of the two conductors as a result of the application of voltage. At one connection, heat is emitted, and at another, where cooling takes place, heat is absorbed. When conventional cooling techniques are not appropriate, TEC offers a wide range of advantages. TEC is also more ecologically friendly than other cooling systems on the market. Gökçek et al. [1] investigated the thermoelectric refrigerator's thermal performance using a mini-channel heat sink integrated on the hot side of TEC. The effectiveness of TEC for air heating and cooling applications was investigated experimentally and numerically at various fan speeds and operating voltages by Yilmazoglu et al. [2]. Cosnier et al. [3] worked to develop the thermoelectric cooling and heating system for indoor use. The performance of the system was experimentally tested, and results were validated through a numerical study. Gull et al. [4] developed a thermoelectric cooling module via a combined series and the parallel connection of Peltier modules. The module was numerically simulated, and the results were validated with experimental findings. Siahmargoi et al. [5] proposed a mathematical and thermodynamic model to investigate the performance of TEC. The effect of changing the voltage and the thermal resistance of cold and hot side heat sink on the performance of a TEC was tested. In the present study, a thermoelectric system is developed for water-cooling purposes, and the effect of varying flowrate on the cooling effect of water is investigated.

2. Methodology

Heat is absorbed at the cold-side junction and released at the hot-side junction when the DC current is transmitted through the thermocouples of the Peltier module. The methodology involves cooling the water using the Peltier module. The biggest issue is that if the Peltier module becomes too hot, it could become damaged. When the heat produced on the hot side of the Peltier module is effectively removed, the thermoelectric modules operate at their best. The issue of overheating can be overcome using heat sinks. In this study, a water block was placed on the hot side of the Peltier module, and generated heat was removed via different flowrates of water.

2.1. Experimental Setup

The experimental setup is shown in Figure 1. The Peltier module used in the experimentation is TEC1-12706, which is operated by a 12 V power supply. The water to be cooled down is kept in a small water tank that is mounted on the cold side of the Peltier module, and the water flowing via a water block dissipates the heat produced on the hot side of the Peltier module. The water block is of 40×40 mm^2 cross-section attached by thermal paste on the hot side of the module. A 12 V water pump was placed inside a tub full of water to ensure the flow of water through a pipe. Water flows through pipes and passes through water blocks attached to remove the heat. A flow meter was attached at the end of the pipe and connected through a computer initially to check the flowrate using an Arduino code. A waterproof temperature sensor LM35 was used to determine a change in the temperature of the water that passes through the water block and in a water tank on the top of the Peltier module with the help of an Arduino code and PC setup.

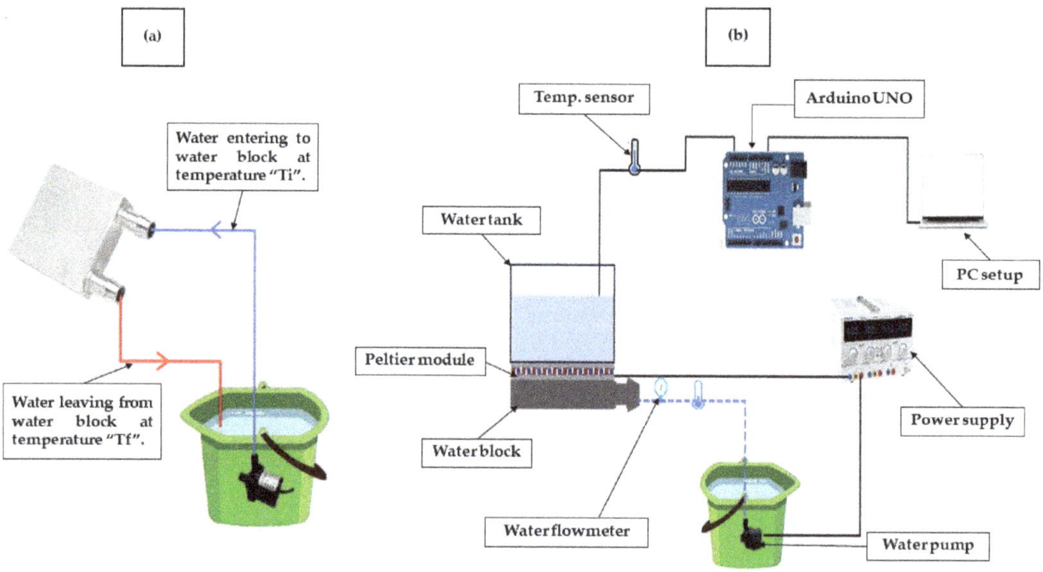

Figure 1. (**a**) Water flow path from water block to water bucket. (**b**) Schematic of experimental setup.

2.2. Performance Parameters

A flow sensor and flow control valve are used to measure and control the flow of water. The temperature of the water can be measured by the LM35 sensor. A mass flowrate of water can be determined using Equation (1),

$$\dot{m} = \dot{V} \times \rho \quad (1)$$

The cooling effect in the present study has been computed using the rate of heat transfer of the cold side by assuming an energy balance between the hot and cold sides of the fluid.

$$\text{Heat transfer rate} = \dot{m}C_p\Delta T \tag{2}$$

The coefficient of performance (COP) of this water-cooling system can be measured by the following relation,

$$\text{COP} = \frac{\text{Heat trasnfer rate}}{\text{Peltier power} + \text{pump power}} \tag{3}$$

Here, the mass flowrate of flowing water through the water block is shown by $\dot{m}$ and $\dot{V}$ is the volumetric flowrate, ρ is the density of the water, C_p is the specific heat at constant pressure, and ΔT is the difference between T_i and T_f. T_i is the temperature of the water before entering the water block, and T_f is the temperature of the water leaving the water block after absorbing the heat.

3. Results and Discussion

The experiments were carried out to determine the effect of varying flowrates of water flowing on the hot side of the module through the water block. Three flowrates of water, 7.5 mL/s, 20.1 mL/s and 43 mL/s, were selected to test the cooling effect, COP, and temperature difference of this water-cooling system. Temperature was directly measured via the temperature sensor. Equations (2) and (3) can be used to determine the cooling effect and COP of the system, respectively. The measured temperature of the water, cooling effect, and COP of the system are shown in Table 1.

Table 1. Experimental results at different flowrates.

Sr. No.	Volumetric Flowrate (ml/s)	Initial Water Temperature (°C)	Final Water Temperature (°C)	Rate of Heat Transfer (W)	Peltier Power (W)	Pump Power (W)	COP
1	7.5	27.82	24.40	226	92	0.9	2.72
2	20.1	27.82	24.97	635	92	1.96	3.94
3	43.0	27.82	25.10	1360	92	3.52	3.99

Figure 2a shows the variation in the cooling effect of the system by varying the flowrate. Equation (2) is used for the calculation of the cooling effect. The effect of flowrate on cooling effect is noticed for three different values of flowrate and linear relation is obtained between them. By increasing the flowrate of water, the cooling effect produced can also be increased. The maximum value of the cooling effect obtained is 1360 W, which is at 43 mL/s. Initially, water to be cooled is at 27.82 °C and at a flowrate of 7.5 mL/s, water temperature decreases to 24.40 °C. At 20.1 mL/s, the temperature difference between the final water temperature and initial water temperature before operating the cooling system decreases to 2.85 °C. Similarly, at 43 mL/s, this temperature difference is 2.72 °C.

Figure 2b shows the variation in COP of the system with flowrate. Similar to the cooling effect, there is an increasing trend of COP due to the varying flowrates. At a low flowrate, the change in the value of COP is significant; the reason for this variation is less pumping power at a low flowrate since the COP is the ratio of cooling effect to power input. At higher flowrate values, the pumping power increases, which results in a slight increase in the value of COP. The maximum value of COP is 3.99 at a flowrate of 43 mL/s.

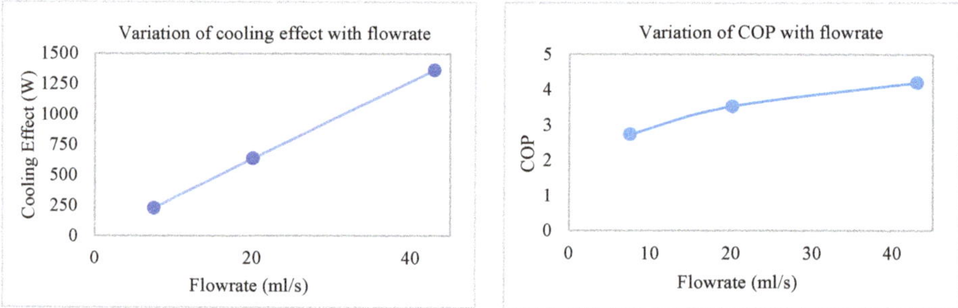

(**a**) (**b**)

Figure 2. Effect of flowrate on (**a**) cooling effect, and (**b**) COP.

4. Conclusions

In this study, the cooling effect produced in water is investigated, by removing the generated heat on Peltier's hotter side through water flowing at different flowrates. The cooling effect, COP and temperature drop of the water are determined at different flowrates. From the results, by increasing the flowrate of water, there is an increase in the cooling effect produced. Variation in flowrate also shows an increasing trend with COP. It is noticed that at the higher flowrate, there is a slight increase in the COP of the system, due to the increase in pumping power.

Author Contributions: Conceptualization, F.I. and M.N.G.; methodology, D.A. and F.I.; validation, D.A., A.S. and S.U.H.; formal analysis, M.N.G. and H.M.R.; investigation, F.I.; resources, D.A.; writing—original draft preparation, M.N.G. and H.M.R.; writing—review and editing, M.N.G. and T.A.C.; supervision, T.A.C.; project administration, T.A.C.; funding acquisition, T.A.C. All authors have read and agreed to the published version of the manuscript.

Funding: The authors acknowledge the support of the Interdisciplinary Engineering, Modelling and Simulation Research Group (IEMSRG) of GIK Institute, Topi, 23460, Pakistan.

Institutional Review Board Statement: Not applicable.

Informed Consent Statement: Not applicable.

Data Availability Statement: Data described in this study will be provided upon request.

Conflicts of Interest: The authors declare no conflict of interest.

References

1. Gökçek, M.; Şahin, F. Experimental performance investigation of minichannel water cooled-thermoelectric refrigerator. *Case Stud. Therm. Eng.* **2017**, *10*, 54–62. [CrossRef]
2. Yilmazoglu, M.Z. Experimental and numerical investigation of a prototype thermoelectric heating and cooling unit. *Energy Build.* **2016**, *113*, 51–60. [CrossRef]
3. Cosnier, M.; Fraisse, G.; Luo, L. An experimental and numerical study of a thermoelectric air-cooling and air-heating system. *Int. J. Refrig.* **2008**, *31*, 1051–1062. [CrossRef]
4. Gull, M.N.; Khan, S.; Rizwan, H.M.; Muzaffar, A.; Cheema, T.A. Experimental and Numerical Investigation of Thermoelectric Cooling Module. In Proceedings of the International Conference on Emerging Power Technologies (ICEPT), Topi, Pakistan, 6–7 May 2023; pp. 1–4. [CrossRef]
5. Siahmargoi, M.; Rahbar, N.; Kargarsharifabad, H.; Sadati, S.E.; Asadi, A. An Experimental Study on the Performance Evaluation and Thermodynamic Modeling of a Thermoelectric Cooler Combined with Two Heatsinks. *Sci. Rep.* **2019**, *9*, 20336. [CrossRef] [PubMed]

engineering proceedings

Proceeding Paper

Refrigeration Potential Investigation of Half-Cycle Refrigeration System Using Liquefied Petroleum Gas †

Ghazanfar Ali [1], Atif Muzaffar [1,*], Taqi Ahmad Cheema [1,*] and Syed Farhad Shah [2]

1 Faculty of Mechanical Engineering, GIK Institute of Engineering Sciences and Technology, Topi 23460, Pakistan; ghazanfar2816@gmail.com

2 Department of Mechanical, Robotics and Energy Engineering, Dongguk University, Seoul 04620, Republic of Korea; sfshah72@gmail.com

* Correspondence: atif.muzaffar@giki.edu.pk (A.M.); tacheema@giki.edu.pk (T.A.C.)

† Presented at the Third International Conference on Advances in Mechanical Engineering 2023 (ICAME-23), Islamabad, Pakistan, 24 August 2023.

Abstract: This study explores the potential of Liquefied Petroleum Gas (LPG) with a half-cycle refrigeration process before combustion. The results show significant cooling efficiency, with a 30 L compartment's temperature dropping from 14.7 °C to 0 °C in 30 min. When using 600 mL of water, its temperature reduced from 12.3 °C to 4.4 °C. The total LPG consumption was 469 g. By extrapolating to a standard 14 kg LPG cylinder, the process could last 14.94 h, cooling 30 drinks from 12.3 °C to 4.4 °C. This study highlights practicality in smaller setups, which is a promising option for energy-efficient cooling using LPG.

Keywords: LPG; half-cycle refrigeration; combustion

Citation: Ali, G.; Muzaffar, A.; Cheema, T.A.; Shah, S.F. Refrigeration Potential Investigation of Half-Cycle Refrigeration System Using Liquefied Petroleum Gas. *Eng. Proc.* **2023**, *45*, 44. https://doi.org/10.3390/engproc2023045044

Academic Editors: Mohammad Javed Hyder, Muhammad Mahabat Khan, Muhammad Irfan and Manzar Masud

Published: 15 September 2023

1. Introduction

Half-cycle LPG refrigeration deviates from the conventional full-cycle vapor compression cycle by employing only two distinct processes instead of the usual four. These processes involve constant-pressure heat addition and isenthalpic expansion, effectively eliminating the need for constant-pressure heat rejection in the condenser and adiabatic compression in the mechanical compressor. In this system, LPG is utilized once and then transferred to an evaporator to absorb heat, after which it is directly burned in a hob or combustion chamber, rendering a condenser unnecessary. Moreover, as the LPG intake originates from a high-pressure cylinder, there is no requirement for a compressor. Consequently, LPG refrigeration only requires half the usual refrigeration components.

Numerous studies have explored the use of LPG as a refrigerant. For example, reference [1] compared the coefficient of performance (COP) of household refrigerators using both R134a and LPG in an international publication. Similarly, another study [2] designed an LPG refrigeration system and conducted an energy comparison with a residential refrigerator. The term "zero-cost refrigeration system" is often associated with half-cycle LPG refrigeration.

A recent study presented the rapid cooling ability of liquid LPG on water temperature, achieved by leveraging its refrigerating properties, even under typical environmental conditions, followed by its eventual evaporation [3]. Another study investigated the performance of a half-cycle air-conditioning system utilizing liquefied petroleum gas (LPG). The study thoroughly assessed the system's efficiency and effectiveness, offering valuable insights for potential applications in cooling technologies [4]. In this setup, the LPG undergoes a transformation into saturated vapors as it passes through the evaporator. However, reverting gaseous LPG back to a liquid state proves costly, making this configuration feasible only when the outgoing LPG is continuously consumed in the burner. The main goal of this project is to use LPG to its best extent, which includes passing it through a refrigerator to produce cooling effects, in addition to using it for burning.

2. Materials and Methods

2.1. Design Methodology

To attain the set objectives and goals, conducting experiments was deemed the most suitable research method, ensuring the effective and precise resolution of the problem at hand. The focus of this study was to fully exploit the potential of LPG found in cylinders for combustion purposes in various daily applications. The complete LPG-based refrigeration system is shown in Figure 1.

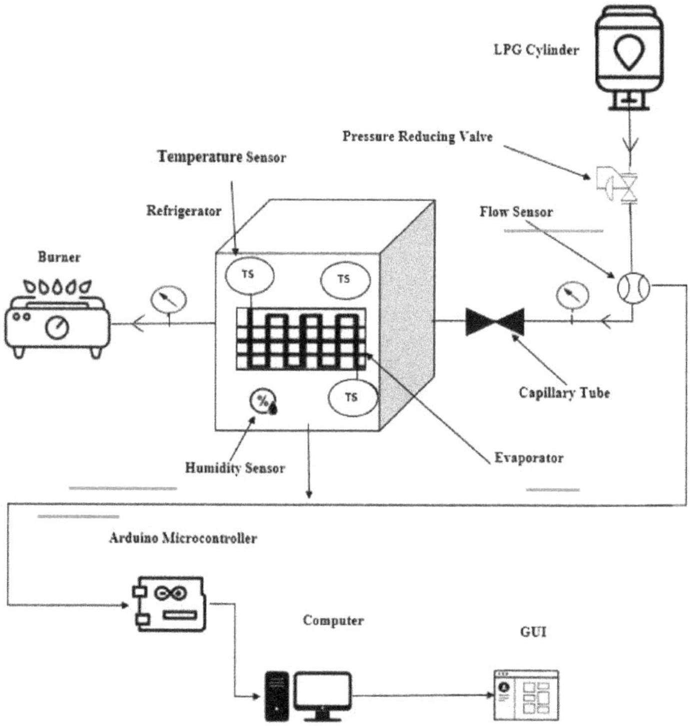

Figure 1. The schematic of the final setup (including automation part).

The cooling effect generated by LPG was determined using the equation as follows:

$$Q = m\, c_p\, (T_2 - T_1) \tag{1}$$

where c_p denotes the specific heat capacity and T_2 and T_1 are the final and initial temperature of the refrigerator compartment, respectively.

2.2. Description of Refrigerator

A small refrigerator with the specifications provided in the table below was chosen for the experiments.

3. Design of Experiment

Two types of experiments were performed using domestic refrigerator refer to Table 1: first without a load in the compartment and the second with a load in the compartment. Each type of experiment was completed in two stages (30 min duration each): the first stage was the cooling of the compartment, and the second stage was the warming of the compartment with a zero-input pressure of LPG. Both types of experiments were performed for different input pressures, ranging from 3 bar up to 5 bar. The temperatures of the compartment and evaporator plate were measured using a digital thermometer and an LM-35, respectively.

Table 1. Refrigerator specification used for experiments.

S. No	Quantity	Value
1	Operating Voltage	100 V
2	Operating Current	1.20 A
3	Operating Frequency	50 Hz/60 Hz
4	Power Input	44 W/60 W
5	Power Consumption	0.34 KWh/24 h
6	Mass of refrigerator	18.5 kg

4. Results and Discussion

A number of experiments were performed by changing the condition of the input pressure. The compartment temperature variation was plotted against time at different input pressures, as shown in Figure 2a–c. Figure 2a shows the results of the experiment with an input pressure of LPG of 4.5 bar and 396 g of consumed LPG. The temperature of the compartment changed from 15.7 °C to −7 °C and that of the evaporator plate temperature changed from 15 °C to −26.6 °C in a 30 min duration of regular supply of LPG. As shown in Figure 2b, during warming, that is, when the supply valve of the LPG is closed, the temperature of the evaporator plate changes rapidly, whereas that of the compartment changes slowly. The temperature of the evaporator plate increased from −26.6 to 0 °C, whereas that of the compartment increased from −7 °C to 1 °C in a duration of 30 min.

Figure 2c shows the experiment with an input pressure of LPG as 4 bar, and 1.5 L of water was taken as a load in the compartment. The temperature of water decreased from 12.8 °C to 9.2 °C in a duration of 30 min. At the same time, the temperature of the compartment decreased from 13 °C to 1 °C, whereas that of the evaporator plate decreased from 12.6 °C to −15.3 °C. From the results and plots, an important pattern is observed between the input pressure and cooling rate of the compartment. In addition to the benefit of the refrigeration effect, the combustion efficiency of LPG increases and CO emissions decrease during burning because of the preheating of LPG before burner. The preheating of LPG takes place in the compartment as a result of heat from the cold drinks and other food stuffs. This study was validated using a standard research article [5].

Based on the data presented in Figure 2 from the two types of experiments, it can be deduced that the temperature of the compartment exhibits a faster rate of change in the first case (without load) compared to the second case (with a load). In the first case, where only air was present in the compartment, the LPG solely absorbed heat from the air. Conversely, in the second case, with both air and water present, the LPG absorbed heat from both sources, resulting in a slower temperature change in the compartment.

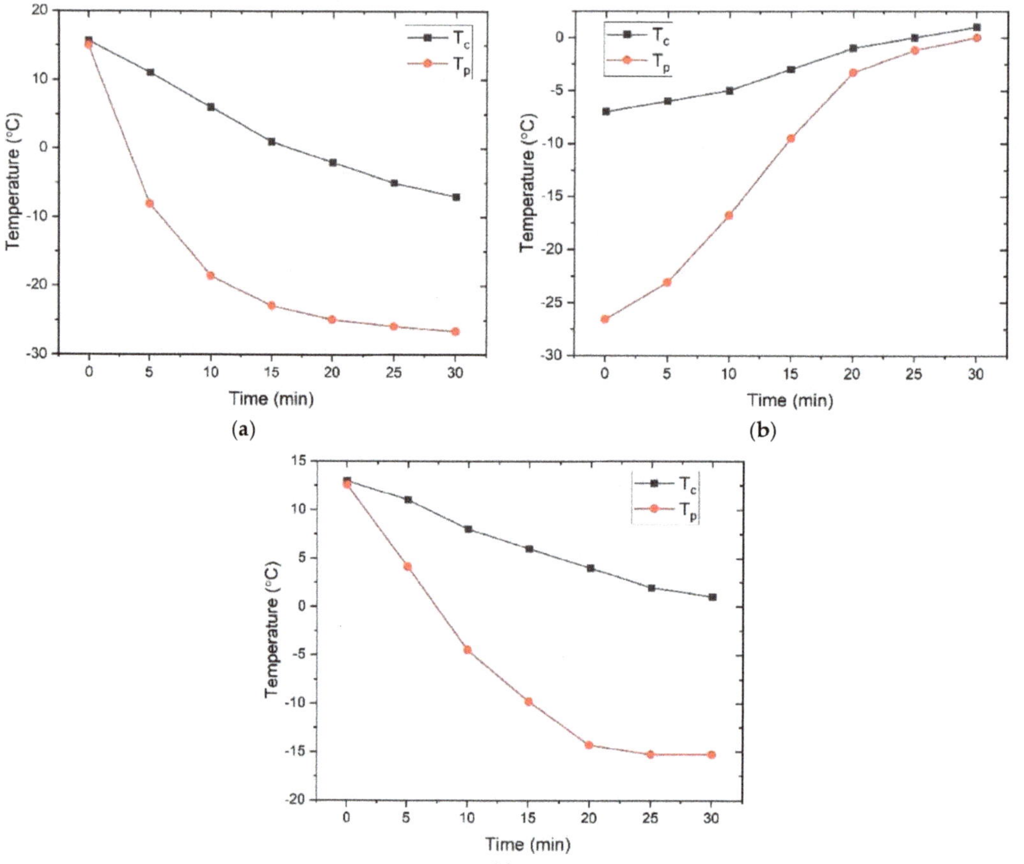

Figure 2. Compartment temperature variations during (**a**) cooling and (**b**) warming with no load. (**c**) Compartment temperature variations during cooling with a load of 1.5 L of water. T_c and T_p represent compartment and evaporator plate temperatures, respectively.

5. Conclusions

In addition to its role as a fuel, LPG can be harnessed for refrigeration purposes due to its compressed nature. This study demonstrates the utilization of LPG as a refrigerant in domestic refrigerators, which eliminates the need for compressors and condensers (half-cycle refrigeration).

The system's COP was 13.4, significantly surpassing that of typical domestic refrigerators. As a result, this innovative approach leads to electricity-free refrigeration, fully tapping into the potential of LPG for cooling applications.

Author Contributions: Conceptualization T.A.C.; methodology, G.A. and A.M.; software, S.F.S.; validation, T.A.C.; writing—original draft preparation, G.A.; writing—review and editing, A.M. and S.F.S.; experiments conducted, G.A. and A.M.; supervision, T.A.C. All authors have read and agreed to the published version of the manuscript.

Funding: This research received no external funding.

Institutional Review Board Statement: Not applicable.

Informed Consent Statement: Not applicable.

Data Availability Statement: Data described in this study will be provided upon request.

Acknowledgments: The authors express their gratitude for the technical and financial support provided by the GIK Institute of Engineering Sciences and Technology, Topi 23460, Pakistan.

Conflicts of Interest: The corresponding author affirms that all authors have no conflict of interest to declare in relation to this submission.

References

1. Kumar Rai, A.; Alabd Mohamed, S. Study of Performance Evaluation of Domestic Refrigerator with Mixture of Propane, Butane and Isobutene Refrigerant (LPG). *Int. J. Mech. Eng. Technol. (IJMET)* **2016**, *7*, 161–169.
2. Hussain Shah, I.; Gupta, K. Design of LPG Refrigeration System and Comparative Energy Analysis with Domestic. *Int. J. Eng. Sci. Res. Technol.* **2014**, *3*, 206–213.
3. Muzaffar, A.; Tariq, M.H.; Abbas, A.; Tayyab, M.; Cheema, T.A. Refrigeration Potential Investigation of Liquefied Petroleum Gas under Atmospheric Conditions. *Eng. Proc.* **2022**, *23*, 32.
4. Muzaffar, A.; Cheema, T.A.; Abbas, A.; Tayyab, M.; Ilyas, M.; Park, C.W. Performance Analysis of Liquified Petroleum Gas (LPG) Driven Half-Cycle Air Conditioning System. *Heat Mass Transf./Waerme-Und Stoffuebertragung* **2020**, *56*, 3177–3197. [CrossRef]
5. Aisyah, L.; Rulianto, D.; Setyo Wibowo, C. Analysis of the Effect of Preheating System to Improve Efficiency in LPG-Fuelled Small Industrial Burner. *Energy Procedia* **2015**, *65*, 180–185. [CrossRef]

engineering proceedings

MDPI

Proceeding Paper

Heat Transfer Enhancement in Louvered Fin Flat Tube Radiator Using Hybrid Nanofluids †

Waqar Ali *, Abid Hussain, Ahmed Usman, Khalid Mahmood, Muhammad Mubashir Iqbal and Haris Khan

Mechanical Engineering Department, University of Engineering and Technology, Taxila 47050, Pakistan; abid.hussain@uettaxila.edu.pk (A.H.); muhammadmubashar928@gmail.com (M.M.I.)
* Correspondence: waqaralibalti1996@gmail.com
† Presented at the Third International Conference on Advances in Mechanical Engineering 2023 (ICAME-23), Islamabad, Pakistan, 24 August 2023.

Abstract: This research explores the potential of hybrid nanofluids to improve the thermal efficiency of a car's louvered fin flat-tube radiator. Hybrid nanofluids were prepared by combining distilled water with a 0.1% vol. concentration of SiO_2 and MWCNT nanoparticles, using different ratios of nanoparticles: 80:20, 50:50, and 20:80. The experimental analysis focused on examining the heat-transfer performance of the radiator. The results clearly demonstrated a significant improvement in the radiator's thermal performance when using hybrid nanofluids. These nanofluids effectively enhanced the rate and coefficient of heat transfer. Notably, an increase of 15.6% in the Nusselt number was observed with the SiO_2–MWCNT 20:80 water containing a 0.1% volumetric concentration of nanoparticles. Overall, the findings highlight the promising application of hybrid nanofluids in boosting the thermal efficacy of car radiators.

Keywords: hybrid nanofluids; flat tube radiator; CFD; thermal performance; heat-transfer coefficient

Citation: Ali, W.; Hussain, A.; Usman, A.; Mahmood, K.; Iqbal, M.M.; Khan, H. Heat Transfer Enhancement in Louvered Fin Flat Tube Radiator Using Hybrid Nanofluids. *Eng. Proc.* **2023**, *45*, 51. https://doi.org/10.3390/engproc2023045051

Academic Editors: Mohammad Javed Hyder, Muhammad Mahabat Khan, Muhammad Irfan and Manzar Masud

Published: 19 September 2023

1. Introduction

The development of efficient cooling systems is crucial for optimizing the performance and longevity of automotive engines. The effectiveness of traditional coolants, such as water or ethylene glycol, is limited. However, recent advancements in nanotechnology have led to the emergence of hybrid nanofluids as promising alternatives for car radiator systems [1]. Hybrid nanofluids, consisting of a base fluid and nanoscale additives, offer improved thermal properties and enhanced heat-transfer capabilities [2]. Hybrid nanofluids exhibit significantly enhanced heat-transfer properties compared to conventional coolants [3].

This research presents an innovative method to improve car radiator efficiency using a mix of SiO_2 and MWCNT nanoparticles. The findings have important implications for making cars and other cooling systems work better, saving energy and being more eco-friendly.

2. Experimental Setup and Procedure

SiO_2 and MWCNT nanoparticles were dispersed in distilled water to create hybrid nanofluids. In this experiment, nanoparticle concentrations were set at 0.1 vol% with varying nanoparticle ratios of 80:20, 50:50, and 20:80. The formulation process involved magnetic stirring to achieve a uniform dispersion, followed by sonication to further enhance dispersion and break down agglomerates. Sodium carbonate was added as a surfactant to enhance stability. The resulting nanofluid exhibited improved stability and reduced agglomeration. In the experimental setup, a Suzuki Mehran car radiator made of aluminum was used. The setup included a pump for coolant circulation, a hot fluid reservoir with a heater, a flow meter, and valves for flow control. A total of five K-type thermocouples were placed at the radiator's inlet, outlet, and various points on its wall to measure the radiator's temperature. A data-acquisition system and laptop were utilized for real-time temperature monitoring and analysis. A constant 65 °C coolant temperature was maintained while the ambient temperature was 27 °C. We analyzed the performance of the radiator under different fluid flow rates to assess its efficiency. Figure 1 represents the experimental arrangement.

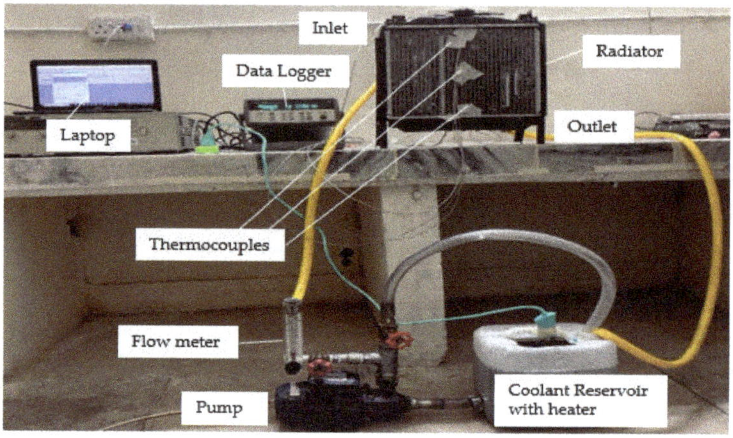

Figure 1. View of experimental arrangement.

The thermophysical properties of nanofluids were determined using correlations proposed by different scientists [4–6]. These properties are calculated using Equations (1)–(4).

$$\rho_{hnf} = (\varphi_{nf1} * \rho_{nf1}) + (\varphi_{nf2} * \rho_{nf2}) + (1 - \varphi) * \rho_{bf} \tag{1}$$

$$C_{p_{hnf}} = \frac{\varphi_{nf1}\rho_{nf1}C_{p_{nf1}} + \varphi_{nf2}\rho_{nf2}C_{p_{nf2}} + (1-\varphi)\rho_{bf}C_{p_{bf}}}{\rho_{hnf}} \tag{2}$$

$$\mu_{hnf} = (1 + 7.3\varphi + 123\varphi^2)\mu_{bf} \tag{3}$$

$$K_{hnf} = \frac{[(\varphi_{nf1}K_{nf1} + \varphi_{nf2}K_{nf2})/(\varphi_{nf1} + \varphi_{nf2}) + 2K_{bf} + 2(\varphi_{nf1}K_{nf1} + \varphi_{nf2}K_{nf2}) - 2\varphi K_{bf}]}{[(\varphi_{nf1}K_{nf1} + \varphi_{nf2}K_{nf2})/(\varphi_{nf1} + \varphi_{nf2}) + 2K_{bf} - 2(\varphi_{nf1}K_{nf1} + \varphi_{nf2}K_{nf2}) - 2\varphi K_{bf}]}K_{bf} \tag{4}$$

The rate and overall coefficient of heat transfer were determined using Equations (5) and (6):

$$Q = \dot{m}C_p(T_{in} - T_{out}) \tag{5}$$

$$U = \frac{Q}{nA_s(LMTD)} \tag{6}$$

In Equation (5), Q represents the rate of heat transfer, $\dot{m}$ represents the mass flow rate (kg/s), C_p represents the specific heat capacity (J/kg-K), and T_{in} and T_{out} represent the inlet and outlet temperatures (K). In Equation (6), U represents the overall heat transfer coefficient, n represents the tube count of the radiator, and A_s represents the radiator tube surface area. Equation (7) was used to calculate the logarithmic mean temperature difference (LMTD):

$$LMTD = \frac{T_{in} - T_{out}}{\ln\frac{T_{in}}{T_{out}}} \tag{7}$$

The Nusselt number, denoted as Nu, can be calculated using Equation (8):

$$Nu = \frac{h_{avg}D_h}{k} \tag{8}$$

In this equation, h_{avg} represents the average coefficient of heat transfer, D_h represents radiator's hydraulic diameter, and k represents coolant's thermal conductivity.

The average coefficient of heat transfer denoted as h_{avg}, was determined using Equation (9):

$$h_{avg} = \frac{Q}{nA_s(T_b - T_{wall})} \tag{9}$$

where, Q represents the rate of heat transfer, n represents the number of tubes of the radiator, A_s represents the surface area of the radiator, T_b represents the bulk temperature, which is mean of the inlet and outlet temperatures of the coolant, and T_{wall} is the radiator's wall temperature.

The Prandtl number, denoted as Pr, was determined using Equation (10):

$$Pr = \frac{\mu C_p}{k} \tag{10}$$

3. Result and Discussion

In this study, the Prandtl number decreased as the SiO_2–MWCNT nanoparticles were added, with the lowest value observed in the 20:80 ratio nanofluid. The nanofluid with a 20:80 ratio of SiO_2–MWCNT exhibited the most effective heat-dissipation capabilities. Hence, nanoparticles increased the thermal diffusivity of the nanofluid compared to its kinematic viscosity, enhancing its heat-transfer properties.

The graphs in Figure 2 compare the heat-transfer rates and coefficients for distilled water and the SiO_2–MWCNT hybrid nanofluids flowing at various ratios. The results show that hybrid nanofluids outperformed distilled water in terms of heat-transfer rates and coefficients. Both the volume flow rate and hybrid nanofluids contributed to the increased heat transfer.

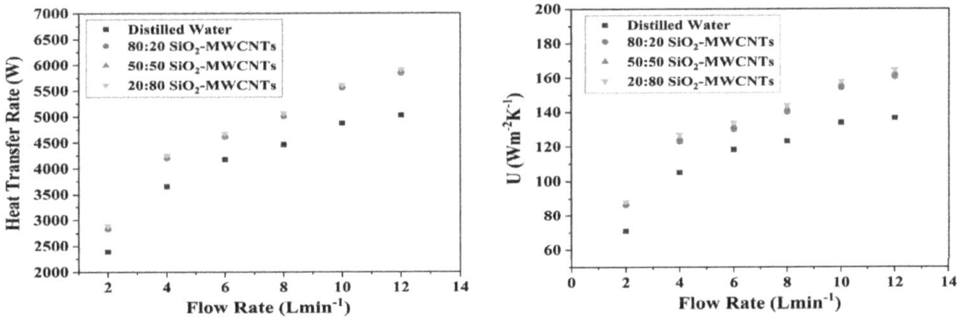

Figure 2. Rate and coefficient of heat transfer with varying flow rates of coolant.

In Figure 3, the graph compares the Nusselt number and Reynolds number for distilled water and the SiO_2–MWCNT hybrid nanofluids with different ratios of nanoparticles. The outcomes demonstrated that for both the distilled water and nanofluids, raising the Reynolds number resulted in a rise in the Nusselt number. In comparison to distilled water, nanofluids consistently exhibited higher Nusselt numbers. Among the nanofluids, the SiO_2–MWCNT 20:80 water nanofluid with a 0.1% concentration showed the highest enhancement in the Nusselt number, i.e., 15.63%.

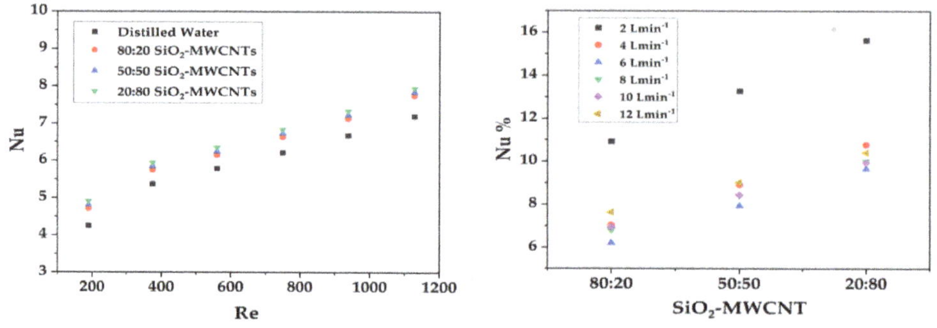

Figure 3. Relationship between Nusselt number and Reynold's number, and relationship between Nusselt number enhancement and different nanoparticle ratios.

4. Conclusions

This study examined the effectiveness of hybrid nanofluids in improving radiator thermal efficiency. Experimental analysis revealed that the combination of SiO_2 and MWCNT nanoparticles with distilled water enhanced its heat-transfer performance. Using the SiO_2–MWCNT 20:80 nanofluid, a 15.6% rise in Nusselt number was seen. These findings demonstrate the potential of hybrid nanofluids for optimizing cooling systems in automobiles and provide valuable insights for the design of efficient radiators. Overall, the study confirms the efficacy of hybrid nanofluids in enhancing thermal performance of car radiators.

Author Contributions: Writing, review, and editing: W.A., A.H., A.U., K.M., M.M.I. and H.K.; supervision: A.H.; project administration: A.H.; conceptualization: W.A.; methodology: W.A.; software: W.A. and H.K. All authors have read and agreed to the published version of the manuscript.

Funding: This research was self-funded and did not receive any external financial support.

Institutional Review Board Statement: Not applicable.

Informed Consent Statement: Not applicable.

Data Availability Statement: Not applicable.

Conflicts of Interest: The authors declare no conflict of interest.

References

1. Jamil, F.; Ali, H.M. Applications of hybrid nanofluids in different fields. In *Hybrid Nanofluids for Convection heat Transfer*; Elsevier: Amsterdam, The Netherlands, 2020; pp. 215–254.
2. Sarkar, J.; Ghosh, P.; Adil, A. A review on hybrid nanofluids: Recent research, development and applications. *Renew. Sustain. Energy Rev.* **2015**, *43*, 164–177. [CrossRef]
3. Sahoo, R.R.; Sarkar, J. Heat transfer performance characteristics of hybrid nanofluids as coolant in louvered fin automotive radiator. *Heat Mass Transf.* **2017**, *53*, 1923–1931. [CrossRef]
4. Pak, B.C.; Cho, Y.I. Hydrodynamic and heat transfer study of dispersed fluids with submicron metallic oxide particles. *Exp. Heat Transf. Int. J.* **1998**, *11*, 151–170. [CrossRef]
5. Qasim, M.; Sajid Kamran, M.; Ammar, M.; Ali Jamal, M.; Yasar Javaid, M. Heat transfer enhancement of an automobile engine radiator using ZnO water base nanofluids. *J. Therm. Sci.* **2020**, *29*, 1010–1024. [CrossRef]
6. Xuan, Y.; Roetzel, W. Conceptions for heat transfer correlation of nanofluids. *Int. J. Heat Mass Transf.* **2000**, *43*, 3701–3707. [CrossRef]

engineering proceedings

Proceeding Paper

Enhancing Energy Efficiency: Geothermal Heat Pumps Utilizing Existing Water Boreholes in Pakistan †

Saif Ullah *, Muhammad Irfan and Muhammad Mahabat Khan

Department of Mechanical Engineering, Capital University of Science and Technology, Islamabad 44000, Pakistan
* Correspondence: saifullah@cust.edu.pk

† Presented at the Third International Conference on Advances in Mechanical Engineering 2023 (ICAME-23), Islamabad, Pakistan, 24 August 2023.

Abstract: Vertical borehole geothermal heat pumps (GHPs) offer eco-friendly and energy-efficient heating and cooling. This paper proposes enhancing the coefficient of performance (COP) by converting air source heat pumps (ASHPs) into ground source heat pumps (GSHPs). This study presents a cost-effective approach to utilizing existing water boreholes in Pakistani buildings, reducing drilling and piping expenses. The results show that GSHPs achieved average COP values of 5.92, 33.33% higher than ASHPs. GSHPs demonstrated an average energy efficiency ratio (EER) of 20.12, exceeding ASHPs by 33.9%. These findings highlight the enhanced COP and energy efficiency of GSHPs, making them an attractive option for sustainable HVAC systems.

Keywords: ground source heat pump (GSHP); air source heat pump (ASHP); sustainable heating and cooling

1. Introduction

Geothermal heat pumps (GTHPs) have emerged as a promising alternative to conventional heating and cooling systems, demonstrating great potential for sustainable development and contributing to the attainment of sustainable development goals (SDGs). The energy-intensive nature of heating, ventilating, and air-conditioning (HVAC) systems in residential buildings raises concerns about their reliance on fossil fuels and their associated environmental impacts. Consequently, there is a growing initiative to replace conventional HVAC systems with environmentally friendly alternatives. Ground source heat pumps (GSHPs) have gained considerable attention as a sustainable solution, leveraging the Earth's stable thermal properties to provide energy-efficient heating and cooling. By tapping into renewable energy stored in the ground, GSHPs offer advantages such as higher energy efficiency and reduced greenhouse gas emissions.

S. Zhang et al. conducted a study on the performance evaluation of five residential GSHP systems in a region that experienced hot summers in China. The study revealed that the heat pumps exhibited an average coefficient of performance (COP) ranging from 3.36 to 5.94, while the system's energy efficiency ratio (EER) varied between 1.95 and 4.35 [1]. In another investigation, J. Luo et al. continuously monitored a GSHP system that had been installed in an office building in Germany for four years. The results indicate an annual increase of 8.7% in the system's seasonal energy efficiency Ratio (SEER), while the seasonal COP exhibited a decrease of 4.0% [2].

Y. Hwang et al. presented a study focused on the cooling performance of a closed vertical-type ground heat exchanger with 24 boreholes, each 175 m deep, installed in a school building in Korea. The study reported an average cooling COP of 8.3 and an overall COP of 5.9. In comparison, the air source heat pump (ASHP) system, with an equivalent capacity, exhibited an average COP of 3.9 and an overall COP of 3.4. These findings suggest that the GSHP system outperforms the ASHP system due to the Earth's stable and lower temperatures [3].

Citation: Ullah, S.; Irfan, M.; Khan, M.M. Enhancing Energy Efficiency: Geothermal Heat Pumps Utilizing Existing Water Boreholes in Pakistan. *Eng. Proc.* **2023**, *45*, 6. https://doi.org/10.3390/engproc2023045006

Academic Editors: Mohammad Javed Hyder and Manzar Masud

Published: 7 September 2023

This research paper focuses on investigating the performance of GSHPs in cooling applications in Pakistan. The study aims to compare COP and EER of a conventional ASHP with a heat pump that has been converted to utilize a vertical bore ground-cooled condenser.

2. Experimental Methodology

The experimental methodology employed in this study was used to assess the performance and feasibility of a GSHP system. The first step involved determining the undisturbed ground temperature at a depth of 30–50 feet, as this temperature is crucial for GSHP system design. In this study, the measured ground temperature at the selected site in Sector I-14, Islamabad, Pakistan, was found to be approximately 20 °C.

Soil properties, including thermal conductivity and thermal diffusivity, were examined using the ASTM D5334-14 standard test method. The moisture content of the soil samples was measured to approximate the thermal properties. Gravimetric and sieving methods were employed to determine the grain size and moisture content of the soil as shown in Table 1.

Table 1. Bore hole soil samples test results.

Sample No.	Wet Soil Mass (kg)	Dry Soil Mass (kg)	Moisture Content (gm)
1	6.96	5.75	21.043
2	7.10	5.89	20.543
3	6.85	5.63	21.669
4	7.00	5.74	21.951
5	6.68	5.53	20.796

For the design of the ground connection, the method proposed by Ingersoll et al. [4] was applied, considering a closed-loop ground heat exchanger design. The length of the ground connection was calculated based on the heat transfer equation per unit length. To reduce drilling costs and utilize existing boreholes, a spiral-type ground connection was implemented with a length of approximately 80 feet.

The heat pump selection involved choosing a DC inverter heat pump with a capacity of 1.5 tons operating with refrigerant R-410A, which was then converted into a GSHP. The selection was based on the space heating and cooling loads.

Grouting, an essential aspect of the system, involved filling the space between the ground and the bore casing GI pipe with an appropriate grout material. In this study, the local clay-based mud was identified as the optimal grout material, with an approximate moisture content of 20% and a density of 1600 kg/m^3. The determined thermal diffusivity was approximately 0.086 m^2/day, while the thermal conductivity of the grout material measured as 2.68 W/m·°C.

The experimental setup included the installation of the ground connection, with winding of 250 ft, 3/8 inch diameter copper tubing on the 5 inch diameter GI bore casing for the fabrication of ground heat exchanger. Temperature sensors were strategically positioned to gather data for the performance evaluation, as illustrated in Figure 1.

Data analysis involved operating the original air source heat pump (ASHP) in cooling mode with an indoor set point of 17 °C, collecting temperature, pressure, electric current, and voltage readings at regular intervals of 30 min. The ASHP was then converted into a GSHP by replacing the air-cooled condenser with a water-cooled condenser connected to the ground connection spiral heat exchanger. The GSHP was tested under similar conditions, and data were collected for analysis.

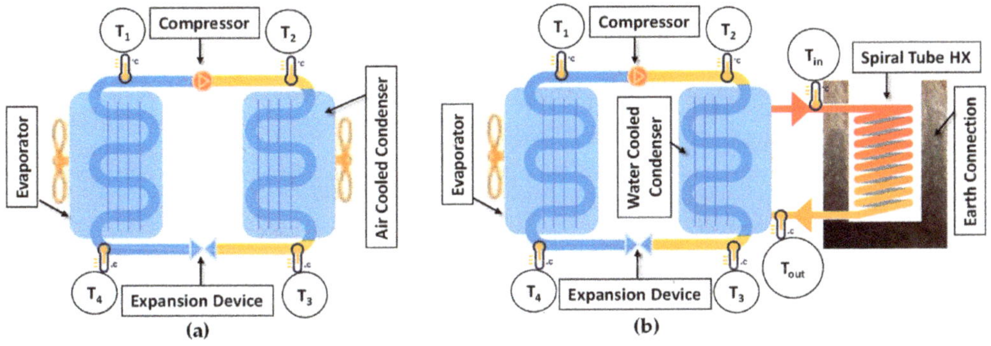

Figure 1. Layout of the experimental setup with sensors placement: (**a**) ASHP and (**b**) GSHP.

3. Results

The collected data were analyzed to evaluate the performance parameters of the GSHP system, providing valuable insights into its effectiveness compared to ASHP.

Figure 2 presents the temperature results obtained from the ground connection. It is evident that circulating water at eight different inlet temperatures through the ground connection experienced a significant temperature decrease. The temperature difference increased with higher inlet water temperatures, indicating the effectiveness of the heat exchanger.

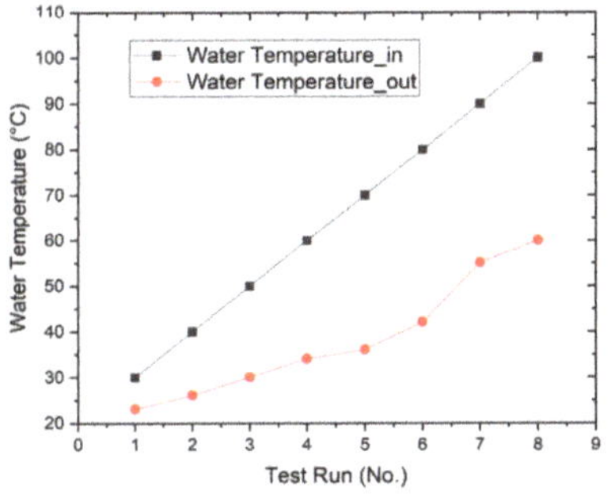

Figure 2. Ground connection heat exchanger temperatures for eight test runs in summer.

Figure 3a shows COP of the GSHP and ASHP under identical indoor and outdoor conditions throughout different times of the day during summer in Pakistan. The graph clearly illustrates that the GSHP exhibited a superior COP compared to the ASHP when operating under the same conditions, validating the enhanced performance of the GSHP.

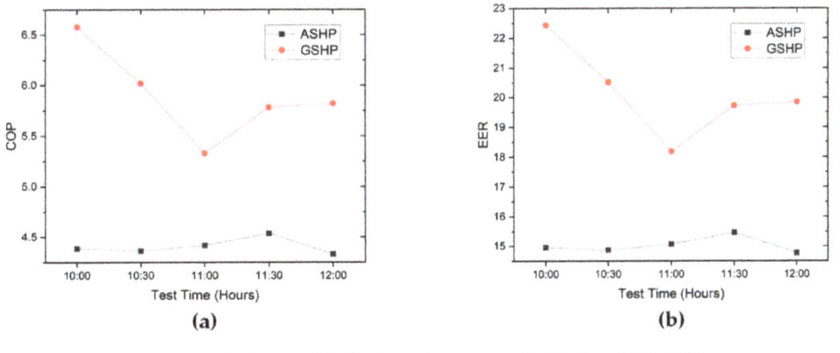

Figure 3. Comparison of ASHP and GSHP performance: (**a**) COP and (**b**) EER.

Similarly, Figure 3b shows a comparison of the EER of both the GSHP and ASHP. The graph demonstrates that the GSHP outperformed the ASHP, mirroring the advantages observed in the COP analysis. The comparable electrical requirements of the ASHP condenser fan and GSHP recirculating pump affirm the analogous trends observed between COP and EER. Since the GSHP utilizes water as a heat carrier, which is more efficient than air, and transfers heat to the ground at approximately 20 °C, the condenser temperatures are lower compared to the ASHP case. Consequently, this results in higher COP and EER values.

4. Conclusions

The experimental results of this study demonstrate significant improvements in the performance of a GSHP compared to an ASHP. The average COP achieved for the GSHP was approximately 5.92, whereas the average COP for the ASHP was found to be 4.44 during operation. This indicates that the GSHP exhibited a COP that was approximately 33.33% higher than ASHP.

Furthermore, the average energy efficiency ratio (EER) for the ASHP was approximately 15.02, whereas the GSHP achieved an average EER of approximately 20.12. Consequently, the EER of the GSHP exceeded that of the ordinary heat pump by approximately 33.9%.

In conclusion, this research highlights the superior performance of geothermal heat pumps in terms of COP and EER compared to conventional heat pumps. The findings emphasize the substantial potential of GSHPs to enhance energy efficiency. Moreover, utilizing an existing water borehole can lead to reduced operating and installation costs.

Future research should focus on assessing GSHPs' heating capabilities in diverse winter climates and investigating the impact of cooling fluids on ground connections. Additionally, long-term ground temperature monitoring and protective measures against the degradation of the buried heat exchanger are essential.

Author Contributions: Conceptualization, methodology, investigation, data curation, writing—original draft preparation, S.U.; editing, M.I.; review, M.M.K. All authors have read and agreed to the published version of the manuscript.

Funding: This research was funded by ASHRAE, USA, under an undergraduate program equipment grant titled Geothermal Heat Pump.

Institutional Review Board Statement: Not applicable.

Informed Consent Statement: Not applicable.

Data Availability Statement: The data presented in this study are available on request.

Conflicts of Interest: The authors declare no conflict of interest.

References

1. Zhang, S.; Zhang, L.; Wei, H.; Jing, J.; Zhou, X.; Zhang, X. Field Testing and Performance Analyses of Ground Source Heat Pump Systems for Residential Applications in Hot Summer and Cold Winter Area in China. *Energy Build.* **2016**, *133*, 615–627. [CrossRef]
2. Luo, J.; Rohn, J.; Bayer, M.; Priess, A.; Wilkmann, L.; Xiang, W. Heating and Cooling Performance Analysis of a Ground Source Heat Pump System in Southern Germany. *Geothermics* **2015**, *53*, 57–66. [CrossRef]
3. Hwang, Y.; Lee, J.K.; Jeong, Y.M.; Koo, K.M.; Lee, D.H.; Kim, I.K.; Jin, S.W.; Kim, S.H. Cooling Performance of a Vertical Ground-Coupled Heat Pump System Installed in a School Building. *Renew. Energy* **2009**, *34*, 578–582. [CrossRef]
4. Kavanaugh, S.; Rafferty, K. *Geothermal Heating and Cooling Design of Ground-Source Heat Pump Systems*; ASHRAE: Atlanta, GA, USA, 2014.

engineering proceedings

MDPI

Proceeding Paper

Numerical Analysis of a Super-Insulated Pipe for the Transportation of Liquid Nitrogen (LN_2) †

Danish Bin Nisar, Maaz Ahmed, Ali Mohsin Hussain, Muzaffar Ali * and Hafiz Sohaib Muhammad

Mechanical Engineering Department, University of Engineering and Technology Taxila, Taxila 47080, Pakistan; danish.nisar@students.uettaxila.edu.pk (D.B.N.); maaza641@gmail.com (M.A.); alimohsinhussain3@gmail.com (A.M.H.); sohaib.me@must.edu.pk (H.S.M.)

* Correspondence: muzaffar.ali@uettaxila.edu.pk

† Presented at the Third International Conference on Advances in Mechanical Engineering 2023 (ICAME-23), Islamabad, Pakistan, 24 August 2023.

Abstract: In this study, a super-insulated pipe incorporating Multi-Layer Insulation (MLI) and vacuum is numerically analyzed to overcome the challenges faced during the transportation of cryogenic fluids like nitrogen. A super-insulated pipe incorporating an inner process pipe of SS 304 L insulated by twenty-four consecutive layers has been used. Each layer consists of aluminized mylar (as a radiation shield) and dacron netting (as a spacer material). High vacuum (10^{-9} torr) is applied and numerically analyzed at multiple flow rates, i.e., 250 LPH, 500 LPH and 1000 LPH. The results show a gradual increase in temperature along the flow direction from 77 K to 79 K at the most. Moreover, the temperature increases with the increase in the length of the pipe and decreases with the increase in the flow rate of LN_2.

Keywords: Multi-Layer Insulation; cryogenic transfer line; super-insulated pipe; process pipe; radiation shield

Citation: Nisar, D.B.; Ahmed, M.; Hussain, A.M.; Ali, M.; Muhammad, H.S. Numerical Analysis of a Super-Insulated Pipe for the Transportation of Liquid Nitrogen (LN_2). *Eng. Proc.* **2023**, *45*, 55. https://doi.org/10.3390/engproc2023045055

Academic Editors: Mohammad Javed Hyder, Muhammad Mahabat Khan, Muhammad Irfan and Manzar Masud

Published: 11 October 2023

1. Introduction

Liquid nitrogen is a cryogenic fluid as its temperature is below −150 °C (123 K), i.e., −196 °C (77 K). It should be managed with caution as it could result in extreme frost bites and other health hazards. It finds its application in multiple industries, for example, in space exploration for rocket propulsion, in metal working to shrink-fit components, in the food industry to preserve food products and so on.

In 2009, Sun et al. [1] experimentally investigated the effect of different spacer gases over a wide range of temperature (77 K–300 K) and pressure (10^{-3} Pa–10^{5} Pa) and concluded that Ar has the least apparent thermal conductivity among all the gases assessed. Similarly, Wei et al. [2] experimentally investigated heat transfer across a Perforated Multi-Layer Insulation Blanket (PMLIB) and inferred that the shape and structure of a PMLIB play a significant role in the accuracy of its performance test. Bapat et al. [3] also predicted the performance of MLI incorporating aluminized mylar (12 μm) and glass fabric (76.2 μm). The effective thermal conductivity of MLI seems to increase with the increase in the number of layers due to gas conduction. Later, Johnson et al. [4] experimentally evaluated several specimens of MLI in order to calculate their optimal layer density. Then, Funke et al. [5] conducted a comparative study between multiple insulations, taking liquid hydrogen and liquified natural gas (LNG) as the process fluid. They found that the heat flux (q) and effective thermal conductivity (k_e) were the lowest for the MLI systems under consideration. In another study, Fesmire et al. [6] conducted extensive experimentation on six different MLI configurations.

In the current study, a super-insulated pipe incorporating Multi-Layer Insulation (MLI) and vacuum was designed and numerically analyzed on ANSYS to observe its thermal behavior at multiple flow rates, i.e., 250, 500 and 1000 LPH. The aim of this study is to determine heat flow through temperature distribution. The results obtained were then validated through Experimental Fluid Dynamics (EFD) data compiled by Lim et al. [7].

2. Materials and Methods

MLI is most effective for extremely large temperature gradients as considered in this study. MLI consists of aluminized mylar (0.08 mm) and dacron netting (0.02 mm) due to its lower heat transfer rates. Aluminized mylar acts as a reflector for radiation while dacron netting minimizes the thermal contact area between adjacent aluminized mylar layers. SS 304 L-grade pipes were used owing to their better strength at extremely low temperatures. Argon was introduced as vacuum due to its inert nature. A high-density polyethylene (HDPE) spacer was used due to its lower thermal conductivity and high strength. A single layer comprises a combination of one aluminized mylar layer and one dacron netting layer.

The flow was turbulent in this study. The total thermal resistance is determined by the following:

$$R_{Total} = \frac{1}{h_{LN2} \cdot A_{ip}} + \frac{1}{2\pi k_{ss} \cdot L}\left(ln\frac{r_2}{r_1}\right) + \frac{1}{2\pi k_{ins} \cdot L}\left(ln\frac{r_3}{r_2}\right) + \frac{1}{2\pi k_{ss} \cdot L}\left(ln\frac{r_5}{r_4}\right) + \frac{1}{h_{Air} \cdot A_{op}} \tag{1}$$

Here, h and k are convective and conductive heat transfer coefficients, L is the length of the pipe and r refers to the radius of either the pipe or insulation. A_{op} and A_{ip} is outer and inner pipe surface area respectively. The convective heat transfer coefficient is calculated using the Nusselt number (Nu) [8].

$$Nu = \frac{h \cdot L}{k} = 0.0023 \cdot (Re)^{0.8} (\mathrm{Pr})^{0.3} \tag{2}$$

3. Modelling and Simulation

The insulated pipe model was developed in SOLIDWORKS. The length of the pipe is 1 m and its inner diameter is 25.4 mm. The surface contact area of the spacer is kept to a minimum to reduce the heat conduction between layers and the outer jacketed pipe.

The mesh size in Figure 1a was kept to 10 mm. Figure 1b shows sectional view of the super insulated pipe developed for numerical analysis. The SOLIDWORKS model was imported into ANSYS for thermal analysis. k-epsilon turbulence model and SIMPLE (Semi-Implicit technique for Pressure Linked Equations) method was implemented. The solution was converged after 85 iterations (Table 1).

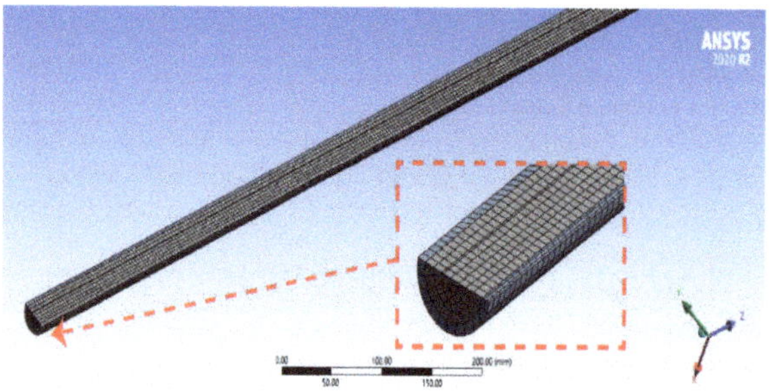

(a)

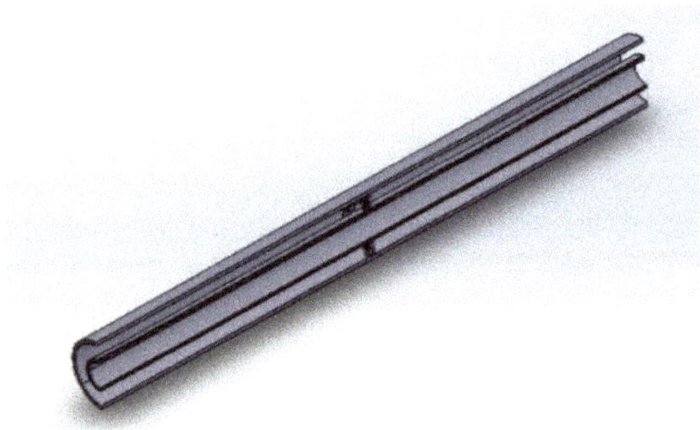

(b)

Figure 1. Modelling and meshing. (**a**) Mesh size. (**b**) Section view of the model.

Table 1. Boundary conditions and layers data.

Number of Layers 'n'	Layer Density (Lyrs/mm)	Cold Body Temperature (K)	Hot Body Temperature (K)	Cold Vacuum Pressure (Torr)
25	10	77	300	10^{-9}

4. Results

The following results were obtained after the completion of analysis (Table 2).

Table 2. Temperature variations with respect to volume flow rate.

Inlet Mass Flow Rate (kg/s)	0.0561	0.1123	0.2245
Max Wall Line Temperature (K)	78.6	77.9	77.6
Min Wall Line temperature (K)	77	77	77
Change in Temperature (K)	1.6	0.9	0.6

Inlet mass flow rate values of 0.0561, 0.1123 and 0.2245 kg/s are equivalent to 250, 500 and 1000 LPH, respectively.

Figure 2a shows that the temperature of LN_2 increases along the direction of flow. The temperature contour is set between 77 K to 77.8 K to compensate minor deviation in temperature. This rise is quite abrupt in the beginning due to the large temperature gradient, while this rise in temperature becomes gradual along the length as the temperature gradient becomes small. Moreover, it can be seen that the boundary wall experiences a higher temperature than the asymmetric line due to frictional effect as it delays flow along the boundary.

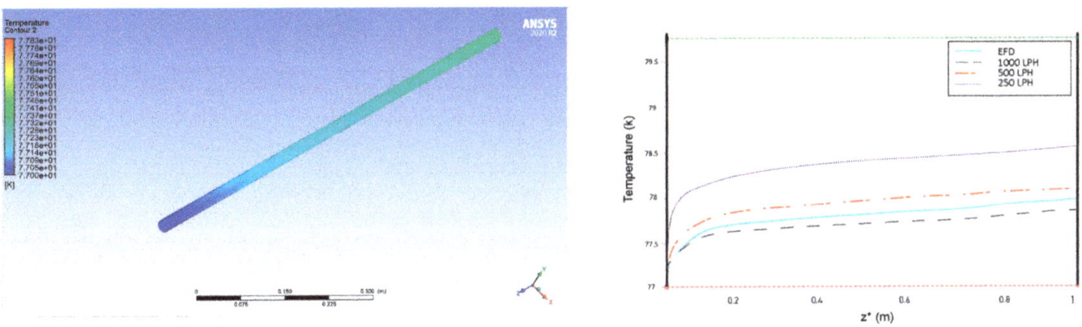

(a) (b)

Figure 2. Temperature profile of super-insulated pipe. (**a**) Temperature contour at 1000 LPH. (**b**) Temperature variation with respect to length.

Figure 2b shows a comparison between current analysis with the EFD trend for 1000 LPH developed in previous studies [7]. There is a minor deviation (i.e., less than 5%) in the analysis results from the Experimental Fluid Dynamics (EFD) trend. The lower flow rates seem to follow the same trend but with a slightly higher temperature gradient. This proves that lower flow rates are more prone to temperature variations than higher flow rates. Lower flow rates increase interaction time between boundary wall and the LN_2 resulting in higher temperature variations. As the temperature gradient between ambient (i.e., 300 K) and LN_2 (i.e., 77 K) is quite high.

5. Conclusions

It has been concluded from this study that the temperature gradient along the flow reduces with the increase in the flow rate. Moreover, the boundary wall experiences a higher temperature than the asymmetric line. Temperature distributions are within the range of 77 K to 79 K for inlet volume flow rates of 250 LPH, 500 LPH and 1000 LPH. The temperature profile follows a hyperbolic trend for the LN_2 flow inside the super-insulated pipe.

Author Contributions: D.B.N. conceived the idea and developed the model. D.B.N. and M.A. (Maaz Ahmed) conducted numerical analysis and drafted the manuscript. H.S.M. provided the necessary guidance on numerical analysis. A.M.H. conducted analytical calculations. M.A. (Muzaffar Ali) reviewed the analysis results and supervised the project. All authors have read and agreed to the published version of the manuscript.

Funding: This research received no external funding.

Institutional Review Board Statement: Not applicable.

Informed Consent Statement: Not applicable.

Data Availability Statement: The data is not available online to avoid any misuse of data.

Conflicts of Interest: The authors declare no conflict of interest.

References

1. Sun, P.J.; Wu, J.Y.; Zhang, P.; Xu, L.; Jiang, M.L. Experimental study of the influences of degraded vacuum on multilayer insulation blankets. *Cryogenics* **2009**, *49*, 719–726. [CrossRef]
2. Wei, W.; Li, X.; Wang, R.; Li, Y. Effects of structure and shape on thermal performance of Perforated Multi-Layer Insulation Blankets. *Appl. Therm. Eng.* **2009**, *29*, 1264–1266. [CrossRef]
3. Bapat, S.L.; Narayankhedkar, K.G.; Lukose, T.P. Performance prediction of multilayer insulation. *Cryogenics* **1990**, *30*, 700–710. [CrossRef]
4. Johnson, W.L. Thermal analysis of low layer density multilayer insulation test results. In Proceedings of the ADVANCES IN CRYOGENIC ENGINEERING: Transactions of the Cryogenic Engineering Conference–CEC, Spokane, WA, USA, 13–17 June 2011; pp. 1519–1526. [CrossRef]
5. Funke, T.; Haberstroh, C. A calorimeter for measurements of multilayer insulation at variable cold temperature. *Phys. Procedia* **2015**, *67*, 1062–1067. [CrossRef]
6. Fesmire, J.E.; Johnson, W.L. Cylindrical cryogenic calorimeter testing of six types of multilayer insulation systems. *Cryogenics* **2018**, *89*, 58–75. [CrossRef]
7. Lim, C.L.; Adam, N.M.; Ahmad, K.A. Cryogenic pipe flow simulation for liquid nitrogen with vacuum insulated pipe (VIP) and Polyurethane (PU) foam insulation under steady-state conditions. *Therm. Sci. Eng. Prog.* **2018**, *7*, 302–310. [CrossRef]
8. Cengel, Y.A.; Ghajar, A.J. *"Fundamentals of Convection." Heat and Mass Transfer: A Practical Approach*; McGraw-Hill Education: New York, NY, USA, 2019; pp. 333–358.

engineering proceedings

Proceeding Paper

Techno-Economic Investigation of Standalone Photovoltaic Energy Systems for Rural Areas of Quetta †

Riaz Ahmed [1], Abdul Basit [1], Qamar ud Din Abid [1,2], Muhammad Haroon [2,3,*], Fasihullah Kakar [1], Najeeb Ullah [1] and Danish Ahmed Khan [1]

1 Department of Mechanical Engineering, Balochistan University of Information Technology, Engineering and Management Sciences, Quetta 87300, Pakistan; 50397@student.buitms.edu.pk (R.A.); 48955@student.buitms.edu.pk (A.B.); qamaruddin.abid@buitms.edu.pk (Q.u.D.A.); fasihullah.kakar@buitms.edu.pk (F.K.); 51413@student.buitms.edu.pk (N.U.); 51320@student.buitms.edu.pk (D.A.K.)

2 Department of Mechanical Engineering, International Islamic University, Islamabad 44000, Pakistan

3 Department of Mechanical Engineering, Capital University of Science and Technology (CUST), Islamabad 44000, Pakistan

* Correspondence: muhammad.haroon@cust.edu.pk

† Presented at the Third International Conference on Advances in Mechanical Engineering 2023 (ICAME-23), Islamabad, Pakistan, 24 August 2023.

Abstract: The purpose of this techno-economic study is to investigate the potential of the implementation of standalone photovoltaic (PV) energy systems in rural areas of Quetta, Pakistan. This study focuses on alleviating the region's energy crisis and giving the local population access to clean and sustainable energy. The technique used entails data collection, load demand analysis, technical and economic evaluation, site selection, and data interpretation. The findings and discussion provide an estimate of PV energy systems' lifetime costs, annual maintenance expenses, and energy prices. The results show that standalone PV energy systems are a feasible and financially viable option for electricity generation in Quetta's rural areas that operates via monitoring the economic development and environmental sustainability of the region. Three plants are proposed for generating 26.8 KW, 15 KW, and 6.8 KW of power, and it is demonstrated that green and clean energy can be provided for domestic consumers at reasonable unit costs of PKR. 21.33, PKR. 21.9, and PKR. 23.89, respectively.

Keywords: renewable energy; solar energy; standalone PV system; sustainable energy; levelized cost of electricity; breakeven point

Citation: Ahmed, R.; Basit, A.; Abid, Q.u.D.; Haroon, M.; Kakar, F.; Ullah, N.; Khan, D.A. Techno-Economic Investigation of Standalone Photovoltaic Energy Systems for Rural Areas of Quetta. *Eng. Proc.* **2023**, *45*, 30. https://doi.org/10.3390/engproc2023045030

Academic Editors: Mohammad Javed Hyder, Muhammad Mahabat Khan, Muhammad Irfan and Manzar Masud

Published: 12 September 2023

1. Introduction

Pakistan's energy crisis is a growing concern for all stakeholders due to the nation's incremental population growth and decremental economic growth. It is important to explore the potential of renewables in Pakistan to generate cleaner and sustainable energy. The most important globally employed renewable energy resources are solar energy, wind energy, hydropower, geothermal energy, bioenergy, and ocean thermal energy conversion cycles (OTEC) [1]. Balochistan province is particularly affected by this energy and economic crisis. Pakistan needs an effective, long-term energy plan to bridge the supply and demand gap, which has been precipitated by obsolete power plants, inadequate financial resources, and poor transmission and distribution (T&D) infrastructure. Domestic energy resources in Pakistan's Balochistan province have a combined capacity of 500.041 TWh, which can be used to address energy crises. Pakistan's energy consumption from 2018 to 2030 is expected to be in the range from 312 TWh to 399 TWh [2]. To fulfill this demand, serious efforts are required from researchers, government officials, higher authorities, and decision makers.

Studies have been carried out on different locations of Balochistan to explore the potential of photovoltaic energy. Gohram et al. [3] carried out a techno-economic investigation of a solar power plant in Quetta with a 50 MW capacity and concluded that the production of 91.980 GWh of energy (per year), with a USD 59.689 million capital cost and a USD 0.9 million (per year) O&M cost, can be achieved. Noman et al. [4] conducted an economic analysis of photovoltaic (PV) installations in big cities in Pakistan (Karachi, Quetta, Multan, Lahore, and Peshawar) and concluded that PV is a suitable replacement for non-renewable energy sources that can decrease the load on the national grid by 60%, if installed at higher scales. Pakistan has installed a 100 MW solar system in Bahawalpur that has five modules, each having a capacity of 20 MW [5].

The current study was carried out to render the rural areas of Balochistan energy-independent and provide some comfort for the local population. The location selected for this study is Chashma Achozai Garden Town. Hardware, including PV panels, batteries, and invertors, have been chosen such that the best output can be attained at low capital and maintenance costs. The cost of electricity per kWh has been calculated, and the levelized cost of electricity has been evaluated. Furthermore, break-even and payback times were calculated for three power plants with capacities of 6.8 KW, 15 KW, and 26.8 KW.

2. Materials and Methods

The important information used in the current study was obtained using a pyranometer installed at BUITEMS Quetta by the World Bank, information from the energy info website [6], software programs like RET Screen and PVsyst, and information from Meteonorm and NASA. The mentioned pyranometer was used to measure the study area's sun radiation levels on a regular basis. Software programs like RET Screen and PVsyst provided functionalities for simulating and analyzing energy systems, including solar energy. Meteonorm and NASA also provided information on temperature, solar radiation, and other climatic variables. The power-consumption-related data of Chashma Achozai Garden Town was collected using a questionnaire, which provided estimates of monthly/daily electricity consumption.

Technical and economic analyses were conducted to evaluate the feasibility and financial viability of the analyzed solar energy projects. Solar radiation data were used to estimate solar energy potential, and statistical analysis and modelling approaches were used to evaluate variability and availability.

Financial modelling tools were used to assess the solar energy projects' financial viability and economic viability, considering capital costs, operation and maintenance costs, energy tariff rates, financing alternatives, and social and environmental aspects.

It is important to carry out data collection for solar energy projects in order to determine a project's viability and system development characteristics and perform a corresponding performance evaluation. Solar resource data include global horizontal irradiance (GHI), diffuse horizontal irradiance (DHI) levels, and direct normal irradiance (DNI). Historical weather information includes past temperatures, cloud cover, and precipitation levels that have affected the production of solar energy. Site-specific data include latitude and longitude, shading analysis information, topography, economic and financial data, regulations and permits, local laws, environmental impact assessments, performance and monitoring data, and reliability reports. Radiation statistics of Quetta city were gathered for 24 months, with the greatest values of global horizontal irradiance (GHI), incidence angle modifier (IAM) losses, and shadings recorded in May, June, and July, amounting to 228.5, 231.7, and 223.7 KWh/m^2, respectively.

The load profile divides a load into three phases. This method considers the load demand and load calculations to obtain the daily maximum consumption for

- 3 to 4 houses (6.8 KW);
- 10 to 15 houses (15 KW);
- 20 to 25 houses (26.5 KW).

Levelized cost of electricity (LCOE), annual operation and maintenance (O&M) cost, and total cost were calculated using the following equations:

$$\text{LCOE} = \text{Total Cost}/\text{Total Electricity Produced} \quad (1)$$

$$\text{Annual unit O\&M cost} = \text{Annual O\&M cost/Plant size} \quad (2)$$

$$\text{Total cost} = \text{Installation cost} + \text{Maintenance cost} \quad (3)$$

3. Results and Discussion

The details of the hardware suggested in this work along with the installation and maintenance costs (excluding consumables) are tabulated in Table 1.

Table 1. Quantity and Specifications of the Hardware suggested along with costs.

Load (KW)	No. of Panels	Panels' Specifications	No. of Batteries	Batteries' Specifications	No. of Inverters	Inverters' Specifications	Installation Cost (PKR)	Maintenance Cost per Year (PKR)
6.8	20	Mono 340W-Power Generic Panels	128	ANTBatt 12 V-101an Li-ion	1	Generic MPPT Convertor	6,803,800	1–2% of initial cost
15	44	Mono 340W-Power Generic Panels	272	Abwatt 12 V-101an Li-ion	1	Generic MPPT Convertor	14,073,400	1–2% of initial cost
26.5	80	Mono 340W-Power Generic Panels	496	ANTBatt 12 V-101an Li-ion	1	Generic MPPT Convertor	25,431,000	1–2% of initial cost

The selected PV panels, batteries, and inverters were chosen based on their efficiency and cost-effectiveness. Three different power plants, with loads of 26.8 kW, 15 kW, and 6.8 kW, were simulated using PVsyst software. This analysis included both economic and technical aspects. For the 26.8 kW load, the software estimated an installation cost of PKR 25,431,000. The annual maintenance cost was PKR 30,000, and the energy cost was PKR 22.35/kWh. A daily input/output diagram for this plant is shown in Figure 1a.

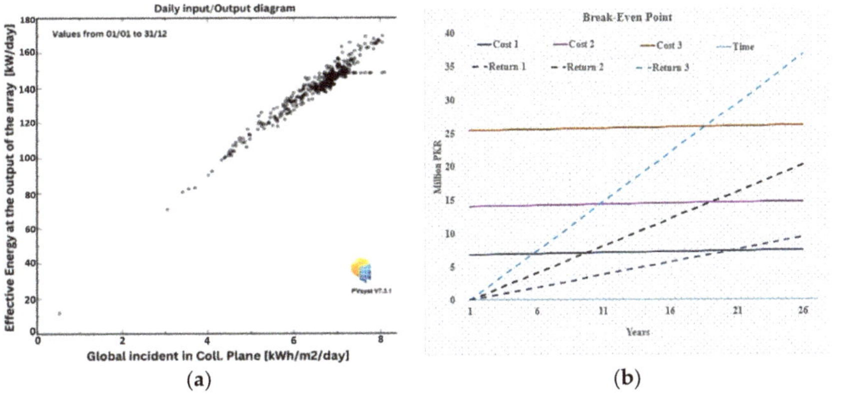

(a) (b)

Figure 1. (**a**) Daily input/output diagram for the 26.8 KW plant generated in PVsyst; (**b**) break-even points for all three powerplants.

For the 15 KW load, the installation cost was estimated to be PKR 14,073,400, with an energy cost of PKR 21.922/kWh. For the 6.8 kW load, the installation cost was projected to be PKR 6,803,800, with an energy cost of PKR 19.98/kWh. The maintenance costs (excluding consumables) of all three plants were approximated to account for 1–2% of initial cost. The estimated lifespan of the PV panels is 25 years, while the batteries have a lifespan of 3–5 years.

The cost of electricity calculated in this study is very economical compared to the rates of electric power supplied by the national grid, which will increase even further due to the current financial crisis the country is experiencing; hence, the suggested plants are the need of the hour.

The proposed plants have shown great promise, and their investment return is quite high. Figure 1b presents the break-even points of the three proposed plants. In Figure 1b, Cost 1, Cost 2, and Cost 3 represent the total cost of the plants with a capacity of 6.8 KW, 15 KW, and 26.8 KW, respectively, while Return 1, 2, and 3 represent the total outcomes for the respective plants. The unit price of the national grid is assumed to be PKR 26/kWh for the calculation of the break-even point, and the prices per unit for the proposed plants of 6.8 KW, 15 KW, and 26.8 KW in PKR are 23.89, 21.9, and 21.33, respectively.

4. Conclusions

The primary purpose of this study is to highlight the techno-economic feasibility of standalone PV plants for rural areas in Quetta. This study reveals that 140 kW/day energy can be received by the analyzed PV plates. Energy can be produced for as cheap as PKR 21.33/kWh for 20 years. Investments can give high returns, as the payback time is about 17 years. Massive projects can have even better returns, provided that hardware is chosen wisely. This energy-deprived region can become energy-efficient, and the local population can gain access to a clean and green power supply, which is economical and reliable.

Author Contributions: Conceptualization; R.A. and A.B.; methodology, F.K. and N.U.; software; R.A. and D.A.K.; formal analysis; M.H. and Q.u.D.A.; investigation; Q.u.D.A. and R.A.; writing—original draft preparation, R.A. and A.B.; writing—review and editing Q.u.D.A. and M.H.; supervision, Q.u.D.A. and F.K. All authors have read and agreed to the published version of the manuscript.

Funding: This research received no external funding.

Institutional Review Board Statement: Not applicable.

Informed Consent Statement: Not applicable.

Data Availability Statement: Not applicable.

Conflicts of Interest: The authors declare no conflict of interest.

References

1. Haroon, M.; Ayub, A.; Sheikh, N.A.; Ahmed, M.; Shalaby, A.-B. Harnessing Ocean Thermal Energy from Offshore Locations in Pakistan Using an Organic Rankine Cycle. *Eng. Proc.* **2022**, *23*, 24. [CrossRef]
2. Raza, M.A.; Khatri, K.L.; Israr, A.; Haque MI, U.; Ahmed, M.; Rafique, K.; Saand, A.S. Energy demand and production forecasting in Pakistan. *Energy Strategy Rev.* **2022**, *39*, 100788. [CrossRef]
3. Khan, G.; Hassan, M.; Anwar, M.; Waqas, A.; Shakir, S.; Sajid, J. Technical and Economic Evaluation of a 50 MW Solar Power Plant in Quetta. *Eng. Proc.* **2022**, *20*, 46.
4. Shabbir, N.; Usman, M.; Jawad, M.; Zafar, M.H.; Iqbal, M.N.; Kütt, L. Economic analysis and impact on national grid by domestic photovoltaic system installations in Pakistan. *Renew. Energy* **2020**, *153*, 509–521. [CrossRef]
5. Khaliq, A.; Ikram, A.; Salman, M. Quaid-e-Azam Solar Power park: Prospects and challenges. In Proceedings of the 2015 Power Generation System and Renewable Energy Technologies (PGSRET), Islamabad, Pakistan, 10–11 June 2015; pp. 1–6. [CrossRef]
6. Welcome to ENERGYDATA.INFO. Available online: https://energydata.info/ (accessed on 17 May 2023).

engineering proceedings

MDPI

Proceeding Paper

A Thermal Analysis of Atmospheric Balloons Using Different Coating Materials †

Nadeem Aslam [1], Muhammad Haroon [1,2,*], Shummaila Rasheed [2], Fazal Umer Rai [1] and Mohsin Iftikhar [1]

1 Department of Mechanical Engineering, International Islamic University, Islamabad 44000, Pakistan; nadeem.phdme49@iiu.edu.pk (N.A.); fazal.phdme50@iiu.edu.pk (F.U.R.); mohsin.iftikhar@iiu.edu.pk (M.I.)

2 Department of Mechanical Engineering, Capital University of Science and Technology (CUST), Islamabad 44000, Pakistan; shummaila@cust.edu.pk

* Correspondence: muhammad.haroon@cust.edu.pk

† Presented at the Third International Conference on Advances in Mechanical Engineering 2023 (ICAME-23), Islamabad, Pakistan, 24 August 2023.

Abstract: Atmospheric balloons are the cheapest source utilized by scientists to investigate various research areas. Thermal analyses of atmospheric balloons is usually performed to attain the required thermal equilibrium in the shortest timeframe. In this paper, a mathematical model was developed that numerically solved material properties (i.e., ε (emissivity of balloon surface), α_V (balloon absorptivity averaged over the visible and near infrared band), ε_{IR} (absorptivity averaged over infrared band), α_{IR} (balloon absorptivity averaged over infrared band), etc.) as well as geometric properties (i.e., A_C (cross-sectional area), A_S (surface area), M (balloon mass), etc.) by incorporating these values in the heat balance equation. An optimized code was engendered in MATLAB, which simultaneously solved input parameters and delivered the required thermal equilibrium in the shortest timeframes. This may serve as a guide to new material generation.

Keywords: atmospheric balloon; transient thermal behavior; optimization process

Citation: Aslam, N.; Haroon, M.; Rasheed, S.; Rai, F.U.; Iftikhar, M. A Thermal Analysis of Atmospheric Balloons Using Different Coating Materials. *Eng. Proc.* **2023**, *45*, 42. https://doi.org/10.3390/engproc2023045042

Academic Editors: Mohammad Javed Hyder, Muhammad Mahabat Khan, Muhammad Irfan and Manzar Masud

Published: 14 September 2023

1. Introduction

Balloon technology is considered one of the low-cost options for carrying different types of scientific payloads. The studies on balloon and airship systems have carried vital importance for many years in the fields of control system, flight testing, structural analysis and others. Payloads with long durations and massive area/mass have the capacity to fly over nearby space conditions that provide charming opportunities for the development of actual science, e.g., for many of NASA's top-priority zones for future and current missions. Missions involving balloon-borne technology carry less cost than concerned satellite missions with minimum timescales. Atmospheric balloons designed to analyze the "in situ" data and air mass tracking intelligence information at high altitudes have been studied around the globe for many years.

In the thermal analysis of a balloon, solar and ground radiation are the source of temperature elevation, while the thermal emissivity, surface reflection and convection heat of the balloon results in temperature deterioration [1]. Infrared (IR) radiation is an important candidate for affecting the thermal performance of balloons at high altitudes [2]. The rate of the temperature change on the surface of a balloon is much faster compared with other infrared targets.

This study is focused on the thermal analysis of atmospheric balloons by varying mass, size, initial deployed temperature and coating materials. The main goal of this study is to produce an optimized transient thermal model to numerically investigate the transient thermal behavior of spherical atmospheric balloons at various geometric and material properties and their acquiring of peak thermal equilibriums. It can be noticed that the heating effect produced by the external heating element is not only linked with the thermo-physical properties and surface structure of the thermal balloon but also linked with the shape and initial deployed temperature of the balloon.

2. Methodology

To study the thermal equilibrium effect of atmospheric balloons placed in space, a numerical code was developed in MATLAB (9.7 R2019b version), which describes the complete optimization behavior of thermal atmospheric objects. To obtain maximum and minimum thermal equilibrium ranges, a thermo-physical model could be generated and solved numerically for objects in daylight and at night. But, in this paper, only objects in daylight are discussed. The thermo-physical model for daylight is described below:

For a balloon illuminated by different fluxes and at a temperature T,

$$P_E = P_A + P_I \tag{1}$$

where P_E is the power emitted by the object, P_A is the power absorbed by the object and P_I is the internal power generated by the object, which is neglected (gives $P_E = P_A$).

$$P_E = A_S \times \varepsilon_{IR} \times \sigma \times T_{eq}^4 \tag{2}$$

$$P_A = A_C \cdot [(S + S_R) \cdot \alpha_V + \alpha_{IR} \cdot E] \tag{3}$$

where σ is the Stefan–Boltzmann constant (5.67×10^{-12} W/cm^2-K^4), T_{eq} is the equilibrium temperature, S is the solar flux (1360 W/m^2), S_R is the albedo flux (0.3 S), E is the earth infrared flux (240 W/m^2) and C_P represents the specific heat capacity of the balloon.

Using the fact that $\alpha_{IR} = \varepsilon_{IR}$, for an object in thermal equilibrium,

$$P_A = A_C \cdot [(S + S_R) \cdot \alpha_V + \varepsilon_{IR} \cdot E] \tag{4}$$

$$T_{eq} = \left[\frac{(A_C/A_S) \cdot \left[(S + S_R) \cdot \frac{\alpha_V}{\varepsilon_{IR}} + E \right]}{\sigma} \right]^{\frac{1}{4}} \tag{5}$$

where T_{eq} is independent of the balloon radius and for a spherical balloon, $A_C/A_S = \frac{1}{4}$.

Various parameters like mass, initial temperature, size and coating materials depend on thermal equilibrium, and the effect of these parameters is incorporated in the heat balance Equation (6) below:

$$M \times C_P \times \frac{dT}{dt} = -A_S \cdot \sigma \cdot \varepsilon \cdot T^4 + A_C \cdot [(S + S_R) \cdot \alpha_V + \varepsilon_{IR} \cdot E] \tag{6}$$

The equation is solved numerically by incorporating material and geometric properties, as shown in Table 1 below. All parameters and equations are collected from [3].

Table 1. Material properties of various coatings and geometric properties of an atmospheric balloon [3].

Material Properties of Various Coatings					Geometric Properties of the Atmospheric Balloon	
Surface Coating	α_V	ε_{IR}	$\frac{\alpha_V}{\varepsilon_{IR}}$	T_{eq}	Mass (kg)	0.3, 0.4, 0.5
White TiO_2 Paint	0.19	0.94	0.20	227	Size (m)	2.0, 3.0, 4.0
Aluminum Foil Shiny Side	0.192	0.036	5.33	454	Initial temperature (K)	300, 350, 400, 600
Black Paint	0.975	0.874	1.12	314	Cp (J/kg.K)	1090
Aluminum paint	0.54	0.45	1.20	320	Ac (m^2)	$\pi \times r^2$
					As (m^2)	$4 \times \pi \times r^2$

The optimization code simultaneously solves the various material and geometric properties of a thermal balloon to attain the desired ranges of maximum and minimum thermal equilibriums.

3. Results and Discussion

3.1. Results Validation

The validation results of the data received through the numerical process show great symmetry with the results drawn from the literature [3], as shown in Figure 1a. A simple model of a balloon was used to illustrate both the range of temperatures and equilibrium time that could be achieved for balloons. After the validation of the numerical code, an optimization process was carried out to obtain the maximum as well as minimum values of thermal equilibrium for the atmospheric balloon with corresponding time for various scientific and engineering applications. The combination of input values was 36, which led toward the optimization process.

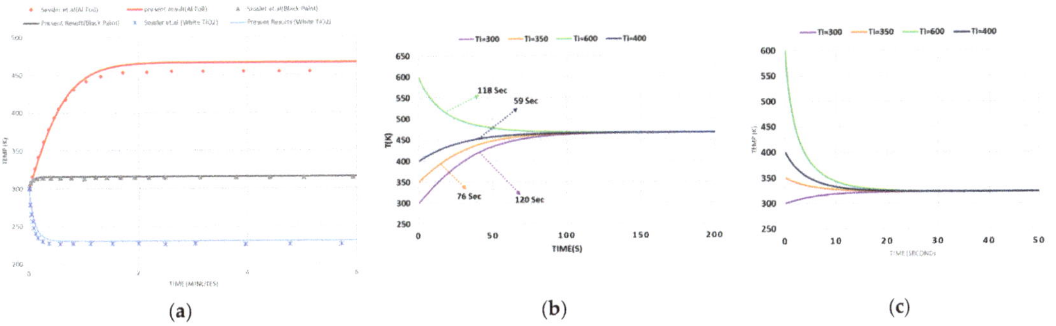

Figure 1. (**a**) Validation results for different coating materials (**b**) Temperature analysis of Aluminum foil shiny side and (**c**) Aluminum paint.

3.2. Initial Temperature Change

When the initial temperature of coated material was changed, the balloon thermal equilibrium achievement time also changed, depending upon its size, material and mass. Therefore, the main objective of this analysis was to attain thermal equilibrium in a significantly short interval of time. The rate of thermal equilibrium procurement was highly dependent upon the initial temperature of the material. The thermal equilibrium of Aluminum Foil Shiny Side was 454 K and the maximum time taken by the balloon to reach a thermal equilibrium state was 120 s when it was launched to the atmosphere at 600 K. Similarly, the balloon stretched very quickly to its final equilibrium state when it was launched at 400 K (very near to the thermal equilibrium, i.e., 59 s), as shown in Figure 1b. The same type of analyses were performed for black paint, white TIO_2 paint and aluminum paint (Figure 1c) coatings with a fixed diameter of 3 m and mass of 0.5 kg.

3.3. Thermal Mass Change

The thermal mass of the balloon also affected the speed needed to achieve thermal equilibrium for various material coatings. Three thermal masses (0.3 kg, 0.4 kg, 0.5 kg) were selected for four material coatings and launched in the atmosphere at a fixed initial temperature (300 K) and balloon size (3.0 m). It was perceived that the balloon with Aluminum Foil Shiny side paint attained thermal equilibriums as 53.5 s for 0.3 kg, 71.4 s for 0.4 kg and 120 s for 0.5 kg (Figure 2a).

3.4. The Effect of Size Change and Various Coating Materials

The size of the balloon was the most important parameter for the absorption and repulsion of radiation in space. The size variation analysis was carried out at a fixed mass (i.e., 0.5 kg) and an initial temperature of 300 K for all four types of coatings. It was observed that the magnitude of thermal equilibrium changes for the same material and the balloon with the largest size had a minimum value of thermal equilibrium but the shortest time required to attain the thermal equilibrium. The results of black and White TiO_2 paint coatings for varying diameters are enclosed in Figure 2b,c, respectively.

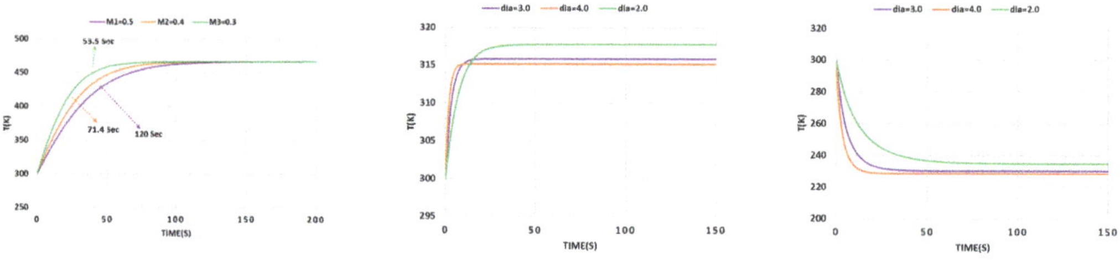

(a) (b) (c)

Figure 2. (**a**) Mass analysis of Aluminum foil shiny side, (**b**) Size analysis of black and (**c**) White TiO_2 paint coatings.

4. Conclusions

It is concluded that (1) thermal equilibrium achievement time depends on the deviation of the initial temperature from the equilibrium temperature. (2) For fixed initial temperature and balloon size, all materials with lesser thermal mass achieved thermal equilibrium at a faster rate than heavier balloons. (3) The size change of atmospheric balloons is inversely proportional to the thermal equilibrium achievement time, i.e., larger balloons reach their thermal equilibriums in a very short time interval and vice versa.

Author Contributions: Conceptualization; N.A. and M.H.; methodology; S.R. and N.A.; formal analysis; M.H. and F.U.R.; investigation; M.I. and S.R.; data curation; F.U.R. and M.I.; writing—original draft preparation; N.A.; writing—review and editing; M.H. and S.R.; supervision; N.A. All authors have read and agreed to the published version of the manuscript.

Funding: This research received no external funding.

Institutional Review Board Statement: Not applicable.

Informed Consent Statement: Not applicable.

Data Availability Statement: Not applicable.

Conflicts of Interest: The authors declare no conflict of interest.

References

1. Wei, J.; Ma, R.; Hou, X.; Zhen, D.; Tan, H. Analysis of the thermodynamical property of super-pressure balloons. *Acta Mech.* **2019**, *230*, 1355–1366. [CrossRef]
2. Dai, Q.; Xing, D.; Fang, X.; Zhao, Y. Numerical research on the thermal performance of high altitude scientific balloons. *Appl. Therm. Eng.* **2017**, *114*, 51–57. [CrossRef]
3. Sessler, A.M.; Cornwall, J.M.; Dietz, B.; Fetter, S.; Frankel, S.; Garwin, R.L.; Gottfried, K.; Gronlund, L.; Lewis, G.N.; Postol, T.A.; et al. *Countermeasures, a Technical Evaluation of the Operational Effectiveness of the Planned US National Missile Defense System*; Union of Concerned Scientists: Cambridge, MA, USA, 2000; p. 200.

engineering proceedings

MDPI

Proceeding Paper

Investigating Energy-Saving Strategies: A Numerical Study of Translucent Insulation and Phase Change Materials in Windows †

Ubaid ur Rehman [1,*], Muhammad Ashiq [2], Muhammad Ahmad Rafi [3], Usama Malik [3], Waqas Javid [1] and Mehmood ul Hassan Amjad [1]

1 Department of Mechanical Engineering, Wah Engineering College, Wah Cantt 47010, Pakistan
2 Department of Mechanical Engineering, College of Engineering and Technology, University of Sargodha, Sargodha 40100, Pakistan
3 Department of Mechanical Engineering, University of Engineering and Technology Taxila, Taxila 47080, Pakistan
* Correspondence: ubaid.rehman@wecuw.edu.pk; Tel.: +92-309-1705973
† Presented at the Third International Conference on Advances in Mechanical Engineering 2023 (ICAME-23), Islamabad, Pakistan, 24 August 2023.

Abstract: In Pakistan, residential energy consumption is predominantly devoted to ensuring thermal comfort, making energy reduction a significant task in building load management. Windows, which are notorious for having poor thermal barriers, contribute considerably to energy losses under harsh weather conditions. Incorporating high thermal inertia materials in windows, such as transparent insulation materials (TIM) and phase change materials (PCM), offers the potential for energy reduction. Using numerical simulations in ANSYS Fluent, this study compares three window types and explores their influence on interior temperature. The findings show that PCM-based windows have a low temperature increase during the melting phase, indicating their great energy-saving potential. Furthermore, PCM absorbs almost 90% of exposed heat, emphasizing its usefulness for energy saving in the Pakistani building industry.

Keywords: PCM; TIM; energy; buildings; fluent; numerical simulation

Citation: Rehman, U.u.; Ashiq, M.; Rafi, M.A.; Malik, U.; Javid, W.; Amjad, M.u.H. Investigating Energy-Saving Strategies: A Numerical Study of Translucent Insulation and Phase Change Materials in Windows. *Eng. Proc.* **2023**, *45*, 12. https://doi.org/10.3390/engproc2023045012

Academic Editors: Mohammad Javed Hyder, Muhammad Mahabat Khan, Muhammad Irfan and Manzar Masud

Published: 8 September 2023

1. Introduction

Growing worldwide worries about energy consumption, CO_2 emissions, the effects of climate change, and energy crises have generated a critical need for efficient energy reduction methods in both developed and developing countries. The construction industry stands out as a key contributor to the spectrum of energy-consuming industries, owing to the large energy needs associated with heating and cooling requirements [1]. As a result, it is critical to investigate novel techniques that might handle these energy concerns comprehensively and lead to more sustainable construction practices.

The use of phase change materials (PCMs) in the building envelope is one possible path for improving energy efficiency and lowering dependency on traditional energy sources. PCMs have an extraordinary capacity to absorb and release significant quantities of thermal energy during phase transitions such as melting and solidification while maintaining essentially constant temperatures [2–6]. Buildings can benefit from utilizing the latent heat capabilities of PCMs in a variety of ways, including greater thermal comfort, improved energy performance, lower energy consumption, and a reduction in peak temperature loads.

Traditional coated windows, which have long been used in building construction, have drawbacks such as high total heat transfer coefficients and insufficient thermal insulation [7]. As a result, researchers have concentrated on developing alternative approaches to overcome these flaws and optimize the thermal properties of glazed windows. Various solutions, such as the use of laminated glass, multilayered glass, evacuated glass, intelligent glass, and the insertion of elements, such as gases, aerogels, and phase change materials within window cavities, have been examined [8–13]. Filling the spaces between windowpanes with PCM has emerged as a particularly promising option for improving the thermal efficiency of glazed windows among these approaches. Solar thermal energy may be successfully captured during the day and gradually released at night by using PCM, such as high thermal storage density paraffin wax, effectively managing interior temperature conditions.

In the context of Pakistan's climatic conditions, especially in Islamabad, this study focuses on rigorous numerical simulations using ANSYS Fluent to investigate the possible advantages and energy-saving implications of adding PCM to double-glazed windows. We intend to assess the usefulness of PCM integration in minimizing excessive heat absorption via windows by analyzing the performance of various window layouts. Previous research in comparable fields has shown that numerical simulations and experimental findings match well, providing confidence in the use of numerical models in analyzing the performance of PCM-based systems [14].

This study addresses two key issues: Firstly, most research on phase change materials and energy-efficient windows has focused on climates outside of Pakistan. Secondly, there is limited knowledge of the energy-saving potential of different passive window variants in Pakistan's climate conditions. Therefore, this study investigates various passive window options specifically for Pakistan's climate, utilizing numerical analysis to assess their energy-saving capabilities. The establishment of a numerical model facilitates further exploration for future research.

2. Model Description

This research study delves into the intricate process of heat transfer through windows, focusing on the utilization of Phase Change Materials (PCM) and Translucent Insulation Materials (TIM). The objective is to compare and analyze the thermal behavior of three distinct types of double-glazing windows.

Solar radiation takes center stage as the primary driver of heat transfer, with conduction and convection playing secondary roles. The window prototypes shown in Figure 1 comprise a 5 mm thick glass layer, accompanied by an air gap, TIM, and PCM. Notably, the PCM utilized in this study is RT25 paraffin wax, chosen for its exceptional thermal properties, while the TIM employed is silica aerogel, renowned for its translucent insulation capabilities.

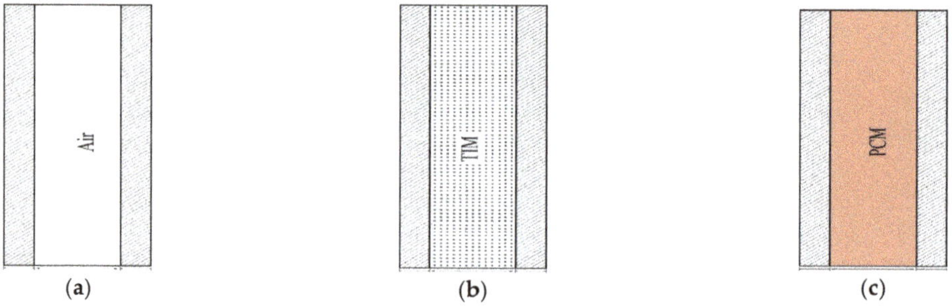

Figure 1. Schematic of (**a**) air-based, (**b**) TIM-based, and (**c**) PCM-based windows.

The model incorporates the thermophysical properties of the materials involved, including RT25 PCM and silica aerogel TIM. These properties are essential for accurately simulating the heat transfer characteristics and assessing the thermal performance of the windows as given in Table 1.

Table 1. Thermo-physical properties of RT25 and TIM.

Parameters	PCM RT25	TIM
Solidus Temperature	27 °C	-
Liquidus Temperature	29 °C	-
Latent Heat of Fusion	230 KJ/Kg	-
Specific Heat Capacity	2 KJ/Kg.K	1500
Density Liquid	0.77 Kg/L	-
Density Solid	0.88 Kg/L	0.1 Kg/L
Thermal Conductivity	0.2 W/m.K	0.018 W/m.K

3. Governing Equations and Model Validations

The Fluent software (ANSYS 2022) is used for simulating the melting process within the enclosure. The enthalpy–porosity approach is employed, calculating the melt fraction instead of tracking the melt interface. Assumptions include laminar and transient PCM melting flow, constant thermophysical properties, except for density variations with temperature, and the utilization of Boussinesq approximations for handling density changes during natural convection. In this numerical simulation, the continuity and energy equations are provided as follows:

$$\frac{\partial}{\partial x_i}(\rho u_i) = 0, \quad (1)$$

$$\frac{\partial(\rho H)}{\partial t} + \nabla.\left(\rho\vec{V}H\right) = \nabla.(k\nabla T) + S, \quad (2)$$

where ρ is the density of PCM, $\vec{V}$ is the velocity of the fluid, and S is the heat generated within PCM, which will be zero. k is the thermal conductivity of PCM and H is the total enthalpy of the system given by (Equation (3)):

$$H = h + \Delta H, \quad (3)$$

The numerical model was initially validated by solving and comparing the same problem as reported by Ravindra D. Jilte et al. [15]. The results of the present model demonstrated strong agreement with Ravindra D. Jilte et al. [15] shown in in Figure 2 with variations of less than 5%. This validation confirms the reliability of the present model in terms of formulation and grid independence. Subsequently, the validated model was utilized for additional simulations.

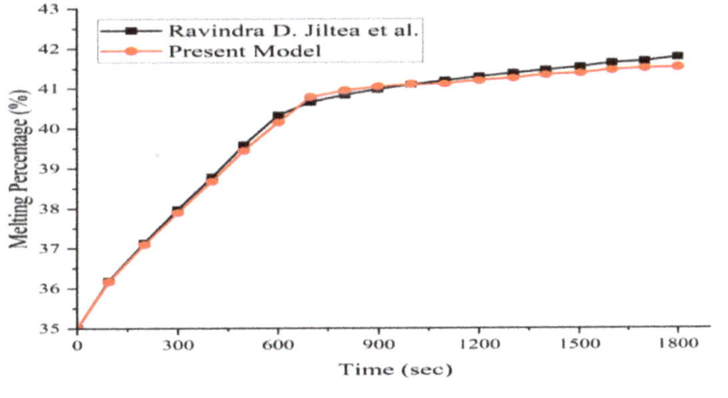

Figure 2. Validation of numerical model [15].

4. Results and Discussions

The numerical results from this study reveal important insights into the performance of different window types in controlling room temperature. With a peak ambient temperature of 45 °C in Islamabad, the desired room temperature was set at 26 °C, which led to the selection of a suitable phase change material (PCM).

Using ANSYS Fluent, the PCM's solidification and melting behavior were simulated under transient conditions. A constant heat flux of 900 W/m^2 was applied to the windows' outer glass surface. The simulations demonstrated the PCM's ability to absorb and store heat, regulating room temperature by capturing excess heat from the environment.

Comparing different window types, the PCM-based window exhibited the lowest heat flow as shown in Figure 3c, 98% lower than the air-based window. The heat flow remained constant over time for the PCM-based window, while it increased for the air-based and translucent insulation material (TIM)-based windows. After 5400 s, the TIM-based window's heat flow was 31.51% lower than the air-based window, indicating the TIM's role in reducing heat transfer.

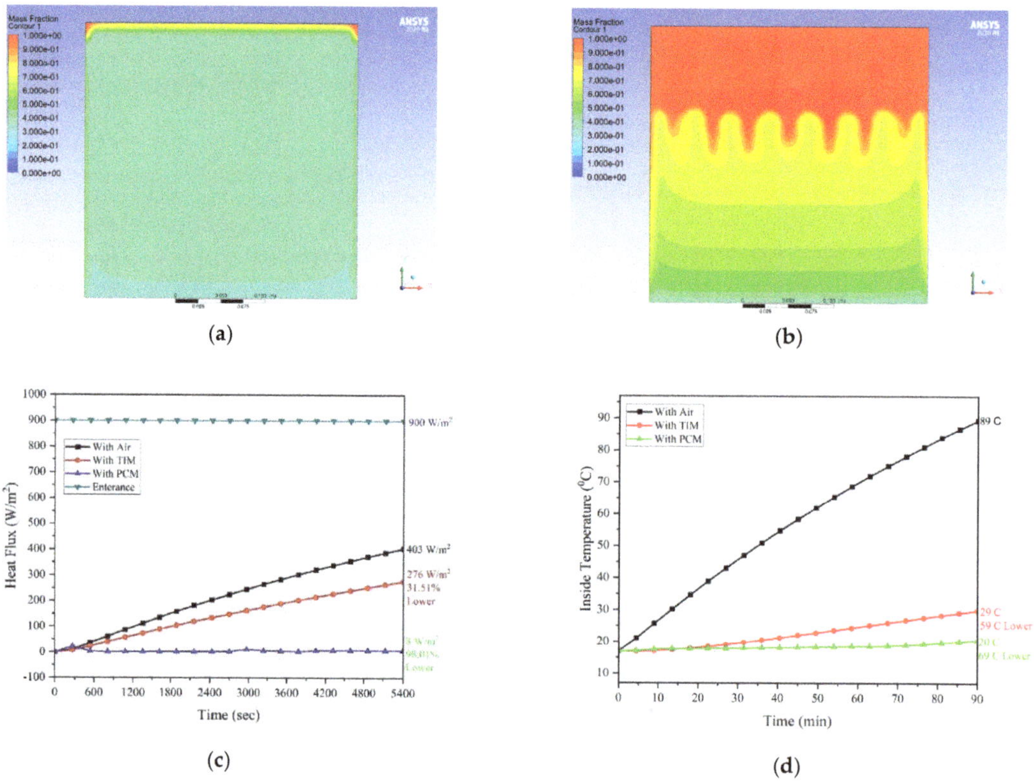

Figure 3. (**a**) Melting contour at 500 s, (**b**) melting contour at 1500 s, (**c**) comparison of heat flux, and (**d**) comparison of indoor temperature.

Regarding indoor temperatures, the PCM-based window maintained the lowest temperature shown in Figure 3d, 69 °C lower than the air-based window. The PCM-based window's indoor temperature remained relatively constant, while it rose rapidly for the air-based window and at a slower rate for the TIM-based window. The TIM-based window showed an indoor temperature 59 °C lower than the air-based window, demonstrating its effectiveness in reducing heat transfer. Additionally, there was a 10 °C temperature difference between the TIM-based and PCM-based windows, highlighting the PCM-based window's superior ability to block solar radiation.

Overall, these findings underscore the significance of window type selection in controlling indoor temperature. The PCM-based window proved to be the most efficient solution, offering potential benefits for energy efficiency and thermal comfort in buildings.

5. Conclusions

This study provides valuable insights into the heat transfer characteristics and performance of different window types in controlling the thermal conditions within a room. By carefully selecting a suitable phase change material (PCM) and employing numerical simulations using the pressure-based solver in ANSYS Fluent, we were able to investigate the solidification and melting behavior of the PCM and analyze its impact on heat transfer.

The results demonstrate that the PCM-based window exhibits the lowest heat flow among the three window types studied. With a reduction of 98% compared to the air-based window, the PCM-based window effectively absorbs and stores heat during the melting process, preventing the majority of it from entering the room. Only a small amount of 8 W/m^2 of heat is transferred to the room, resulting in improved thermal insulation.

Furthermore, the PCM-based window demonstrates temperature stability, maintaining a relatively constant indoor temperature. In contrast, the air-based window experiences a rapid temperature rise, while the TIM-based window shows a slower temperature increase. The TIM-based window also provides a notable temperature reduction of 59 °C compared to the air-based window, indicating the effectiveness of thermal interface materials in reducing heat transfer.

The observed temperature difference of 10 °C between the TIM-based window and the PCM-based window highlights the superior ability of the PCM-based window to block solar radiation. This finding underscores the potential energy efficiency benefits associated with PCM-based windows.

Overall, the findings of this study emphasize the significant impact that different window types have on indoor temperature control. The PCM-based window emerges as a highly effective solution for maintaining lower indoor temperatures and minimizing heat transfer from the external environment. These results contribute to our understanding of energy-efficient building design and highlight the potential advantages of incorporating PCM-based windows in real-world applications.

Future research could further investigate the long-term performance and durability of PCM-based windows, as well as explore the potential for optimizing PCM properties to enhance their heat transfer characteristics. Such advancements would contribute to the development of sustainable and energy-efficient building solutions.

Author Contributions: All authors contributed significantly to the conception, design, and implementation of this research study. U.u.R. played a key role in the conceptualization of the study, conducted the numerical simulations using ANSYS Fluent, and analyzed and interpreted the simulation results. U.M. and M.u.H.A. assisted in the design of the study, conducted literature reviews, contributed to the development of the numerical model, and analyzed the heat transfer dynamics and performance of different window types. M.A.R. provided expertise on the selection and properties of the phase change material (PCM), contributed to the interpretation of the results, and participated in discussions on thermal behavior and indoor temperature control. M.A. and W.J. contributed to the research design, reviewed the methodology and data analysis, and provided guidance throughout the study. This research represents a collaborative effort that draws upon the expertise and insights of all authors. All authors have read and agreed to the published version of the manuscript.

Funding: This research received no external funding.

Institutional Review Board Statement: Not applicable.

Informed Consent Statement: Not applicable.

Data Availability Statement: Not applicable.

Conflicts of Interest: The authors declare no conflict of interest.

References

1. International Energy Agency. Energy Efficiency Indicators. Available online: https://www.iea.org/topics/energy-efficiency/indicators (accessed on 4 June 2023).
2. Cui, Y.; Xie, J.; Liu, J.; Pan, S. Review of Phase Change Materials Integrated in Building Walls for Energy Saving. *Procedia Eng.* **2015**, *121*, 763–770. [CrossRef]
3. MFarid, M.M.; Khudhair, A.M.; Razack, S.A.K.; Al-Hallaj, S. A review on phase change energy storage: Materials and applications. *Energy Convers. Manag.* **2004**, *45*, 1597–1615. [CrossRef]
4. Du, K.; Calautit, J.; Wang, Z.; Wu, Y.; Liu, H. A review of the applications of phase change materials in cooling, heating and power generation in different temperature ranges. *Appl. Energy* **2018**, *220*, 242–273. [CrossRef]
5. Madad, A.; Mouhib, T.; Mouhsen, A. Phase Change Materials for Building Applications: A Thorough Review and New Perspectives. *Buildings* **2018**, *8*, 63. [CrossRef]
6. Kishore, R.A.; Bianchi, M.V.; Booten, C.; Vidal, J.; Jackson, R. Enhancing building energy performance by effectively using phase change material and dynamic insulation in walls. *Appl. Energy* **2021**, *283*, 116306. [CrossRef]
7. Cannavale, A.; Ayr, U.; Fiorito, F.; Martellotta, F. Smart Electrochromic Windows to Enhance Building Energy Efficiency and Visual Comfort. *Energies* **2020**, *13*, 1449. [CrossRef]
8. Meng, Y.; Tan, Y.; Li, X.; Cai, Y.; Peng, J.; Long, Y. Building-integrated photovoltaic smart window with energy generation and conservation. *Appl. Energy* **2022**, *324*, 119676. [CrossRef]
9. Ali, H.; Hayat, N.; Farukh, F.; Imran, S.; Kamran, M.S.; Ali, H.M. Key design features of multi-vacuum glazing for windows: A review. *Therm. Sci.* **2017**, *21*, 2673–2687. [CrossRef]
10. Allard, F.; Santamouris, M.; Alvarez, S. European Commission Directorate-General for Energy; ALTENER Programme. In *Natural Ventilation in Buildings: A Design Handbook*; James and James (Science Publishers) Limited: London, UK, 1998.
11. Balaji, D.; Sivalingam, S.; Bhuvaneswari, V.; Amarnath, V.; Adithya, J.; Balavignesh, V.; Surya, R.G. Aerogels as alternatives for thermal insulation in buildings–A comparative teeny review. *Mater. Today Proc.* **2022**, *62*, 5371–5377. [CrossRef]
12. *ISO-13786:2017*; Thermal Performance of Building Components—Dynamic Thermal Characteristics—Calculation Methods. International Organization for Standardization: Geneva, Switzerland, 2018.
13. Ashrae. Energy Standard for Buildings Except Low-Rise Residential Buildings. 2021. Available online: www.ashrae.org (accessed on 4 March 2023).
14. Al-Yasiri, Q.; Szabó, M. Numerical analysis of thin building envelope-integrated phase change material towards energy-efficient buildings in severe hot location. *Sustain. Cities Soc.* **2023**, *89*, 104365. [CrossRef]
15. Jilte, R.D.; Kumar, R.; Ahmadi, M.H.; Chen, L. Battery thermal management system employing phase change material with cell-to-cell air cooling. *Appl. Therm. Eng.* **2019**, *161*, 114199. [CrossRef]

engineering proceedings

MDPI

Proceeding Paper

Reduction in Specific Energy Consumption in Desalination through Hybrid Desalination Techniques †

Faisal Iqbal and Muhammad Asif *

Faculty of Mechanical Engineering, Ghulam Ishaq Khan Institute of Engineering Sciences and Technology, Topi 23460, Pakistan; gme2262@giki.edu.pk

* Correspondence: masif@giki.edu.pk

† Presented at the Third International Conference on Advances in Mechanical Engineering 2023 (ICAME-23), Islamabad, Pakistan, 24 August 2023.

Abstract: Freshwater scarcity is a growing concern worldwide due to increasing demand and climate change. A promising method that could provide fresh water is seawater desalination. Presently, both thermal- and membrane-based desalination technologies consume a high amount of specific energy (measured in kWh/m^3), which limits the use of seawater for drinking purposes. The development of hybrid desalination technologies has the potential to significantly reduce the energy consumption and cost of desalination, making it a more viable solution to address freshwater scarcity. By integrating the advantages of several desalination techniques and eliminating their drawbacks, hybridization can improve system performance. In the current study, a hybrid desalination system was developed by integrating the vacuum membrane distillation (VMD) output into the multi-effect distillation (MED) input. The results indicated a drop in specific thermal energy consumption (STEC) at various feed flow rates and a decrease in STEC in hybrid mode compared to stand-alone VMD.

Keywords: desalination; vacuum membrane distillation (VMD); hybrid desalination; multi effect distillation (MED); specific energy consumption

Citation: Iqbal, F.; Asif, M. Reduction in Specific Energy Consumption in Desalination through Hybrid Desalination Techniques. *Eng. Proc.* **2023**, *45*, 2. https://doi.org/10.3390/engproc2023045002

Academic Editors: Mohammad Javed Hyder, Muhammad Mahabat Khan, Muhammad Irfan and Manzar Masud

Published: 7 September 2023

1. Introduction

Heat transfer is typically involved in distillation, a thermal process based on natural distillation processes. The major thermal technologies for water desalination are multi-effect distillation (MED) and multi-stage flash (MSF), while membrane processes rely on electricity and membrane filtration. This process-based classification of water desalination is shown in Figure 1 [1]. Each desalination technology offers distinct benefits and drawbacks. Vacuum membrane distillation (VMD) is notable for its operation at lower inlet feed temperatures and smaller hydrostatic pressure [2], allowing it to utilize low-grade waste heat sources like geothermal energy, solar energy, and waste heat from power plants. This attribute gives VMD an advantage over other techniques. Additionally, VMD exhibits relatively higher permeate flux compared to other membrane distillation (MD) technologies, which is attributed to its superior thermal efficiency, making it highly suitable for various MD applications [3]. Despite these advantages, VMD faces challenges, particularly in terms of high specific heat consumption (SHC), which significantly affects its commercialization [4].

To address these challenges, researchers have begun exploring hybrid systems to enhance system performance and overcome these drawbacks. Due to the numerous benefits of hybrid systems, they have gained significant attention as a potential solution to current challenges in sustainable desalination. By combining the advantages of multiple desalination techniques while mitigating their drawbacks, hybridization holds the potential to enhance system performance. The literature demonstrates a promising future, with various reports highlighting the use of different hybrid desalination systems [5]. Conventional hybrid plants that integrate thermal technologies, such as multi-stage flash (MSF) or multi-effect evaporation (MEE) with reverse osmosis (RO), have already been implemented and are widely adopted. However, upon studying the existing literature on hybrid systems, it becomes apparent that no research has specifically examined the hybridization of VMD and MED systems, particularly regarding the specific thermal energy consumption for VMD/MED (without heat recovery) and VMD-HR/MED (with heat recovery) systems at different inlet feed temperatures. The impact of inlet feed temperature was investigated at a practical range of operation (60–70 °C), as mentioned in the majority of the literature.

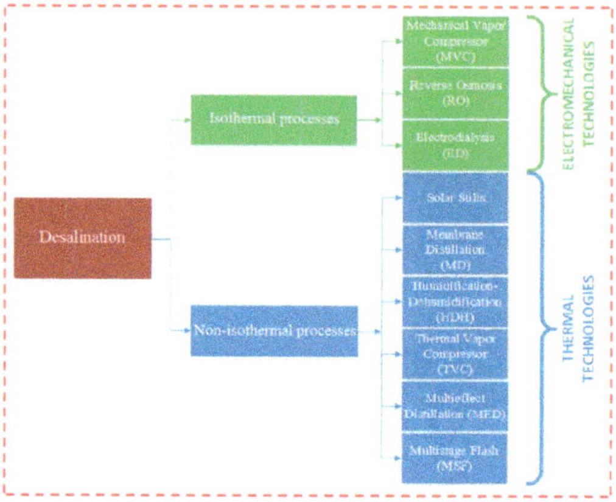

Figure 1. Classification of water desalination based on the type of process.

Thus, the main objective of this research work is to present a novel hybrid distillation model and to study the specific energy consumption of these hybrid systems. By integrating VMD with MED, there has been a significant reduction in specific thermal energy consumption. The findings of this research contribute to a deeper understanding of hybrid systems and their potential for energy efficiency improvements.

2. Modelling Procedure

To minimize the energy consumption of desalination technologies, the integration of VMD with MED has been proposed. In this model, the brine output $(m_{b,\ vmd})$ from VMD, which carries a significant amount of heat potential, is utilized as a heat source in the first effect of MED. Subsequently, the vapor generated in each preceding effect is used as a heat source for the subsequent effects. The total distillate is the combined distillate from both VMD $(m_{d,\ vmd})$ and MED $(m_{d,\ med})$, collected in the distilled water tank, as depicted in Figure 2. The reject of the VMD is fed into the four effects of MED. The total distillate flux is collected in the tank. To determine the most efficient hybrid system, key performance indicators, such as specific thermal energy consumption, were calculated for each individual system, as well as for the hybrid system, aiming to achieve sustainable desalination.

$$\mathbf{STEC_{VMD}} = \frac{Q_{f,vmd}}{m_{d,vmd}} \tag{1}$$

$$Q_f = m_f \times \mathbf{Cp} \times (\mathbf{T_{f,in}} - \mathbf{T_{f,out}}) \tag{2}$$

$$\mathbf{STEC_{hyb}} = \frac{Q_{f,vmd}}{m_{d,vmd} + m_{d,med}} \tag{3}$$

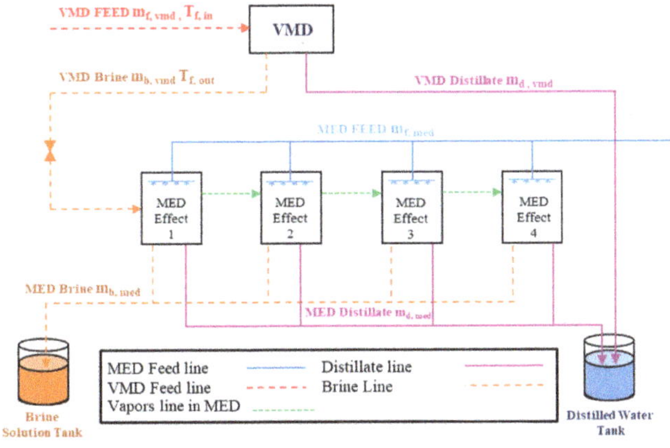

Figure 2. Block diagram of hybrid VMD/MED and VMD-HR/MED desalination system.

3. Results and Discussion

Figure 3 depicts the relationship between the increase in feed temperature and the corresponding reduction in specific thermal energy consumption in the case of VMD alone. This decrease can be attributed to the concurrent rise in the permeating flux of the membrane distillation (MD) process when higher feed temperatures are employed. Additionally, an increase in the feed flow rate results in a decrease in specific thermal energy consumption (STEC). Although there is a slight increase in the temperature drop ($T_{f,in} - T_{f,out}$) across the membrane module as the feed temperature rises, it is considerably smaller compared to the significant enhancement in permeate flux. For example, at a feed flow rate of 3500 L/h, as the feed temperature increases from 65 to 70 °C, the specific thermal energy consumption decreases from 757.45 kWh/m^3 to 734.63 kWh/m^3.

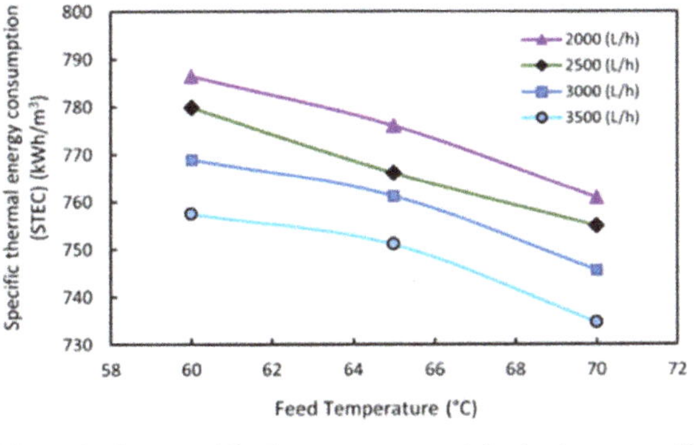

Figure 3. Impact of feed temperature and feed rate on specific thermal energy consumption of VMD alone.

Figure 4 illustrates the variation in Specific Thermal Energy Consumption (STEC) for the hybrid VMD-MED (without heat recovery) and VMD-HR/MED (with heat recovery) systems at different feed temperatures. It is evident that standalone VMD systems consume a significantly higher amount of specific thermal energy compared to hybrid systems, with or without heat recovery. In the hybrid VMD-MED system, the sensible heat from the rejected brine of VMD is fully utilized in MED as a heat source, while the latent heat of vapor in the VMD condenser is used to preheat MED feed water. Consequently, the hybrid system achieves the maximum flux production, resulting in a decrease in specific thermal energy consumption compared to VMD and hybrid systems. By integrating VMD with MED, the specific thermal energy consumption is substantially reduced to

29.96 kWh/m^3 for VMD/MED and 26.38 kWh/m^3 for VMD-HR/MED, in contrast to the initial value of 757.45 kWh/m^3 for VMD alone at a feed temperature of 60 °C. Similar analyses were conducted at inlet feed temperatures of 65 °C and 70 °C, consistently yielding positive results.

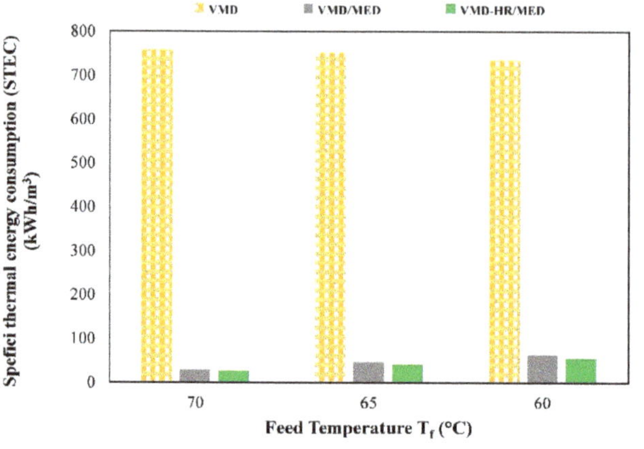

Figure 4. Comparison of STEC for hybrid system with VMD alone at different inlet feed temperatures.

It is important to highlight that there is a correlation between the rising feed temperature and the increasing trend of specific thermal energy consumption. This is due to the fact that higher temperatures raise the transmembrane temperature, leading to a decrease in brine temperature. As a result, when all variables in Equation (3) are taken into consideration, it can be observed that there is an overall escalation in specific thermal energy consumption.

4. Conclusions

In the present study, we explored innovative hybrid desalination systems to evaluate the impact on specific thermal energy consumption (STEC) in comparison to the use of vacuum membrane distillation (VMD) alone. The findings suggest that the hybrid system demonstrates a decrease in STEC at higher feed temperatures and flow rates, which presents significant advantages. The development of this hybrid system contributes to enhancing overall performance by reducing STEC. These results underscore the potential of hybrid systems in achieving enhanced energy efficiency in desalination processes.

Author Contributions: F.I. and M.A. contributed equally to this work. All authors have read and agreed to the published version of the manuscript.

Funding: Higher Education Commission (HEC) Pakistan, NRPU #5550.

Institutional Review Board Statement: Not applicable.

Informed Consent Statement: Informed consent was obtained from all subjects involved in the study.

Data Availability Statement: The data supporting our reported results is available as specified in our article.

Conflicts of Interest: The authors declare no conflict of interest.

References

1. Alam, W.; Asif, M.; Bibi, W.; Rabbi, J. Effect of Multi-staging in Vacuum Membrane Distillation on Productivity and Temperature Polarization. *J. Membr. Sci. Res.* **2022**, *8*, 534548. [CrossRef]
2. Zuo, G.; Wang, R.; Field, R.; Fane, A.G. Energy efficiency evaluation and economic analyses of direct contact membrane distillation system using Aspen Plus. *Desalination* **2011**, *283*, 237–244. [CrossRef]

3. Si, Z.; Han, D.; Gu, J.; Chen, J.; Zheng, M.; Song, Y.; Mao, N. Study on vacuum membrane distillation coupled with mechanical vapor recompression system for the concentration of sulfuric acid solution. *J. Braz. Soc. Mech. Sci. Eng.* **2019**, *41*, 473. [CrossRef]
4. Kim, Y.-D.; Kim, W.-S. Experimental and Theoretical Investigations of a Novel Multi-Stage Direct Contact Membrane Distillation Module. In Proceedings of the 2nd World Congress on New Technologies (NewTech'16), Budapest, Hungary, 18–19 August 2016. [CrossRef]
5. Ahmed, F.E.; Hashaikeh, R.; Hilal, N. Hybrid technologies: The future of energy efficient desalination—A review. *Desalination* **2020**, *495*, 114659. [CrossRef]

engineering proceedings

MDPI

Proceeding Paper

Parametric Optimization of Gravitational Water Vortex Turbines for Enhanced Torque Generation †

Muhammad Mobeen, Saqib Jaweed, Ahmad Abdullah, Shummaila Rasheed and Manzar Masud *

Department of Mechanical Engineering, Capital University of Science and Technology (CUST), Islamabad 44000, Pakistan; muhammadmobeen78692@gmail.com (M.M.); saqibjaweed12@gmail.com (S.J.); ahmadabdullah9815@gmail.com (A.A.); shummaila@cust.edu.pk (S.R.)

* Correspondence: manzar.masud@cust.edu.pk

† Presented at the Third International Conference on Advances in Mechanical Engineering 2023 (ICAME-23), Islamabad, Pakistan, 24 August 2023.

Abstract: Gravitational water vortex turbines (GWVTs) are favored over fuel-based power plants due to its lower head requirements and smaller flow rates, making it an economically viable energy source. Different process parameters such as flow rate, cone angle, blade position, and blade type directly affect the torque generation and ultimately the efficiency of GWVTs. The objective of this research is to analyze the impact of different process parameters on torque generation and to determine the optimal efficiency by optimizing the controlling parameters through Design of Experiments (DOE) in combination with Taguchi analysis. This study involved fabricating an accurate experimental setup and conducting experiments on various parameter combinations. The results obtained showed that the blade type has a major influence on the response factor (torque). This research holds potential for advancements in renewable energy.

Keywords: gravitational water vortex turbine; experimental setup; optimization; torque

1. Introduction

The gravitational water vortex turbine (GWVT) is an invention by Austrian inventor Franz Zotlöterer, which utilizes circular water flow to generate green and renewable electricity. Its low head requirements and smaller flow rates make it suitable for supplying power to millions in Pakistan, addressing the electricity crisis. GWVT systems contribute to sustainable power generation, reducing greenhouse gas emissions and minimizing environmental impacts compared to other renewable energy technologies [1]. The literature contains studies on the performance parameters and their influence on the efficiency of gravitational water vortex turbines. Rizwan ullah et al. [2] performed experimental research on the effects of runner design parameters on vortex turbine performance. Tauqir Khan et al. [3] investigated the effects of the number of blades and different blade parameters along with different design parameters. Ajay et al. [4] carried out an assessment of a gravity hydro vortex hydroelectric power plant in Nepal. They varied the head and flow rates of the gravity vortex turbine to observe the output in kilowatts. The proposed research goes beyond the studies conducted by the aforementioned researchers by conducting a more comprehensive parametric study [5]. In the proposed study, numerous variables were used to thoroughly analyze their influence on torque generation in GWVTs. The main objective of this research was to examine the combined effect of these parameters and identify the optimal combination that maximizes torque production. By conducting extensive experiments and utilizing Minitab software for data analysis, this research aimed to optimize the experimental results and identify the best parameter settings for enhanced torque generation.

Citation: Mobeen, M.; Jaweed, S.; Abdullah, A.; Rasheed, S.; Masud, M. Parametric Optimization of Gravitational Water Vortex Turbines for Enhanced Torque Generation. *Eng. Proc.* **2023**, *45*, 3. https://doi.org/10.3390/engproc2023045003

Academic Editors: Mohammad Javed Hyder, Muhammad Mahabat Khan and Muhammad Irfan

Published: 7 September 2023

2. Methodology

2.1. Experimental Apparatus and Method

Figure 1 shows a view of the manufactured experimental apparatus. Four parameters were selected for the investigation of their effects on the torque generation. The parameters include flow rate, cone angle, blade position, and blade type. The experiments were conducted at three flow rates (35 LPM, 45 LPM, and 45LPM), three cone angles (25°, 40°, and 55°), three blade positions (P1, P2, and P3) and three blade types (B1, B2, and B3). Water was circulated using a 0.50 horsepower water pump manufactured by Super Aisa, Pakistan, and the flow rates were set using the globe valve. The flowrates were measured in liters per minute (LPM) using a panel mound-type flowmeter (manufactured by Dwyer) and has a minimum capacity of 10 LPM and a maximum capacity of 90 LPM. Three different blades runners were fabricated using acrylic material. The blades were designed in such a way that they yielded the best results in terms of torque generation. In this experiment, a mini non-contact tachometer (Lutron DT-2234) and DC generator were installed to measure the torque. A Deek-Robot Wattmeter (0 to 500 watts) was also installed in the experimental setup to measure the power.

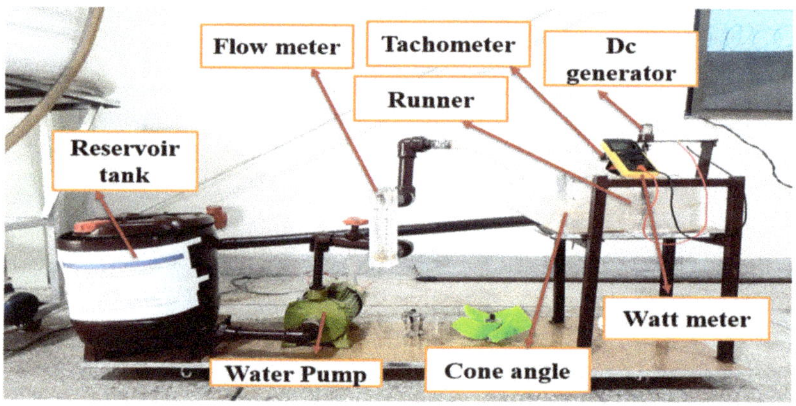

Figure 1. Experimental setup of gravitational water vortex turbine.

2.2. Design of Experiments

Design of Experiments (DOE) is a systematic statistical method used to determine how input factors impact a process. In this study, anticipated alterations to the input variables of a system or process were prepared and tested using structured tests. To assess the process's outcome, four process parameters were chosen. The steps in the DOE encompass factor selection, choosing the response variable, defining factor levels, conducting experiments, optimization, and the confirmatory test. Table 1 presents the parameters (factors) and their corresponding levels utilized in the DOE process. Among the crucial steps in DOE is the selection of the orthogonal array (OA). There were 4 factors with 3 levels for each factor. The Taguchi design was utilized, specifically the 3-level design, for the type of design. Four factors were selected for the Taguchi design, and two options were available: L9 runs and L27 runs. In this research, the decision was made to perform L27 runs due to their potential to yield better results and minimize human error as shown in Table 1. Additionally, Table 1 contains the measured response variable (torque) for the 27 experiments.

Table 1. L27 experimental design and the measured response variable (torque).

Runs	Selected Factors				Response	Runs	Selected Factors				Response
	Flow Rate (LPM)	Cone Angle (Degree)	Blade Position	Blade Type	Torque (Nm)		Flow Rate (LPM)	Cone Angle (Degree)	Blade Position	Blade Type	Torque (Nm)
1	35	25	P1	B1	2.700	15	45	40	P3	B1	4.530
2	35	25	P1	B1	3.220	16	45	55	P1	B2	0.440
3	35	25	P1	B1	3.805	17	45	55	P1	B2	0.527
4	35	40	P2	B2	0.230	18	45	55	P1	B2	0.623
5	35	40	P2	B2	0.260	19	55	25	P3	B2	0.836
6	35	40	P2	B2	0.332	20	55	25	P3	B2	0.681
7	35	55	P3	B3	1.420	21	55	25	P3	B2	1.101
8	35	55	P3	B3	1.300	22	55	40	P1	B3	4.698
9	35	55	P3	B3	1.316	23	55	40	P1	B3	5.159
10	45	25	P2	B3	2.626	24	55	40	P1	B3	5.144
11	45	25	P2	B3	4.048	25	55	55	P2	B1	4.210
12	45	25	P2	B3	3.550	26	55	55	P2	B1	4.560
13	45	10	P3	B1	3.899	27	55	55	P2	B1	5.160
14	45	10	P3	B1	4.270						

3. Results and Discussion

Taguchi analysis was used to improve the process parameters listed in Table 1 in order to identify the factor that has the greatest impact on torque generation. Additionally, Taguchi analysis was performed to forecast the ideal combination of the chosen elements, and then the confirmatory test was utilized to verify the findings. The results of the response with respect to the approach are summarized in Table 2.

Table 2. Torque response table.

Level	Flow Rate	Cone Angle	Blade Position	Blade Type
1	1.6203	2.5074	2.9241	4.0393
2	2.7237	3.1691	2.7751	0.5590
3	3.5054	2.1730	2.1503	3.2512
Delta	1.8851	0.9962	0.7737	3.4804
Rank	2	3	4	1

According to Table 2's delta and rank values, the blade type has the greatest impact on torque generation, followed by flow rate, cone angle, and blade location. Figure 2 displays the mean torque values versus each input parameter as a graph.

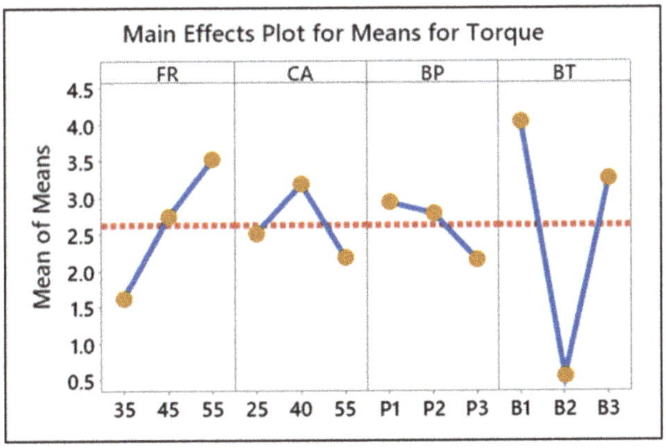

Figure 2. Mean torque vs. the input parameters.

From the analysis of the main effects plots for torque in Figure 2, it was concluded that the optimized parameters are 55 LPM for flow rate, 40° for cone angle, P1 for blade position, and B1 for blade type. Later, the torque results for the ideal combination—5.788 Nm—were predicted using Taguchi analysis. An experimental test was conducted to confirm the validity of the Taguchi analysis findings. As a consequence of the confirmatory test, a torque measurement was made, and a value of 5.76 Nm was recorded. When the results of the confirmatory test and Taguchi analysis were compared, the percentage error was 0.3%, which is noticeably low.

This particular study develops GWVT technology for better torque generation in the production of sustainable energy, boosting confidence in using the ideal combination. The findings demonstrated how crucial careful experimentation and statistical analysis are to maximizing GWVT performance, expanding the possibilities for renewable energy and fostering a more environmentally friendly and sustainable future.

4. Conclusions

A gravitational water vortex turbine was fabricated and optimized using Minitab software to investigate torque. After conducting experiments and analyzing the data, the optimal combination of factor levels for maximum torque was determined to be a 55 LPM for flow rate, 40° for cone angle, curved incline Blade B1as the blade type, and P1 as the blade position. Minitab predicted the torque to be 5.788 Nm for this combination. Subsequent confirmatory tests on an experimental set-up confirmed that the maximum torque was indeed achieved with this combination. The study's findings provide conclusive evidence and valuable guidance for future implementations of these turbines. The systematic approach, including experimentation and data analysis, has successfully enhanced torque generation and improved turbine efficiency. This research contributes to the advancement of water vortex turbines for sustainable energy applications.

Author Contributions: Conceptualization, M.M. (Manzar Masud); methodology, M.M. (Manzar Masud) and S.R.; software, M.M. (Muhammad Mobeen), S.J. and A.A.; validation, M.M. (Manzar Masud) and S.R.; formal analysis, M.M. (Muhammad Mobeen), S.J. and A.A.; investigation, M.M. (Muhammad Mobeen), S.J., A.A. and S.R.; data curation, M.M. (Manzar Masud) and S.R.; writing—original draft preparation, M.M. (Muhammad Mobeen), S.J. and A.A.; writing—review and editing, M.M. (Manzar Masud) and S.R.; supervision, M.M. (Manzar Masud); project administration, M.M. (Manzar Masud). All authors have read and agreed to the published version of the manuscript.

Funding: This research received no external funding.

Institutional Review Board Statement: Not applicable.

Informed Consent Statement: Not applicable.

Data Availability Statement: Not applicable.

Conflicts of Interest: The authors declare no conflict of interest.

References

1. Bajracharya, T.R.; Shakya, S.R.; Timilsina, A.B.; Dhakal, J.; Neupane, S.; Gautam, A.; Sapkota, A. Effects of Geometrical Parameters in Gravitational Water Vortex Turbines with Conical Basin. *J. Renew. Energy* **2020**, *2020*, 5373784. [CrossRef]
2. Ullah, R.; Cheema, T.A. Experimental Investigation of Runner Design Parameters on the Performance of Vortex Turbine. *Eng. Proc.* **2022**, *23*, 14.
3. Khan, T.; Asif, M.M.; Ahmed, H.; Islam, M.; Harun, Z. Design and Development of a Vortex Turbine for the Hilly Regions of Bangladesh. In Proceedings of the 2nd International Seminar of Science and Applied Technology (ISSAT 2021), Video Conference, 12 October 2021.
4. Jha, A.K.; Subedi, D.U.; Timilsina, A.B. Assessment of Gravitational Water Vortex Hydropower Plant in Nepal. *Int. Res. J. Eng. Technol.* **2018**, *5*, 2679–2691.
5. Nishi, Y.; Suzuo, R.; Sukemori, D.; Inagaki, T. Loss Analysis of Gravitation Vortex Type Water Turbine and Influence of Flow Rate on the Turbine's Performance. *Renew. Energy* **2020**, *155*, 1103–1117. [CrossRef]

engineering proceedings

MDPI

Proceeding Paper

A Techno-Economic Viability Analysis of Wind-Powered Power Plants in Panjgur, Balochistan †

Habeeb Ur Rehman [1], Hafeez Ullah [1], Muhammad Zakarya [1], Fasihullah Kakar [1], Qamar ud Din Abid [1,2], Muhammad Haroon [2,3,*], Muhammad Arsalan Munawar [2] and Muhammad Aqil Khan [2]

1 Department of Mechanical Engineering, Balochistan University of Information Technology, Engineering & Management Sciences, Quetta 87300, Pakistan; 51666@student.buitms.edu.pk (H.U.R.); 49793@student.buitms.edu.pk (H.U.); 49314@student.buitms.edu.pk (M.Z.); fasihullah.kakar@buitms.edu.pk (F.K.); qamaruddin.abid@buitms.edu.pk (Q.u.D.A.)
2 Department of Mechanical Engineering, International Islamic University, Islamabad 44000, Pakistan; muhammad.phdme47@iiu.edu.pk (M.A.M.); muhammad.bsme879@iiu.edu.pk (M.A.K.)
3 Department of Mechanical Engineering, Capital University of Science and Technology (CUST), Islamabad 44000, Pakistan
* Correspondence: muhammad.haroon@cust.edu.pk

† Presented at the Third International Conference on Advances in Mechanical Engineering 2023 (ICAME-23), Islamabad, Pakistan, 24 August 2023.

Abstract: This research paper explores the potential of wind turbine power plants as a sustainable energy solution for addressing energy deprivation in the Panjgur region of Balochistan. Balochistan, an area with limited access to traditional energy sources, suffers from chronic electricity shortages, hindering its socio-economic development. This study examines the feasibility of harnessing wind energy through establishing wind turbine power plants. This paper discusses the technical aspects of wind turbine technology, including turbine designs, capacities, and power generation potential. The findings highlight the viability of wind turbine power plants as a renewable energy source to alleviate the energy crisis in the region and promote sustainable development at a very reasonable price.

Keywords: wind energy; sustainable energy; renewable energy; techno-economic analysis

Citation: Rehman, H.U.; Ullah, H.; Zakarya, M.; Kakar, F.; Abid, Q.u.D.; Haroon, M.; Munawar, M.A.; Khan, M.A. A Techno-Economic Viability Analysis of Wind-Powered Power Plants in Panjgur, Balochistan. *Eng. Proc.* **2023**, *45*, 15. https://doi.org/10.3390/engproc2023045015

Academic Editors: Mohammad Javed Hyder, Muhammad Mahabat Khan, Muhammad Irfan and Manzar Masud

Published: 11 September 2023

1. Introduction

Pakistan's energy issue has become a recurring headline, as power usage surges due to electronic devices, advancing technology, and population growth. The country's electricity consumption of 89,361 MW surpasses its generation capacity of 41,557 MW [1], predominantly fueled using coal, natural gas, and petroleum products. However, the environmental risks associated with fossil fuels necessitate a shift towards cleaner and more sustainable energy sources. Pakistan must explore its significant potential for renewable energy to overcome its worst ever crisis and improve the living standards of its citizens.

Among the provinces, Balochistan stands out due to its severe energy deprivation despite being the largest province with minimal industry and population. This study concentrates on the rural areas of Balochistan, aiming to enhance the quality of life by proposing a steady supply of clean, green energy and empowering the local population.

The fuel-wise breakdown of electricity generation in recent times reveals shifts in the energy mix. Hydropower's contribution declined to 24.7%, while RLNG's (Degasified Liquefied Natural Gas) share increased to 23.8%. Coal remained constant, although total installed capacity increased. Natural gas intake decreased from 12.15% to 8.5%, highlighting the transition towards renewable energy [2].

Utilizing renewable energy sources is crucial for several reasons. Firstly, it addresses the urgent need for environmental protection and the fight against global warming caused by greenhouse gas emissions from fossil fuels.

The literature survey indicates the importance of renewable energy generation resources and their incremental share in the energy mix of Pakistan. The most important globally employed renewable energy resources are solar energy, wind energy, hydropower, geothermal energy, bioenergy and ocean thermal energy conversion cycles (OTEC) [3]. This research project aims to evaluate the techno-economic viability of the building wind power plants in Panjgur (a rural area of Balochistan) by using relatively novel tools and an energy consumption data collection methodology.

2. Materials and Methods

This study utilizes several analytic frameworks used by different authors in previous works [4]. Wind speed data were collected from various sources, such as NASA and Globalwindatlas.info [5]. A sample of power consumption-related data of the region was collected through a questionnaire, which included load shedding duration, number of energy savers and 100-watt bulbs, tube lights, mobile phones, celling and table fans, water pumps, computers/laptops, electric geysers, washing machines, refrigerators, etc. This type of data collection provided an estimate of electricity consumption monthly/per day. The data were then analyzed to determine the wind energy potential for the project. Factors like elevation and geographical variables were considered, and wind speed data were analyzed to calculate maximum, minimum, and mean wind speeds. Different locations were assessed, and Panjgur, at a height of 10 m, was found to have the most suitable wind potential based on monthly variations and simulation using Homer Pro software and RET Screen; manual calculations were also performed. Financial modelling tools were used to assess the wind energy project's financial and economic viability, considering capital costs, operational and maintenance costs, energy tariff rates, financing alternatives, and social and environmental aspects.

Qblade software was used to calculate the power produced by different speeds using a NACA 4415 blade profile. This data were then fed to RET screen, which in turn created a power curve and an estimate of the total energy produced by a turbine in KWh. The lifespan of the plant was set at 20 years, and total production in this lifetime was used for economic calculations. The cost of the plant included purchase cost, shipping, and maintenance. The levelized cost of electricity (LCOE) was calculated using Equation (1).

$$\text{LCOE} = \text{Total Cost of the plant/Total Electricity Produced} \quad (1)$$

At various heights of 50, 10, and 80 m above ground level, wind speed measurements were taken. It was discovered after a careful comparison of these observations that wind speed tended to climb with height (Figure 1a). A height of 10 m was determined to be the best option, considering the project's nature and modest size. Figure 1b presents average wind speed variations for Panjgur during the whole year at a 10 m height. This figure is generated through Homer Pro software (version 3.15.2).

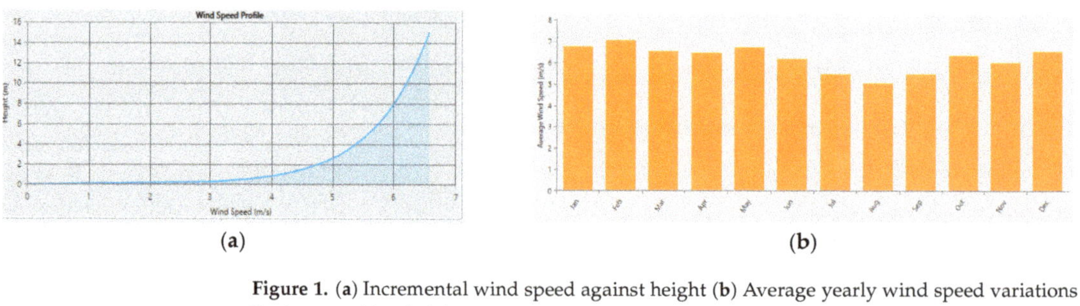

(a) (b)

Figure 1. (a) Incremental wind speed against height (b) Average yearly wind speed variations for Panjgur at a 10 m height.

A load profile divided the load into four phases. A design considered the load demand and load calculation to obtain the daily maximum consumption.

- 2 to 3 houses (3 KW);
- 4 to 6 houses (6 KW);
- 6 to 8 houses (10.5 KW);
- 15 to 20 houses (25.5 KW).

The details of the hardware suggested in this work along with the purchase cost, shipping, installation, and maintenance cost are tabulated in Table 1. Similar calculations were conducted for all plants, but Table 1 presented here is only for the 6 KW plant.

Table 1. Economical and Technical Specifications of a 6 KW plant.

Item	Turbine	Battery	Controller	Invertor
Type	M2-2 (1500) HAWT for Home	Li-ion Battery 48 V 200 ah lifepo4	Automatic MPPT Charge Controller Wind Charge Controller	Pure Sine Wave Power Inverter
Unit Price (PKR)	84,000	72,000	50,700	13,800
Quantity	4	1	1	1
Life Span (Years)	20	6.7	5	10~15
Annual Maintenance cost (Excluding Consumables)	2–3% of initial cost	2–3% of initial cost	2–3% of initial cost	2–3% of initial cost
Shipping Cost	336,000 PKR	432,000 PKR	202,800 PKR	27,600 PKR
Total Cost (20 Years)	806,400 PKR	921,600 PKR	42,5880 PKR	60,720 PKR
Total Cost of the Plant/ Net Present Cost (NPC)	2,214,600 PKR			
Energy Production	13,140 KWh (Annual) and 262,800 KWh (20 years)			
Payback period (N)	9.36 years (Cost/Annual earning)			
LCOE	8.42 PKR			

3. Results and Discussion

Selected turbines, controllers, batteries, and inverters were chosen based on their cost-effectiveness, efficiency, and low maintenance. Shipping charges were calculated based upon the supplier's estimates. All four power plants with loads of 3 kW, 6 kW, 10.5 kW, and 25.5 kW were simulated using QBlade and RET software. The analysis included both economic and technical aspects. For the 6 KW load, calculations are provided in Table 1. Efficiency was calculated at 25% for the power production capacity of the turbines. Prices for different turbines were acquired and the most economical option was chosen. The lifespan of all equipment was not the same. Some equipment needed to be purchased twice or thrice to meet the requirements of twenty years for the whole plant. The power production capability of the turbine against different wind speeds is presented in Figure 2 (for the 6 KW plant only).

It can be observed from Figure 2 that both energy (MWh) and power (KWh) show an incremental trend for incremental wind speed variation (m/s). Power shows steep and smooth variations as compared with energy, which shows the importance of wind speed for power generation. On the other hand, the incremental trend of energy is nonlinear. This includes two types of wind speeds, namely cut-in and cut-off speeds.

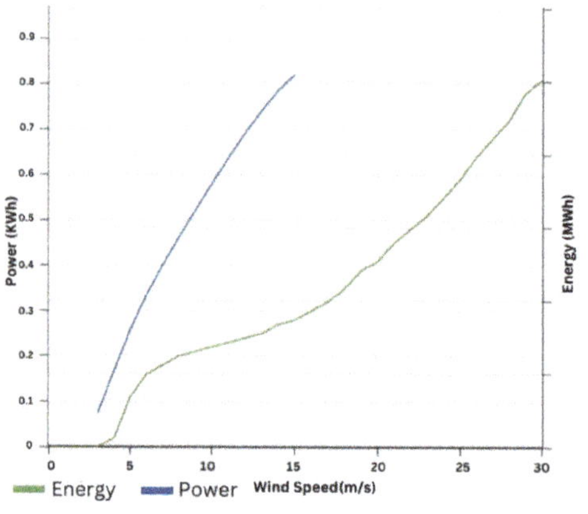

Figure 2. Power and energy variations for varying wind speeds.

4. Conclusions

This study was simulated for four different capacity projects (i.e., 3 KW, 6 KW, 10.5 KW, and 25.5 KW) in Panjgur (Balochistan) using Homer Pro and RET Screen tools. Analysis was performed at different turbine heights, i.e., 10 m, 50 m, and 80 m. A 10 m height measurement was found to have the most suitable wind potential based on monthly variations as per small-scale project standards. Variations in power and energy were also investigated for 0 to 30 m/s wind speed spans. A minimum power of 0.06 KWh and maximum energy of 0.8 MWh was achieved at 2.75 m/s and 30 m/s wind speeds, respectively.

Author Contributions: Conceptualization, F.K., Q.u.D.A. and M.H.; methodology, H.U.R., H.U. and M.Z.; software simulations, H.U., M.Z., M.A.M. and M.A.K.; writing—original draft preparation, H.U.R, H.U. and M.Z.; writing—review and editing Q.u.D.A., M.H. and M.A.M.; supervision, Q.u.D.A., F.K. and M.H. All authors have read and agreed to the published version of the manuscript.

Funding: This research received no external funding.

Institutional Review Board Statement: Not applicable.

Informed Consent Statement: Not applicable.

Data Availability Statement: https://globalwindatlas.info, https://power.larc.nasa.gov/data-access-viewer (assessed on 25 February 2023).

Conflicts of Interest: The authors declare no conflict of interest.

References

1. Pakistan Economic Survey 2021–2022. Available online: https://www.finance.gov.pk/survey/chapter_22/PES14-ENERGY.pdf (accessed on 13 January 2023).
2. Pakistan and Fossil Gas—Global Energy Monitor. Available online: https://www.gem.wiki/Pakistan_and_fossil_gas#cite_note-:0-5 (accessed on 23 June 2023).
3. Haroon, M.; Ayub, A.; Sheikh, N.A.; Ahmed, M.; Shalaby, A.-B. Harnessing Ocean Thermal Energy from Offshore Locations in Pakistan Using an Organic Rankine Cycle. *Eng. Proc.* **2022**, *23*, 24. [CrossRef]
4. Effect of Yaw Angle on the Global Performances of Horizontal Axis Wind Turbine—QBlade Simulation—IOPscience. Available online: https://iopscience.iop.org/article/10.1088/1757-899X/595/1/012047/meta (accessed on 24 June 2023).
5. Global Wind Atlas. Available online: https://globalwindatlas.info (accessed on 24 June 2023).

engineering proceedings

MDPI

Proceeding Paper

Experimental and Computational Analysis of 14 Gauge Rooftop Solar PV Mounting Structure against Wind in Pakistan †

Arhaam Mubarak * and M. Javed Hyder

Department of Mechanical Engineering, Capital University of Science and Technology (CUST), Islamabad 44000, Pakistan; mjavedhyder1@gmail.com or javed.hyder@cust.edu.pk

* Correspondence: arhaamcust@gmail.com

† Presented at the Third International Conference on Advances in Mechanical Engineering 2023 (ICAME-23), Islamabad, Pakistan, 24 August 2023.

Abstract: Fourteen-gauge solar PV mounting structure installations represent 60% of total solar PV installations in Pakistan. At least 34 (1.72%) structures were damaged on 13 June 2021 with an average 56 km/h wind speed on that day. Therefore, a need arose for a thorough analysis of the 14-gauge solar PV mounting structure. Computational analysis was carried out for 14-gauge solar PV mounting structures. The equivalent stress and total deformation in the structure were calculated computationally. Using computational analysis, it was found that 14-gauge solar PV mounting structures deformed at an 80 km/h wind velocity and broke at a 110 km/h wind velocity. Hence, 14-gauge solar PV mounting structures do not fulfil the building code of Pakistan.

Keywords: solar PV mounting structure; equivalent stress; total deformation

Citation: Mubarak, A.; Hyder, M.J. Experimental and Computational Analysis of 14 Gauge Rooftop Solar PV Mounting Structure against Wind in Pakistan. *Eng. Proc.* **2023**, *45*, 14. https://doi.org/10.3390/engproc2023045014

Academic Editors: Muhammad Mahabat Khan, Muhammad Irfan and Manzar Masud

Published: 11 September 2023

1. Introduction

The damage percentage of solar PV structures has grown in various locations in Pakistan. On 13 June 2021, at least 34 (1.72%) solar mounting 14-gauge mild steel (MS) galvanized PV structures were damaged by wind in Islamabad and Rawalpindi alone. A total of 8417 Net-Metering Licenses for solar PV on-grid system installation were issued as per the 2020–2021 NEPRA annual report [1]. For the Islamabad and Rawalpindi district, 1976 licenses were issued and at least 34 (1.72%) solar PV mounting structures were damaged in one day [1]. Figure 1 shows a damaged solar PV mounting structure. According to the metrological office of Pakistan, strong wind speeds from 60 km/h to 100 km/h were recorded in past few years. Many solar PV mounting structures were damaged due to high wind speeds. Most damaged solar PV mounting structures were 14-gauge and 16-gauge galvanized steel rooftop structures that did not withstand the wind load and damaged below the safety requirement mentioned in the building code of Pakistan [2].

Figure 1. Damages caused by wind to solar PV mounting structures at different locations. Red mark shows the damaged solar PV mounting structure.

2. Methodology

The 14-gauge solar PV mounting structure with two bolt-mounted solar panels was modelled using Creo Parametric 7.0 and imported into Ansys for FSI computational analysis. The LONGi 545W Solar Panel was modelled according to real-life dimensions [3]. Eight components, including main support with a 2500 mm length and 74 mm width, back support with a 1060 mm length and 79 mm width, upper foot with a 152 mm length and 83 mm width, and lower foot with a 150 mm length and 80 mm width, make up standard 14 gauge (2 mm thickness) solar PV mounting structure, which is extensively used in Pakistan, as shown in Figure 2. The angle of 24° was fixed throughout the study [4].

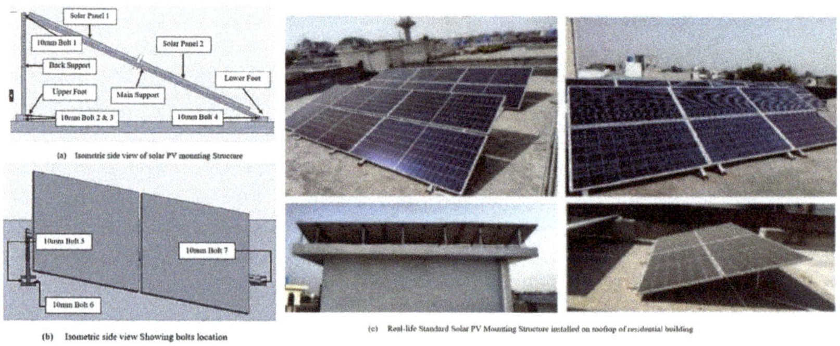

Figure 2. The 14-gauge solar PV mounting structure with 2 solar PV modules.

2.1. Experimental Setup

Single-Width Single-Inlet (SWSI) high-pressure centrifugal fan arrangement was employed for the validation of this study. Using a digital air velocity meter, the velocity at the fan outlet was measured at 31.67 km/h (1732 fpm). The fan and solar PV mounting structure arrangement is shown in Figure 3. The air velocity of 10.3 km/h was constant at the inlet of the solar PV mounting structure. The air velocity at different locations was measured. The height and width of planes perpendicular to the wind direction are 872 mm (2.86 ft) and 2256 mm (7.39 ft), respectively. The air velocity was compared with computational results for validation.

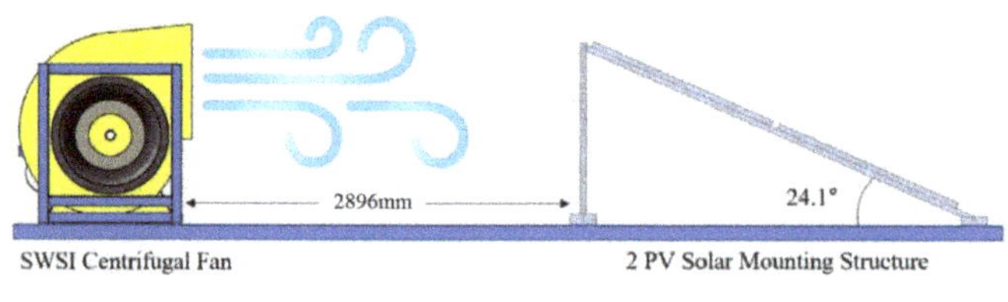

Figure 3. Centrifugal fan setup with solar PV mounting structure.

2.2. Computational Setup

The Fluid Flow (Fluent) package of ANSYS Workbench 2021 R2 was employed in this work to investigate the equivalent stress and deformation in rooftop 14-gauge solar PV mounting structures at various wind speeds. Coarse (176,636 elements), medium (344,321 elements) and fine (487,074 elements) meshes were generated to validate mesh independence using the k$-\omega$ SST model, as shown in Figure 4.

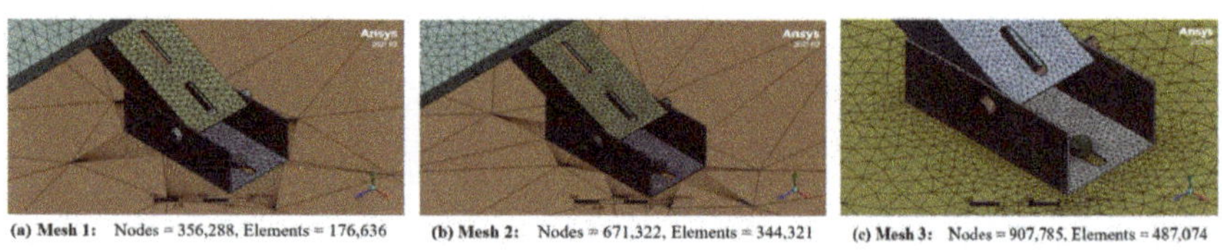

Figure 4. Nodes and elements of 3 tetrahedral meshes.

The difference in results between mesh with 671,322 nodes and one with 907,785 nodes is negligible compared to mesh with 356,288 nodes. The mesh with 907,785 nodes and 487,074 elements was considered sufficient to proceed with the computational analysis of a rooftop solar PV mounting structure subject to wind load. For experimental validation, a 10.3 km/h wind velocity was subjected from the inlet of the fluid domain. The results were validated with less than 10% error in experimental and computational results.

The boundary conditions for ANSYS computation were set up. At the inlet of fluid domain, a uniform wind velocity of 10.3 km/h, 25 km/h, 50 km/h, 75 km/h, 100 km/h, and 120 km/h were applied with a turbulence intensity of 5% and air density of 1.225 kg/m^3.

2.3. Fluid–Structure Interaction (FSI)

After successful Fluent simulation, results from Fluid Flow (Fluent) were imported into a static structural setup. After connecting Fluid Flow (Fluent) with static structural qualities, material properties were initialized. Then, the pressure was imported for all bodies from Fluent to the static structural model. Figure 5 shows the imported pressure from Fluent at a 75 km/h wind velocity. The pressure on top of the solar panels was very low and near atmospheric pressure. The variation in contours is due to vortices of turbulent flow. The pressure at the back of the solar panel is highest because drag and lift forces both act on it [5].

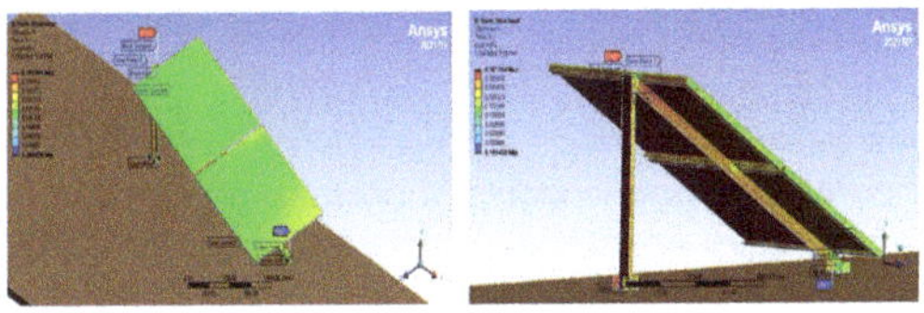

Figure 5. Imported pressure (MPa) from FSI simulation.

3. Results and Discussion

A centrifugal fan was placed and set up 2.896 m (9.5 feet) behind two PV panels fixed on the 14-gauge solar PV mounting structure. The air velocity at the intake plane at a distance of 9 feet varied from 9.9 km/h to 11.2 km/h at various positions. The average air velocity was 10.3 km/h since it was present at the majority of the sites. Throughout the experiment, the wind velocity was measured using a digital anemometer. To record the air velocity value, the digital anemometer was kept stable for five seconds at each location. With 25 points on each plane, a similar procedure was used for each plane. To determine the average air velocity at 25 points on each plane, the experiment was run three times.

Computational Results

The 14-gauge solar mounting structure is stable when the wind velocity is at the start of the strong gale region on the Beaufort wind force scale, which is from 75 km/h to 88 km/h, as shown in Figure 6. In the strong gale region, which starts at a 75 km/h wind velocity, the equivalent stress on the main support of the structure is 221.62 MPa, which is less than the yield strength of steel (250 MPa). The structure failed at a 100 km/h wind velocity. However, structural stability is required up to 120 km/h according to the Building Code of Pakistan [2].

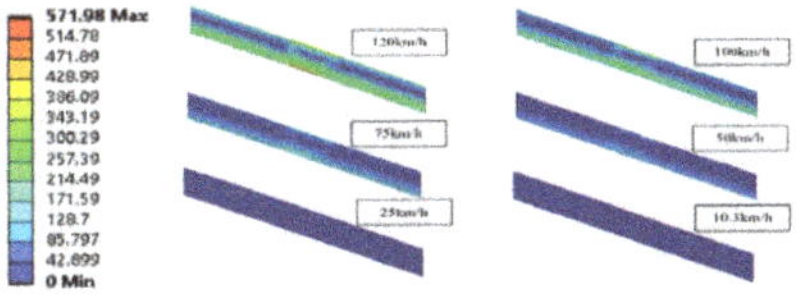

Figure 6. Side view of main support showing equivalent stress (MPa) of 14-gauge solar PV mounting structure at different wind velocities.

The total deformation of the solar PV mounting structure is shown in Figure 7. In this case, the main support plays an important role because the whole structure is in contact with the main support due to the preload on the solar panels. The deflection in the main support is 16.232 mm at a 75 km/h wind velocity. After a 75 km/h wind velocity, the deformation in the main support increased and the structure broke.

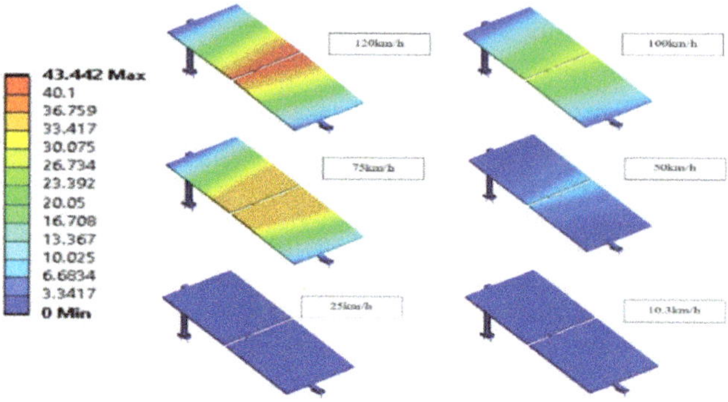

Figure 7. Total deformation (mm) of symmetric 14-gauge structure in isometric view.

4. Conclusions

It was found that 14-gauge solar PV mounting structures installed in Pakistan do not meet the requirements of the Pakistani building code. The 14-gauge structure is being installed on a large scale in Pakistan and, using computational analysis, it was concluded that the structure is not safe at all and exceeds the yield strength at an 80 km/h wind velocity. The main support bears the whole load of the solar PV mounting structure because solar panels are bolted to the main support. It has been observed that most of the damaged solar PV mounting structures were installed with up to 550 W power ratings. The relevant authorities should allow PV modules with a lower area and more power that would bear recommended wind load. The solar PV mounting structure should be designed and built in accordance with the American Society of Civil Engineers (ASCE) worldwide standard.

Author Contributions: Conceptualization, review and supervision of project M.J.H.; Experimentation, investigation, and data processing A.M. All authors have read and agreed to the published version of the manuscript.

Funding: This research received no external funding.

Institutional Review Board Statement: Not applicable.

Informed Consent Statement: Informed consent was obtained from all subjects involved in the study.

Data Availability Statement: The data presented in this study are available on request.

Conflicts of Interest: The authors declare no conflict of interest.

References

1. NEPRA. Annual Report 2020–21. 2021. Available online: https://nepra.org.pk/publications/AnnualReports/AnnualReport2020-21.pdf (accessed on 27 July 2023).
2. Ministry of Housing & Works Experts Committee; PEC; NESPAK; International Code Council (ICC); American Concrete Institute (ACI). Building Code of Pakistan—Seismic Provisions—2007. 2007. Available online: https://www.pec.org.pk/wp-content/uploads/2021/05/Building-Code-of-Pakistan-Seismic-Provisions-2007.zip (accessed on 27 July 2023).
3. LONGi Solar LR5-72HPH-535M-555M (545W) Solar Panel. 2021. Available online: https://static.longi.com/L_Gi_LE_T_TMD_059_107_LR_5_72_HPH_535_555_M_35_35_and_15_V14_895e6db05e.pdf (accessed on 8 September 2023).

4. Shabbir, N.; Usman, M.; Jawad, M.; Zafar, M.H.; Iqbal, M.N.; Kütt, L. Economic analysis and impact on national grid by domestic photovoltaic system installations in Pakistan. *Renew. Energy* **2020**, *153*, 509–521. [CrossRef]
5. Pantua, C.A.J.; Calautit, J.K.; Wu, Y. A fluid-structure interaction (FSI) and energy generation modelling for roof mounted renewable energy installations in buildings for extreme weather and typhoon resilience. *Renew. Energy* **2020**, *160*, 770–787. [CrossRef]

engineering proceedings

Proceeding Paper

Tribological Testing and Analysis of Surface-Textured Metal Surfaces †

Rizwan Siddiqui [1,*], Mian Ashfaq Ali [1], Waqar Arshad [2], Muhammad Shaban [2], Muhammad Usman [2] and Muhammad Zulfiqar [2]

1 School of Mechanical and Manufacturing Engineering (SMME), National University of Science and Technology (NUST), Islamabad 44000, Pakistan; mian.ashfaq@smme.nust.edu.pk
2 Department of Mechanical Engineering, Capital University of Science and Technology (CUST), Islamabad 44000, Pakistan; vickywaqar704@gmail.com (W.A.); shabanqureshi411@gmail.com (M.S.); ukuk81045@gmail.com (M.U.); muhammad.zulfiqar@cust.edu.pk (M.Z.)
* Correspondence: rizwan.siddiqui@cust.edu.pk; Tel.: +92-335-5525093
† Presented at the Third International Conference on Advances in Mechanical Engineering 2023 (ICAME-23), Islamabad, Pakistan, 24 August 2023.

Abstract: This study investigates the friction-reducing potential of "micro surface texturing" on H-13 steel metal surfaces. Tribological tests were performed on a ball-on-plate reciprocating tribometer using SAE30 oil. Circular micro textures with a 0.1 mm diameter and 5–8 μm depth were applied at varying area densities (5%, 10%, and 15%). The results were compared to identify the most effective area density for the micro textures. A friction reduction of 37.95% was observed compared to the untextured sample at 15% area density. The average friction for all the textures is reduced, which shows the effectiveness of laser surface texturing on tribological performance.

Keywords: friction reduction; surface texturing; tribological testing; micro-scale roughness; area density; engineering applications; ball-on-plate system

Citation: Siddiqui, R.; Ali, M.A.; Arshad, W.; Shaban, M.; Usman, M.; Zulfiqar, M. Tribological Testing and Analysis of Surface-Textured Metal Surfaces. *Eng. Proc.* **2023**, *45*, 4. https://doi.org/10.3390/engproc2023045004

Academic Editors: Mohammad Javed Hyder, Muhammad Mahabat Khan, Muhammad Irfan and Manzar Masud

Published: 8 September 2023

1. Introduction

Tribology, the study of friction, wear, and lubrication, is of the utmost importance in engineering applications. The efficient management of friction not only extends the lifespan of machines and components but also improves the overall system efficiency while reducing energy consumption [1]. One effective approach to minimize friction and enhance tribological performance is surface texturing. Surface texturing involves the creation of micro-scale features, such as dimples, ridges, or grooves, on material surfaces. These features have demonstrated the potential to reduce friction between contacting metal surfaces by promoting enhanced lubricant retention, load distribution, and surface roughness [2]. By decreasing the contact area and adhesion between the surfaces, surface texturing contributes to lower friction and wear rates [3].

This study focuses on investigating the impact of micro-surface texturing on the friction reduction between two metal surfaces. We conducted tribological tests using a ball-on-plate reciprocating tribometer, employing SAE30 oil as the lubricant [4,5]. The metal surfaces were modified with circular micro textures, and different area densities (5%, 10%, and 15%) were examined. Through a systematic comparison of the friction coefficients for each area density, our aim is to identify the most effective micro-texture configuration. This research contributes to the growing body of knowledge on surface texturing and its potential to optimize tribological behavior in engineering applications. Surface texturing has emerged as a promising approach to enhance mechanical efficiency, minimize wear, and extend component lifespan. By exploring the influence of textured surfaces on friction and wear characteristics, this study provides valuable insights into designing and implementing tribological systems with improved performance [6,7].

This current research work is focused on the study of the effect of surface texture area density on friction reduction. The surface textures in this work have been produced using a 20 W fiber laser engraver, which required additional iterative processes to create optimal texture design, followed by post-surface cleaning processes.

2. Methodology

H-13 (chromium-molybdenum hot work steel) steel base plate (60 mm × 30 mm × 4 mm) prepared by cutting and filing. Laser engraving with the "X700" laser engraver and "ezcad" software for precise micro-scale patterns. Rigorous calculations determined the number of dimples for each area density (5%, 10%, and 15%). The surface texture diameter is kept less than contact area so that hydrodynamic pressure is created, and surface area densities are selected on the basis of the literature that proved its effectiveness in this range. Texturing is executed with meticulous attention to detail using "ezcad" software. Plate is cleaned using silicon carbide abrasive paper for surface preparation. The laser engraver parameters and required texture details are given in Table 1.

Table 1. Laser engraver parameters and required texture details.

Parameters	Values
Loop count	1
Frequency	20 kHz
Scanning speed	150 mm/s
Power	70 W
Required dimple diameter	100 μm
Required dimple depth	5–8 μm

Diamond slurry and the Metkon Gripo 2 V Grinder-Polisher achieved exceptional surface quality.

The DSX1000 digital microscope provided detailed insights into the texture's unique characteristics shown in Figure 1. The schematic diagram of the LST is shown in Figure 2.

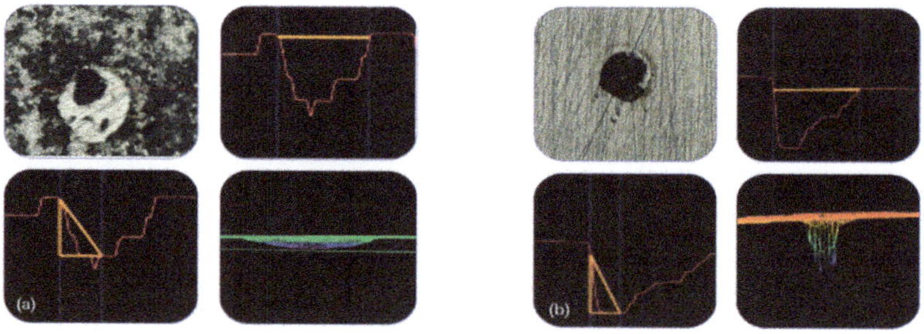

Figure 1. Laser surface texturing: (**a**) surface texture before grinding/polishing; (**b**) surface texture after grinding/polishing.

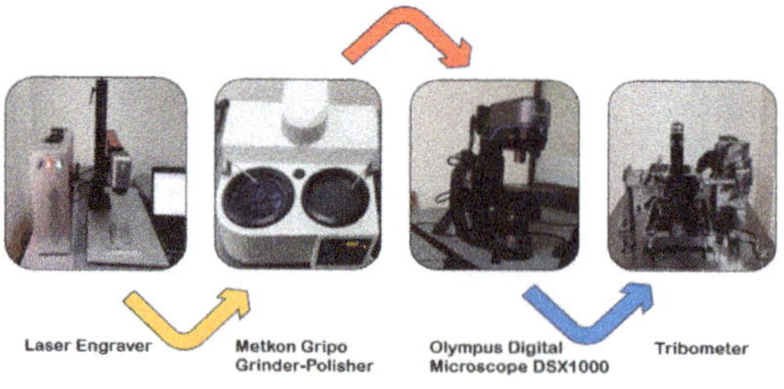

Figure 2. Schematic of experimental setup.

Laser engraving involved fine-tuning key parameters to create textured surfaces with dimple depth of 5–8 μm, dimple diameter of 0.1 mm, and densities of 5%, 10%, and 15%. The textured plate's properties were validated via tribometer testing. Testing parameters are given in Table 2.

Table 2. Testing parameters of tribo-testing.

Parameters	Values
Room temperature	30 °C
Oil used	SAE-30
Type of testing	Ball-on-plate
Normal load	20 N
Roll rate	150 RPM
Area densities	5%, 10%, and 15%
Duration of testing	1 h

The metal plate underwent a ball-on-plate reciprocating test with SAE-30 oil and a 20 N load. The tribometer ensured precise alignment, delivering reliable results for different area densities.

3. Results and Discussion

The untextured sample exhibited a higher friction compared to the textured samples. As the texture density increased from 5% to 10% and then to 15%, the friction was significantly reduced by 23.5%, 41.39%, and 37.95%, respectively. The introduction of micro-scale roughness and enhanced lubricant retention contributed to smoother sliding and minimized adhesion, resulting in the observed friction reduction [8]. With 15% texture density, diminishing returns may occur as closely spaced features limit the lubricant's effectiveness. The ball and textured plates contact visualization is shown in Figure 3.

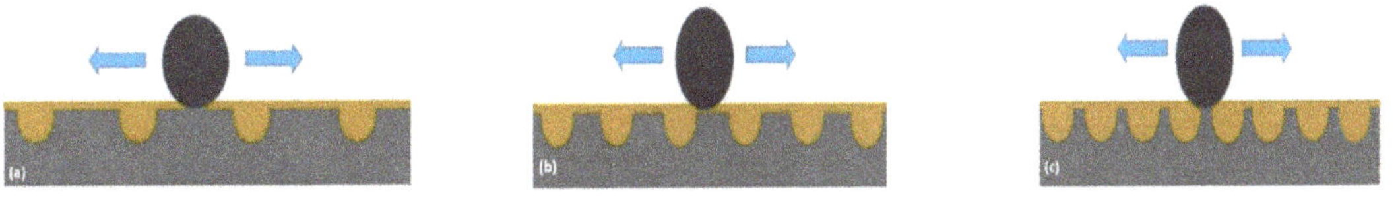

Figure 3. Visualization of contact between ball and textured plate (**a**) 5% area density; (**b**) 10% area density; (**c**) 15% area density.

Comparative Analysis and Discussion

The graph in Figure 4 shows decreasing friction with increasing texture density. A lower density of 5% provides a noticeable reduction, while a higher 10% offers a more substantial decrease. A potential plateau effect of 15% may result from the limitations in hydrodynamic layer formation. The integration of advanced software aids in informed decision making.

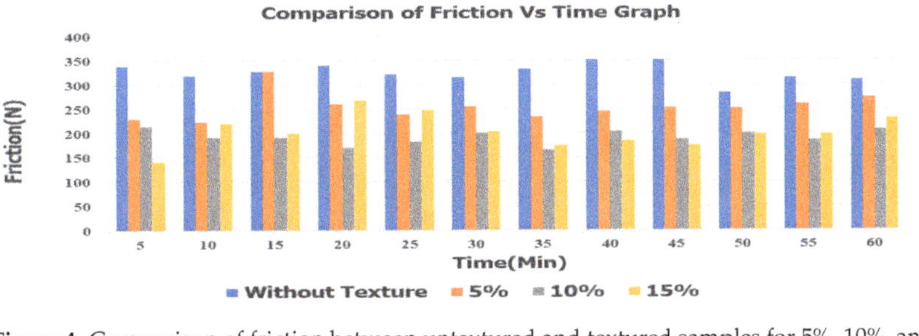

Figure 4. Comparison of friction between untextured and textured samples for 5%, 10%, and 15%. These findings hold promise for engineering applications with reduced friction, increased efficiency, and prolonged component life.

4. Conclusions

Surface texturing proved highly effective in substantially reducing friction within the ball-on-plate reciprocating system. The 5% texture sample achieved a remarkable 23.5% friction reduction, while the 10% texture resulted in approximately 41.39% reduction. At 15% texture density, the reduction was around 37.95%, indicating an optimal range for texture density selection. A careful consideration of the texture parameters is vital for optimizing tribological performance.

Author Contributions: Conceptualization, R.S.; Experimentation and results, W.A., M.S. and M.U.; Technical support, M.A.A.; Proofreading, M.Z. All authors have read and agreed to the published version of the manuscript.

Funding: This research received no external funding.

Institutional Review Board Statement: Not applicable.

Informed Consent Statement: Not applicable.

Data Availability Statement: The data supporting this study's findings are available from the corresponding author upon reasonable request.

Acknowledgments: The authors acknowledge the technical support provided by the National University of Sciences & Technology Islamabad and the administrative support provided by the Capital University of Sciences & Technology Islamabad for the present study.

Conflicts of Interest: The authors declare no conflict of interest.

References

1. Zou, H.; Yan, S.; Shen, T.; Wang, H.; Li, Y.; Chen, J.; Meng, Y.; Men, S.; Zhang, Z.; Sui, T.; et al. Efficiency of surface texturing in the reducing of wear for tests starting with initial point contact. *Wear* **2021**, *482–483*, 203957. [CrossRef]
2. Li, J.; Zeng, S.; Liu, S.; Zhou, N.; Qing, T. Tribological properties of textured stator and PTFE-based material in travelling wave ultrasonic motors. *Friction* **2020**, *8*, 301–310. [CrossRef]
3. Goti, E.; Mazza, L.; Mura, A.; Zhang, B. An early method for the technical diagnosis of pin-on-disk tribometers by reference friction measurements in EHL conditions. *Measurement* **2020**, *166*, 108169. [CrossRef]
4. Kovalchenko, A.M.; Erdemir, A.; Ajayi, O.O.; Etsion, I. Tribological Behavior of Oil-Lubricated Laser Textured Steel Surfaces in Conformal Flat and Non-Conformal Contacts. *Mater. Perform. Charact.* **2017**, *6*, 1–23. [CrossRef]
5. Joshi, G.S.; Putignano, C.; Gaudiuso, C.; Stark, T.; Kiedrowski, T.; Ancona, A.; Carbone, G. Effects of the micro surface texturing in lubricated non-conformal point contacts. *Tribol. Int.* **2018**, *127*, 296–301. [CrossRef]
6. Zhang, T.; Jiang, F.; Yan, L.; Jiang, Z.; Xu, X. A novel ultrahigh-speed ball-on-disc tribometer. *Tribol. Int.* **2021**, *157*, 106901. [CrossRef]
7. Patil, A.S.; Shirsat, U.M. Effect of laser textured dimples on tribological behavior of piston ring and cylinder liner contact at varying load. *Mater. Today Proc.* **2021**, *44*, 1005–1020. [CrossRef]
8. Mao, B.; Siddaiah, A.; Liao, Y.; Menezes, P.L. Laser surface texturing and related techniques for enhancing tribological performance of engineering materials: A review. *J. Manuf. Process.* **2020**, *53*, 153–173. [CrossRef]

engineering proceedings

MDPI

Proceeding Paper

Phase Change Material for the Cooling of Solar Panels—An Experimental Study †

Bilal Hussain [1], Huzaifa Waseem Malik [1], Fakhar Ul Hasnain [2] and Muhammad Irfan [1,*]

1 Department of Mechanical Engineering, Capital University of Science and Technology, Islamabad 45750, Pakistan; bh9143500@gmail.com (B.H.); huzaifa.malik@cust.edu.pk (H.W.M.)
2 Department of Mechanical Engineering, Kyungpook National University, Daegu 37224, Republic of Korea; fakhar948@gmail.com
* Correspondence: m.irfan@cust.edu.pk
† Presented at the Third International Conference on Advances in Mechanical Engineering 2023 (ICAME-23), Islamabad, Pakistan, 24 August 2023.

Abstract: Solar panel efficiency decreases with an increase in the panel surface temperature. This study utilized the Phase Change Material (PCM) based cooling approach along with Aluminum fins to reduce the temperature of the PV panel. The PV panel surface temperature and efficiency are the target parameters we investigated. The results were compared with conventional PV panel values at inclination angles of 30°, 45°, and 60° with respect to the ground. The experimental results show that the PV panel efficiency increased by 6.85%, 6.82%, and 4.2% for the inclination angles of 30°, 45°, and 60°, respectively. The corresponding maximum temperature drops were 13.1 °C, 12.9 °C, and 8.5 °C.

Keywords: phase change material; PV panel cooling; efficiency; solar energy

Citation: Hussain, B.; Malik, H.W.; Hasnain, F.U.; Irfan, M. Phase Change Material for the Cooling of Solar Panels—An Experimental Study. *Eng. Proc.* **2023**, *45*, 43. https://doi.org/10.3390/engproc2023045043

Academic Editors: Mohammad Javed Hyder, Muhammad Mahabat Khan and Manzar Masud

Published: 15 September 2023

1. Introduction

A solar cell, also known as a photovoltaic (PV) cell, converts solar energy into electricity. It consists of a p-n junction diode. A single silicon solar cell can generate a maximum open circuit voltage of approximately 0.5–0.6 Volts. When integrated into a larger solar panel system, it yields significant renewable energy. Roughly 23% of solar irradiance is intercepted by environmental factors like dust, ozone, and water vapor, while approximately 29% is reflected back into space. Only 48% of solar energy traverses the atmosphere and becomes absorbed by the solar panel surface. Consequently, the silicon solar panel converts around 20% of solar energy into electrical energy [1]. An increase in the temperature of the PV panel causes a reduction in the photovoltaic panel's efficiency by approximately 0.5% to 0.6% per degree rise in temperature. Chavan et al.'s study used a 40 W panel with three systems, including conventional PV, PV-PCM fins, and a water circulation system. PV-PCM achieved 4.24% higher electrical efficiency [2]. Prakash et al. used HS36-hydrated salt as a PCM for PV cooling to enhance panel performance. PV-PCM reduced the surface temperature by 25.4% and increased the electrical efficiency by 17.5% [3]. Homlakorn et al. employed organic eutectic PCM in a finned container for PV module cooling. Out of three proposed proportions of acid, a 60:40%wt mixture yielded a 7.06 °C reduction in temperature, a 0.454 W increase in power, and a 4.226% efficiency improvement [4]. This current study utilized a combination of PCM and an Aluminum fin structure to cool down the PV panel in the atmospheric conditions of Islamabad, Pakistan, for various inclination angles with respect to the ground.

2. Experimental Setup

Two identical PV panels, each with a 30 W power rating, were used in the experimental setup, as shown in Figure 1. Monocrystalline solar panels were utilized due to their higher efficiency compared to other panel types. The PV panel optimum operating voltage, open circuit voltage and short circuit current were 19.5 V, 22.9 V and 1.70 A, respectively. The PV panel dimensions were 650 mm × 350 mm × 25 mm.

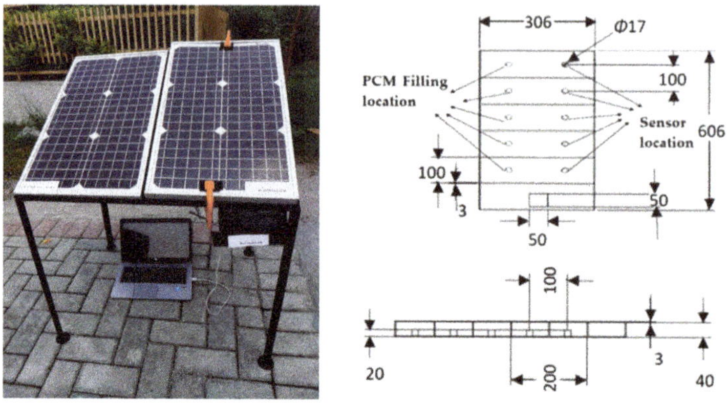

Figure 1. (**Left**) The experimental setup; (**Right**) The schematic of the PCM container installed in the PV-PCM system showing dimensional details (All dimensions are in mm).

One panel called the PV-PCM system, was cooled by the PCM-fin cooling system, while the other panel did not cool but served as the reference for result comparisons. In the PV-PCM system, an Aluminum container containing PCM and fin structures was attached to the back of the PV panel. The PCM chosen for this study was $CaCl_2.6H_2O$ with a density of 1706 kg/m^3 and a melting point of 29 °C, which was suitable according to the environmental conditions of the location. The dimensional details and the container structure are shown in Figure 1. The height of the PCM container was 40 mm, which was dictated by the mass of the PCM and the PV panel dimensions. The mass of the PCM was calculated such that it was sufficient to absorb the total solar energy available throughout the day in the form of latent energy. This helped to maintain the PV panel temperature at the PCM phase change temperature of 29 °C, which was close to the optimum operating temperature of the PV panel (25 °C). K-type thermocouples were used to record the temperature of the PCM, while an infrared thermal gun was used to record the PV panel's surface temperature.

3. Result and Discussion

Each experiment was performed three times, and the average results were used for the purpose of analysis. The efficiency of the conventional PV panel can be calculated as [5]:

$$\eta_{PV} = \eta_{ref}\left[1 - \beta\left(T_{PV} - T_{ref}\right)\right] \quad (1)$$

where η_{ref} is the reference efficiency of the standard PV system and β is the thermal expansion coefficient of the PV system, which is 0.0045/K [5]. Similarly, the efficiency of the PV-PCM system can be calculated as

$$\eta_{PV-PCM} = \eta_{ref}\left[1 - \beta\left(T_{PV-PCM} - T_{ref}\right)\right] \quad (2)$$

Efficiency enhancement is calculated as [5]

$$\text{Enhancement } (\%) = \left|\frac{\eta_{PV} - \eta_{PV-PCM}}{\eta_{PV}}\right| \times 100 \quad (3)$$

Figure 2 presents the trends of the panel surface temperature and panel efficiency of the conventional and PCM-cooled systems for all three angular orientations. From the experimental results plot, it can be seen that the PCM-cooled system exhibited lower surface temperatures and higher efficiencies compared to the conventional PV system for all three angular orientations of the PV panels. At the start of the experiments, both the conventional and the PV-PCM systems were approximately at the same temperatures . As time progressed during the day, the surface temperature of the PV panels increased for both systems; however, the PV-PCM system temperature was lower than the conventional system. This is because the heat was being stored in the PCM filled in the container attached to the PV panel of the PV-PCM system. Consequently, the efficiency of the PV-PCM system was also higher compared to the conventional system. The PCM melted as it stored energy during the process. As time passed, the entire PCM was melted, and no more latent heat could further be stored. Therefore, the surface temperature of PV-PCM started rising at a faster pace and eventually approached that of the conventional panel. Similar trends were also observed for the panel efficiencies. It can be also observed that the panel efficiency for the 30° orientation was highest and continuously decreased as the angle increased to 60°.

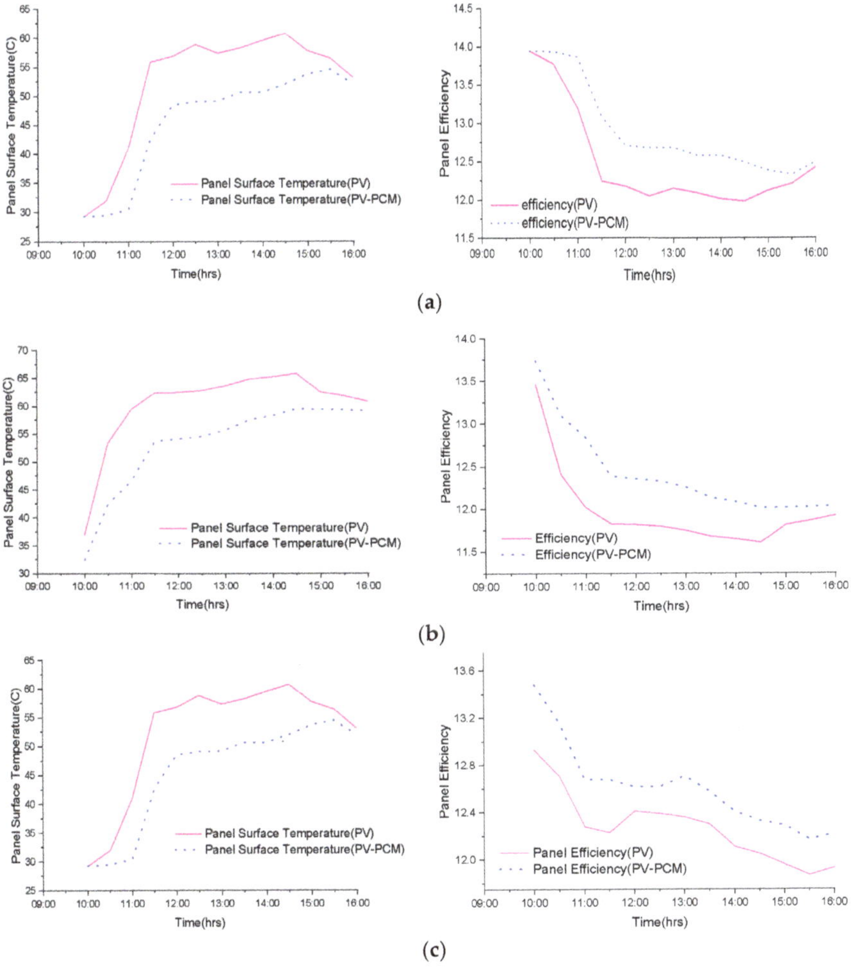

Figure 2. Panel surface temperature and panel efficiency at different angular orientations. (**a**) Angular orientation = 30°; (**b**) Angular orientation = 45°; (**c**) Angular orientation = 60°.

Table 1 presents the maximum values of the temperature difference between the conventional and PV-PCM systems at 11:30 a.m. for all three angular orientations of the PV panel. The corresponding maximum efficiency enhancement values (in percentages) are also tabulated.

Table 1. Performance parameters for various orientations of the PV panel.

Parameters	Orientation Angle		
	30°	45°	60°
Temperature difference, ΔT (°C)	13.1	12.9	8.5
Efficiency Enhancement (%)	6.85	6.82	4.2

4. Conclusions

This study analyzed the performance of the PCM-based cooling system integrated with fins to cool down the PV panel and ultimately enhance electrical efficiency. It was observed that the utilization of PCM decreased the PV panel surface temperature, which consequently increased the efficiency of the panel. The experimental result shows that PCM-based cooling decreased the PV panel's surface temperature by 13.1 °C, 12.9 °C and 8.5 °C compared to the conventional PV panel for the inclination angles of 30°, 45°, and 60°, respectively. The corresponding increase in the PV panel's efficiency were 6.85%, 6.82%, and 4.2%.

Author Contributions: Conceptualization, M.I. and B.H.; methodology, M.I. and B.H.; software, B.H., H.W.M. and F.U.H.; validation, B.H.; formal analysis, B.H., H.W.M., F.U.H. and M.I.; investigation, B.H. and M.I.; data curation, B.H., H.W.M. and F.U.H.; writing—original draft preparation, M.I., B.H., H.W.M. and F.U.H.; supervision, project administration M.I. All authors have read and agreed to the published version of the manuscript.

Funding: This research received no external funding.

Institutional Review Board Statement: Not applicable.

Informed Consent Statement: Not applicable.

Data Availability Statement: The data presented in this study are available on request.

Conflicts of Interest: The authors declare no conflict of interest.

References

1. Uddin, I.; Rangreez, T.A.; Ahamed, M.I.; Chisti, H. Materials for Solar Cell Technologies II; Materials Research Forum LLC, PA USA, 2021.
2. Chavan, S.V.; Devaprakasam, D. Improving the Performance of Solar Photovoltaic Thermal System Using Phase Change Material. *Mater. Today Proc.* **2019**, *46*, 5036–5041. [CrossRef]
3. Prakash, K.B.; Pasupathi, M.K.; Chinnasamy, S.; Saravanakumar, S.; Palaniappan, M.; Alasiri, A.; Chandrasekaran, M. Energy and Exergy Enhancement Study on PV Systems with Phase Change Material. *Sustainability* **2023**, *15*, 3627. [CrossRef]
4. Homlakorn, S.; Suksri, A.; Wongwuttanasatian, T. Efficiency Improvement of PV Module Using a Binary-Organic Eutectic Phase Change Material in a Finned Container. *Energy Rep.* **2022**, *8*, 121–128. [CrossRef]
5. Sark, W.G.J.H.M.V. Feasibility of Photovoltaic—Thermoelectric Hybrid Modules. *Appl. Energy* **2011**, *88*, 2785–2790. [CrossRef]

engineering proceedings

MDPI

Proceeding Paper

Experimental Investigation of Battery Thermal Management System of Lithium-Ion Cells Using PCM †

Sarmad Ali * and Muhammad Mahabat Khan

Department of Mechanical Engineering, Capital University of Science and Technology, Islamabad 44000, Pakistan; drmahabat@cust.edu.pk

* Correspondence: sarmad.ali@cust.edu.pk or sarmadalisa@outlook.com; Tel.: +92-316-5600024

† Presented at the Third International Conference on Advances in Mechanical Engineering 2023 (ICAME-23), Islamabad, Pakistan, 24 August 2023.

Abstract: Experimental analysis of the effect of a lithium-ion battery thermal management system using natural convection and phase change material (PCM) at 3 C discharge rate. The cells are placed in a 2 × 2 square configuration in an acrylic housing with the capability to contain PCM. The experimental setup simulates lithium-ion batteries using ceramic heaters, producing the heat energy per unit volume as produced in a lithium-ion cell. The system placed in natural convection instantly heats up to 45 °C in 305 s. On the other hand, the system, when placed in phase change material, shows that a temperature of 45 °C is achieved in 1490 s.

Keywords: lithium-ion battery; thermal storage; phase change material; battery thermal management mystem; passive cooling

Citation: Ali, S.; Khan, M.M. Experimental Investigation of Battery Thermal Management System of Lithium-Ion Cells Using PCM. *Eng. Proc.* **2023**, *45*, 52. https://doi.org/10.3390/engproc2023045052

Academic Editors: Mohammad Javed Hyder, Muhammad Irfan and Manzar Masud

Published: 19 September 2023

1. Introduction

Lithium-ion batteries have been replacing non-renewable energy sources over the past decade in order to have a clean and green environment with fewer carbon emissions. The internal combustion engines used in vehicles play a major role in the carbon emissions causing global warming [1]. A uniform temperature within the safe limits for lithium-ion batteries should be maintained for optimal operation during the charging and discharging cycle of the cell. There are a number of methods to control these temperatures, some of which are liquid cooling [2], air cooling [3], and phase change materials passive cooling. The phase change material-based BTMS was first proposed by Al-Hallaj and Selman [4]. Their study conducted a performance evaluation of the thermal management system for lithium-ion batteries using both air and PCM. The findings showed that the battery's discharge rate and ambient temperature have a significant impact on the air-cooling management system's performance, with high ambient temperatures and moderate discharge rates making it impossible to maintain a safe battery temperature [2]. Wang et al. conducted an experimental investigation into the impact of utilizing fins on enhancing battery TMS performance. In their technique, paraffin with a melting temperature of 44 °C was used. The average battery temperature was reduced by 8 °C [5].

PCMs with high latent heat storage are a very effective passive cooling method for lithium-ion cells. Battery thermal management systems (TMSs) have been rigorously tested numerically by previous research; however, the numerical analysis of battery TMSs has many simplifications to obtain favorable results. Real-time ambient environment effects with an increase and decrease in ambient temperature cannot be studied numerically. This leads us to the development of an experimental model to simulate the batteries for analysis and investigation of cell temperatures in natural convection, as well as when placed in PCM at a high discharge rate.

2. Methodology

The experimental setup is designed to simulate the heat generation rates of an 18,650 lithium-ion cell using ceramic heaters placed inside an aluminum housing. The aluminum housing and the heater combined have the same dimensions as the 18,650 lithium-ion cell, which is 18 mm in diameter and 65 mm in height. A circuit using a combination of microcontrollers and ICs was designed for pulse wave modulation (PWM), producing similar discharging/charging waves as conducted by the lithium-ion cell controlled by the amount of power supplied in Watts to the ceramic heaters. The circuit has the capability of producing 0.05 W to 6 W with increments of 0.05 W. The circuit, with its current capability, has the ability to simulate 1C, 2C, and 3C discharge rates in a 2 × 2 square configuration of cells. The circuit connected to the cells is shown in Figure 1. The acrylic housing containing the whole system is 72 mm × 72 mm in length and width and has a height of 85 mm.

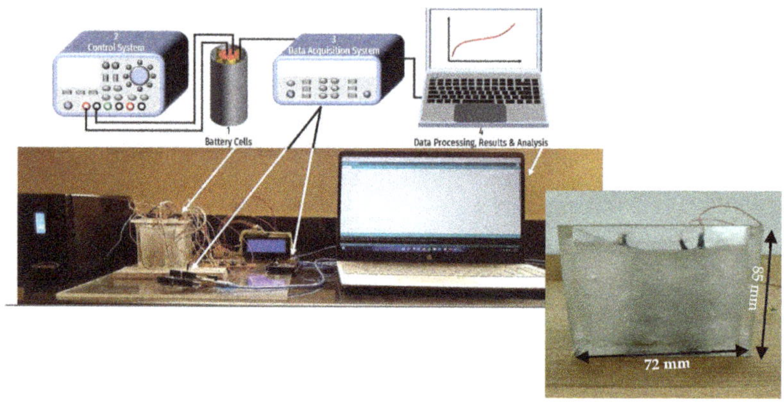

Figure 1. Schematic of the cells in PCM with the data acquisition system.

The data acquisition system includes thermocouples paired with Arduinos for temperature measurement both for the cell and PCM.

The C rating is a crucial performance parameter that describes the discharge and charge capabilities of the battery. It indicates the rate at which a battery can be discharged or charged in relation to its capacity. The C rating is expressed as a multiple of the battery's capacity, typically in units of "C". The general formula to calculate the maximum discharge or charge current (I) based on the C rating [6] is as follows:

$$I\ (mA) = C - rating \times Capacity\ of\ the\ battery\ (mAh) \quad (1)$$

The heat generation rates or the wattage ratings for different discharge rates are taken from Choudhari et al. [7]. The heat generation rates are shown in Table 1.

Table 1. Heat generation rates corresponding to C ratings. Adapted with permission from Ref. [7]. 2020, Elsevier Ltd.

C Rating	Heat Generation Rates (W/m^3)
1 C	10,447
2 C	41,788.37
3 C	94,023.8

The cells are placed with 250 g of RT-42 fully submersed into the PCM. The system is given the heat generation rates of 3C to study the change in the heating curve for the cell. The PCM temperature is measured using 6 K type thermocouples placed at different heights and spacing uniformly, which will average out the PCM temperature, giving the melt fraction. The K-type thermocouple has an accuracy of ±2.2 °C. The experiment time is set to be 6000 s; however, the experiment would be truncated if the cell temperature increases the threshold of 50 °C.

3. Results

In the case of cells placed in natural convection, the rate of change of cell temperature was very high, and it crossed the benchmark temperature of 45 °C at 306 s. The naturally cooled experiment was truncated at 411 s when the temperature exceeded 50 °C. In the case of cells placed in RT-42, the comparison in Figure 2a shows that the benchmark temperature of 45 °C was achieved in 1495 s and the maximum temperature achieved in 6000 s was 56.365 °C; however, the cell in an actual scenario would discharge much sooner, and the temperatures would remain under 50 °C.

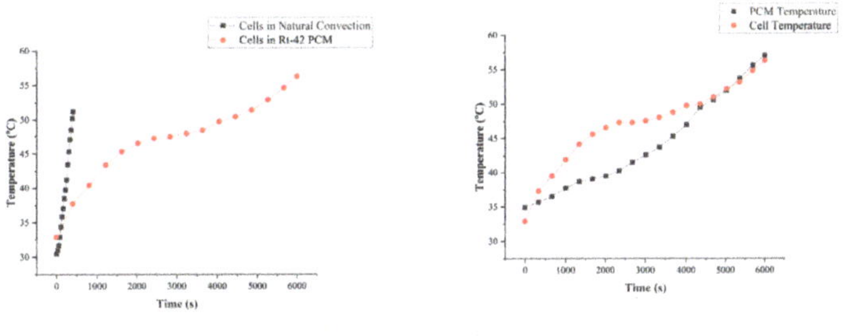

(**a**) (**b**)

Figure 2. (**a**) Temperature–time curve for cells placed in natural convection vs. PCM. (**b**) Temperature–time curve for cell and PCM temperature.

The comparison of PCM and cell temperature is shown in Figure 2b which illustrates that although the PCM prolongs the time to reach the upper limit of the optimal temperature 45 °C, it does not throttle the cell temperature very effectively. The rate of change of cell temperature is more than the rate of change of PCM temperature, which results in a rise in overall cell temperature.

Figure 3 shows the melting fraction between the solidus temperature of 38 °C and the liquidus temperature of 43 °C, as provided by the datasheet from OEM for RT-42. The trends show that the melting slowed down due to the low thermal conductivity of PCM, and the heat started to accumulate. The corresponding cell temperature when melting starts at 1180 s is 43.43 °C, and the melt fraction is 4%. At 2500 s, when the melt fraction is 49.167%, the cell temperature is 47.095 °C, which is above the maximum optimal temperature of 45 °C. This means that due to heat accumulation, although the PCM is melting due to a weak thermal network and no external enhancement for heat transfer from cell to PCM, the cell temperature keeps rising, and the effectiveness of the systems reduces. Finally, a melting fraction of 100% is achieved at 3280 s, where the cell temperature is 48.07 °C.

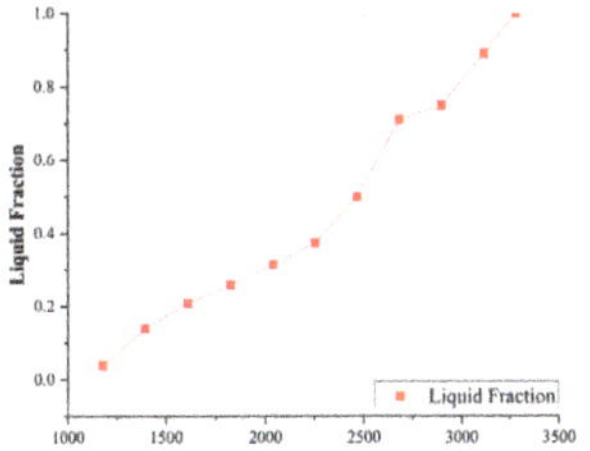

Figure 3. Liquid fraction trends for cells placed in PCM.

4. Discussion

The analysis performed showed that a 79.53% increase in the operating time was observed staying under the maximum optimal temperature for the cells, which was 45 °C. The time range from 1490 s to 6000 s in the case of cells placed in PCM showed that the temperature rise slowed down but crossed 50 °C at 4270 s. The effectiveness of the cooling through PCM dropped when the melting region moved away from the surface of the cell, and the heat started to accumulate around the cell. This is due to the 0.2 W/m-K thermal conductivity value of the PCM. So, beyond a certain point, the rate of change in temperature increased. The solution to this problem is the addition of fins and their optimization while keeping the mass of PCM constant. Another way the thermal conductivity of the PCM can be enhanced is through ongoing research for experimentation.

Author Contributions: Methodology, Experimental Analysis, and Writing—Original Draft: S.A. Editing, Review, and Investigation: M.M.K. All authors have read and agreed to the published version of the manuscript.

Funding: The research received no external funding.

Institutional Review Board Statement: Not applicable.

Informed Consent Statement: Not applicable.

Data Availability Statement: Not applicable.

Conflicts of Interest: The authors declare no conflict of interest.

References

1. Abas, N.; Kalair, A.; Khan, N. Review of fossil fuels and future energy technologies. *Futures* **2015**, *69*, 31–49. [CrossRef]
2. Sabbah, R.; Kizilel, R.; Selman, J.R.; Al-Hallaj, S. Active (air-cooled) vs. passive (phase change material) thermal management of high power lithium-ion packs: Limitation of temperature rise and uniformity of temperature distribution. *J. Power Sources* **2008**, *182*, 630–638. [CrossRef]
3. Wei, Y.; Agelin-Chaab, M. Development and experimental analysis of a hybrid cooling concept for electric vehicle battery packs. *J. Energy Storage* **2019**, *25*, 100906. [CrossRef]
4. Al Hallaj, S.; Selman, J.R. A Novel Thermal Management System for Electric Vehicle Batteries Using Phase-Change Material. *J. Electrochem. Soc.* **2000**, *147*, 3231. [CrossRef]
5. Wang, Z.; Zhang, H.; Xia, X. Experimental investigation on the thermal behavior of cylindrical battery with composite paraffin and fin structure. *Int. J. Heat Mass Transf.* **2017**, *109*, 958–970. [CrossRef]
6. Available online: https://www.power-sonic.com/blog/what-is-a-battery-c-rating/ (accessed on 7 August 2023).
7. Choudhari, V.G.; Dhoble, A.S.; Panchal, S. Numerical analysis of different fin structures in phase change material module for battery thermal management system and its optimization. *Int. J. Heat Mass Transf.* **2020**, *163*, 120434. [CrossRef]

engineering proceedings

Proceeding Paper

The Design and Development of a Solar Dehydrator for Fruits †

Mohammad Javed Hyder, Muhammad Junaid Khan *, Muhammad Abdullah Khan and Salman Saeed

Department of Mechanical Engineering, Capital University of Science and Technology (CUST), Islamabad 44000, Pakistan; javed.hyder@cust.edu.pk or mjavedhyder1@gmail.com (M.J.H.); mabdullahkhan115@yahoo.com (M.A.K.); salman.2000.saeed@gmail.com (S.S.)

* Correspondence: sardarjunaid320@gmail.com or engr.mjunaidkhan@outlook.com

† Presented at the Third International Conference on Advances in Mechanical Engineering 2023 (ICAME-23), Islamabad, Pakistan, 24 August 2023.

Abstract: An active solar dehydrator is designed for dehydrating fruits and vegetables. The solar collector transfers thermal energy to the drying air. The temperature inside the drying chamber is maintained between 45 °C and 50 °C by a feedback control system. The design is based on the December solar irradiance of Islamabad, which is found to be 572 watts per square meter. The energy required to dry one kilogram of apple and banana is 245 watts and 220 watts, respectively. The dried product from the solar dehydrator is better in terms of quality and taste compared to the product produced by ordinary open sun drying.

Keywords: solar dehydrator; drying fruit; dehydrator control; dehydrator temperature

Citation: Hyder, M.J.; Khan, M.J.; Khan, M.A.; Saeed, S. The Design and Development of a Solar Dehydrator for Fruits. *Eng. Proc.* **2023**, *45*, 48. https://doi.org/10.3390/engproc2023045048

Academic Editors: Muhammad Mahabat Khan, Muhammad Irfan and Manzar Masud

Published: 18 September 2023

1. Introduction

Dehydration adds worth to fresh vegetables and fruits by eliminating water, reducing their weight, and increasing the time of usability. It also opens a window for the farmers to sell their surplus produce to local and international markets that would otherwise be wasted due to harsh weather conditions. Drying lessens the burden and quantity of the product, making it easy to keep and use. This will finally lower the cost of storing, packaging, and transportation.

2. Literature Review

The losses of fruit and vegetables are in the order of 40%, and under extremely hindering circumstances, they are as high as 80% [1]. A major percentage of losses is either due to inaccurate or early drying of natural product chips (banana, mango, apple, etc.) [2]. The indirect passive solar drying system was designed and built using convective heat flow principles [3]. For the selected plants, a direct correlation between item temperature and drying rate was postulated. The proposed experimental relationship can depict the drying energy of the picked crops well [4]. The cost of dehydrating chili per kilogram was reduced by 39% in the case of a double-pass solar drier [5]. In the indirect solar dehydrator, the incoming air is heated using solar energy in a separate solar receiver and sent to the drying chamber. A consistent condition of evaporation is maintained with the flow of heat [6].

3. Design Methodology

Andress, Elizabeth L., and Judy A. [7] tabulated the dehydration temperature for fruits and vegetables. The current study is related to banana and apple. Accordingly, the moisture content of apple is 85.9%, and that of banana is 74%. Both dehydration temperature is to be maintained at 45 to 50 °C. Using these data, the moisture amount to be removed from the apple is calculated [8]. In general, the total mass (m) of one standard-size apple is taken as 200 g, and the moisture content is 85.9% by mass (w = 0.859). On a dry basis, the mass of the apple is calculated to be m_o = 107.58 g. Hence, the mass of water present is =200 − 107.58 = 92.42 g. The equilibrium dry mass of an apple with moisture content is 16% by mass [7]. So,

the mass of the apple on a dry basis is calculated as m = 124.79 g. Hence,

the water content present at equilibrium state = 124.79 − 107.58 = 17.21 g

and the mass of water to evaporate per 200 g of sample = 92.42 − 17.21 = 75.21 g

Now, repeating the same process for banana, the mass (m) of one standard-size banana is taken as 120 g, the moisture content is 74 percent by mass, and the equilibrium moisture content in dried banana is 13% [7]. Hence,

the mass of banana on a dry basis = 77.92 g,

the water content present at equilibrium state = 77.92 − 68.96 = 8.96 g,

and the mass of water to evaporate per 120 g of sample = 51.03 − 8.96 = 42.07 g

4. The Volumetric Air Flow Rate Required to Dry One Kilogram of Apple

Unsaturated air is passed over wet material to take away water from the material being dried. This water is evaporated in a drying chamber (Figure 1). The volume of air (V) is cooled in the process of evaporating a mass of water from the fruit sample.

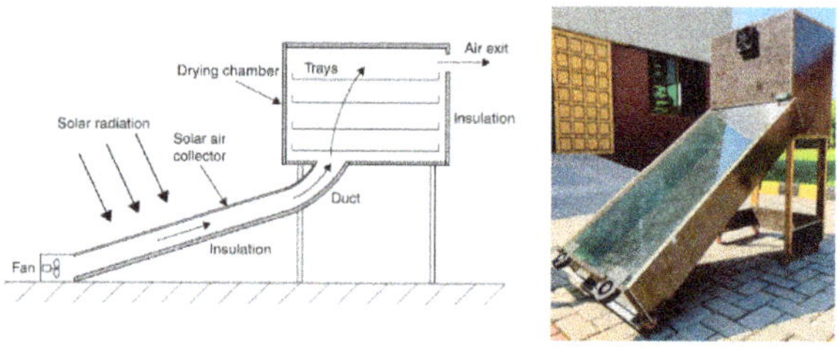

(a) (b)

Figure 1. Active solar dehydrator: (**a**) Schematic diagram; (**b**) Actual system.

In one kilogram of apple, there are approximately five (5) apples. So, for the water content in one kilogram of apple sample, the mass of water removed per kg is m_w = 5 (75.2) = 376 g. So, the volume of air required to dry one kilogram of apple is

$$V = (m_w L_H)/(\rho C_p \Delta T) = ((0.376)(2260))/((1.15)(1.0)(50 - 38)) = 61.57 \text{ m}^3 \quad (1)$$

So, to dehydrate a kilogram of apples in one hour, an air flow rate of 61.57 m^3/h is required.

5. The Heat Energy Required to Dry One Kilogram of Apple

The amount of energy transferred through the air is equal to the amount of energy absorbed by the water. The energy (Q) required to dry the sample is given by

$$Q = m_w L_H \quad (2)$$

where m_w is the mass of water to be removed, and L_H is the latent heat of vaporization. So, the energy required to dry one kilogram of apple is

$$Q = (0.376)\,(2260) = 849 \text{ kJ}$$

To dry it in one hour, the heat rate is given Q = 849.76 kJ/1 h = 849.76 kJ/h = 245 W. So, the energy required for apple is 245 W. Similarly, for banana, the energy required is 220 W.

6. Solar Collector Design

Based on the design methodology discussed by Twidell, J., and Weir, T., the solar collector is designed to meet energy requirements for one kilogram of apple when placed inside a dehydrating chamber. For the collector, the efficiency is η_c = 60%. Total useful power is given by [8]:

$$P_u = \eta_c \, A \, G \tag{3}$$

The solar irradiance on the first June is G_{hmax} = 885 w/m^2, and solar irradiance on the first December is G_{hmax} = 572 w/m^2 [8]. Using irradiance G = 572 w/m^2 and the power required for the apple (calculated above), the energy needed to dry is P_u = 245 watts. So, the area required is A = 0.71 m^2. Hence, the collector is designed for the month of December. The dimensions used for the collector are length 1.2 m and width 0.6 m (A = 0.72 m^2). When the blower is not operating, the flow is stagnated, and the maximum temperature of the plate is observed. The stagnation temperature of the plate in June is 105 °C, and in December, it is 94.6 °C.

7. Experimental Setup and Results

An active solar dehydrator is designed with the wood (Figure 2). The overall dimensions of the active solar dehydrator are length = 1.5118 m, width = 0.6096 m, and height = 1.21 m. The dimensions of the solar collector are 1.206 m long and 0.6069 m wide. The interior of the solar dehydrator is painted black to enhance the absorption of heat energy.

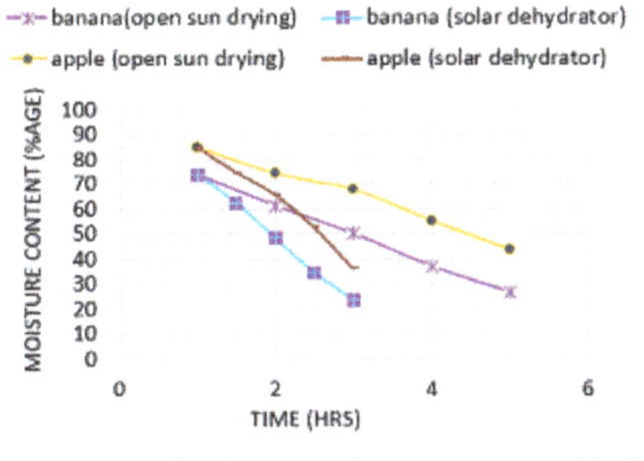

Figure 2. Reduction in moisture content with dehydration time.

The drying chamber maintains a temperature range of 45 to 50 °C, with two temperature sensors: one at the inlet and one inside the chamber. The sensor controls the speed of blowers and the exhaust fan's on/off mechanism. When the chamber temperature is 45 °C or higher, blowers operate at maximum speed, and speed decreases with temperature. The exhaust fan starts working at 50 °C.

In the open sun and dehydrator case, samples are weighed to determine moisture content, impacting color, taste, and overheating, leading to decolorization. Dusty environment also contributes to the end product as shown in Figure 3.

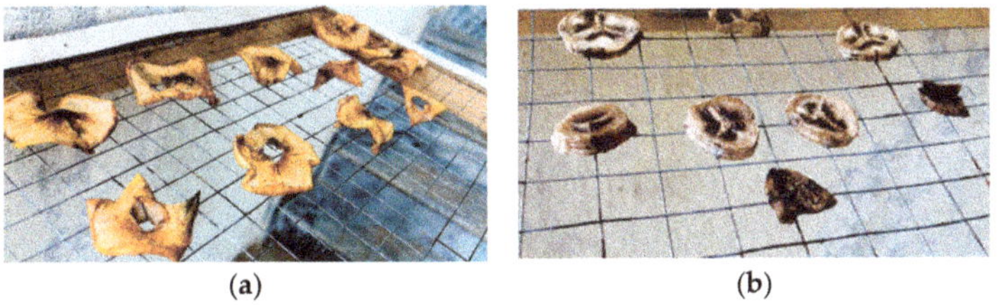

(**a**) (**b**)

Figure 3. Open sun drying: (**a**) Apple; (**b**) Banana.

While in the active solar dehydrator, the drying is faster. The product obtained from an active solar dehydrator is better in terms of quality and hygiene, as shown in Figure 4.

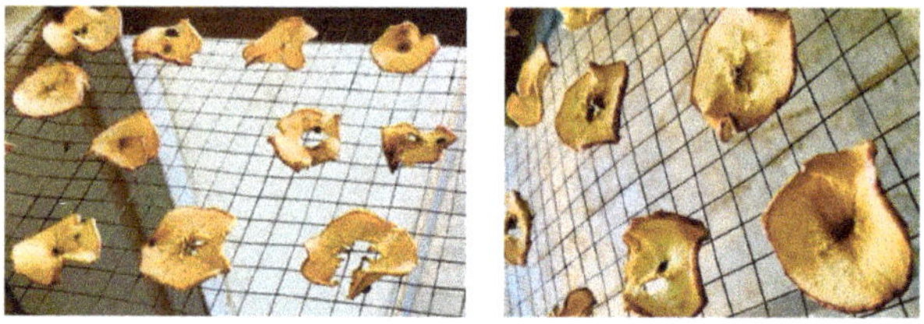

(a) (b)

Figure 4. Drying by active solar dehydrator: (**a**) Apple; (**b**) Banana.

8. Conclusions

Almost 300 watts per kilogram is required to dry one kilogram of sample that has a moisture content of around 80%. In the active solar dehydrator, the drying is faster. The product is dried in 2 h in a solar dehydrator compared to the more than four hours needed for open sun drying.

Author Contributions: Conceptualization, writing of the paper, and supervision of the project, M.J.H.; calculation, M.J.H. and M.J.K.; experimentation, investigation, data processing, and editing, M.J.K., M.A.K., and S.S. All authors have read and agreed to the published version of the manuscript.

Funding: This research received no external funding.

Institutional Review Board Statement: Not applicable.

Informed Consent Statement: Not applicable.

Data Availability Statement: The data presented in this study are available on request.

Conflicts of Interest: The authors declare no conflict of interest.

References

1. Akhtar, J.; Omre, P. Mathematical modeling evaluation for convective hot air drying of poultry meat. *Int. J. Agric. Eng.* **2017**, *10*, 168–178. [CrossRef]
2. Toğrul, I.T.; Pehlivan, D. Modelling of thin layer drying kinetics of some fruits under open-air sun drying process. *J. Food Eng.* **2004**, *65*, 413–425. [CrossRef]
3. Olalusi, A.; Ogunlowo, A.; Bolaji, B. Development and Performance Evaluation of a Mobile Solar Dryer for Cassava Chips. *Energy Environ.* **2012**, *23*, 1261–1272. [CrossRef]
4. El-Sebaii, A.; Aboul-Enein, S.; Ramadan, M.; El-Gohary, H. Experimental investigation of an indirect type natural convection solar dryer. *Energy Convers. Manag.* **2002**, *43*, 2251–2266. [CrossRef]
5. Gallali, Y.; Abujnah, Y.; Bannani, F. Preservation of fruits and vegetables using solar drier: A comparative study of natural and solar drying, III; chemical analysis and sensory evaluation data of the dried samples (grapes, figs, tomatoes and onions). *Renew. Energy* **2000**, *19*, 203–212. [CrossRef]
6. Szabó, D.; Gillet, E.; Vallés, I.; Pereira, J.; Keppens, M.; Krommendijk, P.; Duarte, A.J.; Malheiro, B.; Ribeiro, C.; Justo, J.; et al. Solar Dehydrator. In Proceedings of the Seventh International Conference on Technological Ecosystems for Enhancing Multiculturality, TEEM 2019, León, Spain, 16–18 October 2019.
7. Andress, E.L.; Harrison, J.A. *Preserving Food: Drying Fruit and Vegetables*; University of Georgia: Athens, GA, USA, 2000.
8. Twidell, J.; Weir, T. *Renewable Energy Resources*, 3rd ed.; Routledge, Taylor & Francis Group: New York, NY, USA, 2015; pp. 75–112.

engineering proceedings

MDPI

Proceeding Paper

Design Modification and Prototype Development of a Cross-Flow Turbine for a Low/Zero-Head Water Stream †

Asim Ali, Muhammad Zohaib Azam, Muhammad Faiq and Ghulam Asghar *

Mechanical Engineering Department, Capital University of Science and Technology (CUST), Islamabad 44000, Pakistan; asimaliorakzai502@gmail.com (A.A.); muhammadzohaibazam810@gmail.com (M.Z.A.); faiqr79@gmail.com (M.F.)

* Correspondence: ghulam.asghar@cust.edu.pk

† Presented at the Third International Conference on Advances in Mechanical Engineering 2023 (ICAME-23), Islamabad, Pakistan, 24 August 2023.

Abstract: Low/zero-head cross-flow turbines (LZH-CFT) possess vast potential in the field of renewable energy technology. This research presents a modified design of a cross-flow turbine specifically engineered to harness the untapped hydro-kinetic energy from low/zero-head water resources. Unlike traditional hydro turbines, the LZH-CFT leverages a cross-flow configuration to utilize the water flows of natural streams, making it an ideal solution for flow channels with low/zero head. The design is modified through ANSYS simulations by varying the rotor's blade angles (from 0 to 90°) and 60° angle is found optimum. The prototype model of modified design is prepared and tested, which represents 62.5% RPM efficiency and provides double power as compared to the typical cross-flow turbine.

Keywords: cross-flow turbine; low/zero head; design modification; prototype model; renewable energy

Citation: Ali, A.; Azam, M.Z.; Faiq, M.; Asghar, G. Design Modification and Prototype Development of a Cross-Flow Turbine for a Low/Zero-Head Water Stream. *Eng. Proc.* **2023**, *45*, 49. https://doi.org/10.3390/engproc2023045049

Academic Editors: Mohammad Javed Hyder, Muhammad Mahabat Khan, Muhammad Irfan and Manzar Masud

Published: 18 September 2023

1. Introduction

Hydraulic energy is one of the cleanest and most economical forms of energy, which can be obtained by installing multiple kinds of turbines including Pelton, Kaplan, and Francis turbines according to the required power, available head, and other parameters. The efficiency of the cross-flow turbine depends on various design parameters such as the number of blades, angles of attack, and inner-to-outer diameter ratios. Moreover, the availability of head is a critical parameter to attain and maintain the reasonable efficiency of the cross-flow turbine, which can be achieved through discharge regulator [1].

An attempt is made to utilize the micro-head of urban sewerage discharge water to produce hydro-power by modifying the design of a conventional cross-flow turbine through computational fluid dynamics (CFD) simulation [2]. Another study [3] is focused on exploiting the available micro-heads by modifying the design parameters of the cross-flow turbine to produce power. The design parameters (such as number of blades, runner dimensions, attack angle, etc.) of the cross-flow turbine are evaluated by Nasir [4] to achieve the maximum efficiency of the turbine.

Although numerous studies have been conducted covering various aspects of cross-flow turbines, there is a scarcity of research addressing the utilization of low/zero-head water streams for power generation. Therefore, the aim of this study is to use the flow velocity of a natural water stream and to transfer the kinetic energy of the water flow on the blades to achieve the maximum rotational velocity/revolutions per minute (RPM) of the turbine blades.

2. Methodology and Design

The velocity of the water stream was calculated by using the equation $s = v \times t$ and the noted average velocity of the water was 0.5094 m/s. The prototype model was designed in

CREO parametric and a drawing of the basic design along with dimensions (cm) is shown in Figure 1a, while the wire frame 3D view is displayed in Figure 1b.

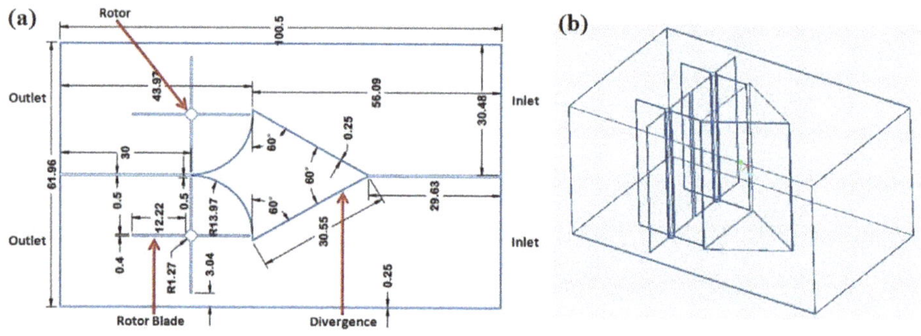

Figure 1. Design of the prototype model: (**a**) 2D design along with dimensions, (**b**) 3D wire frame view.

The inlet channel was separated into two channels through a divergence of 60°, which reduced the inlet dimensions to half at the both edges of the divergence. As a result of the reduction in area by half computed through equation of continuity ($A_1 \times V_1 = A_2 \times V_2$), the inlet velocity (0.5094 m/s) was doubled and became 1.019 m/s at both edges of the divergence. This increased velocity can be used effectively to rotate the vertically placed (two) rotors after the divergence. To achieve maximum moment/torque, the angle of the rotor's blades was varied from 0 to 90° curves as depicted in Figure 2. The dimensions (cm) of the rotor and blades are also displayed in Figure 2. The theoretical value of RPM of the rotor is 72, which was computed by dividing the velocity of the fluid (61.14 m/min) by the circumference of the rotor (0.8463 m).

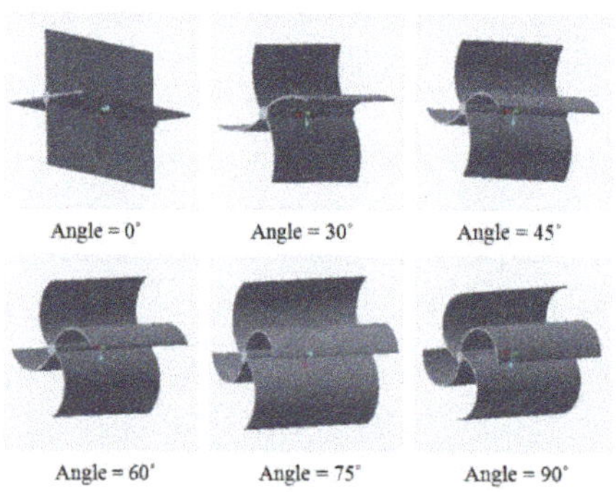

Dimensions of the rotor and blades

Parameter	Dimension (cm)
Height of the rotor	60.96
Thickness of the blade	0.40
Height of the blade	60.20
Width of the blade	12.22
Number of blades	04
Angle between two blades	90°
Diameter of the rotor shaft	2.54
Radius of the rotor shaft + blades' width	13.49

Figure 2. Design of the rotor and blades at various angles along with dimensions (cm).

ANSYS Simulation

The 3D geometry was created in the design modeler ANSYS Fluent and two types of meshing techniques were used (body sizing for the number of elements and edge sizing for the number of divisions between the inner and outer zones) as shown Figure 3b. In the setup, transient and k-epsilon were selected for turbulent flow, water was chosen from the library, and then we set 0.5 m/s as the inlet velocity along with 0 as the absolute pressure of the outlet. Mesh interfaces were generated between the inner and outer zones for the rotation of the rotor. Then, we set the hybrid initialization and 1500 iterations for all simulations to check the effect on rotors. Mesh independence was also checked in order to obtain optimum results.

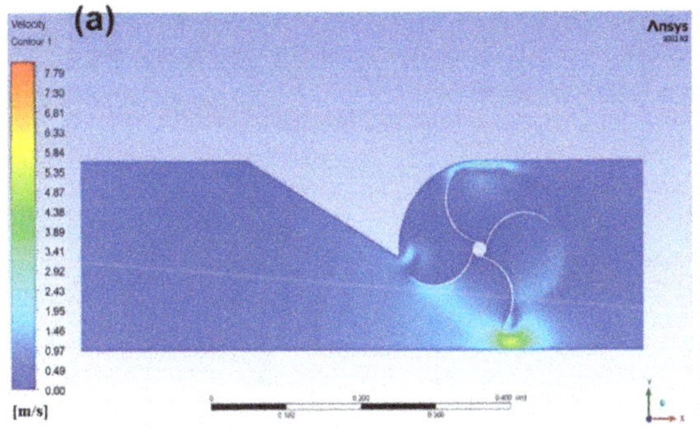

(b) **Meshing Parameters**

Blade angle	Number of elements	Number of nodes	Skewness maximum	Body sizing for number of elements (mm)		Edge sizing number of divisions
				Outer zone	Inner zone	
0°	1194408	1260349	0.59812	6.8	4.8	250
30°	1197306	1263268	0.60585	6.8	4.8	250
45°	1158804	1226433	0.61495	6.8	4.8	250
60°	1105078	1173003	0.59812	6.8	4.8	250
75°	1101421	1033674	0.62004	6.8	4.8	250
90°	1005732	1076136	0.72097	6.8	4.8	250

Figure 3. (**a**) Contour of simulation, (**b**) Meshing parameters and number of elements for different blade angles.

3. Results and Discussion

3.1. Simulation Results

Simulations were performed by varying the blade angle (0°, 30°, 45°, 60°, 75°, and 90°) to identify the blade angle for which the maximum moment can be achieved for preparation of prototype model. Figure 4a represents the simulation results in terms of moment, which shows that the blades at a 60° angle produce the maximum moment (24.5 N-m) among all the examined blade geometries. Figure 3a displayes the velocity contour of the simulation performed for 60° angle blades and demonstrates that the maximum velocity of the water is observed at the edge of the blade. The prototype model is prepared by setting the blades' angle at 60°. The inlet and outlet sections of the prototype model are shown in Figures 4b and 4c, respectively. By using the net moment (obtained from ANSYS simulations for 60° blade angle) and angular velocity, the net power is computed (by using $P = T \times \omega$), which is 128 W for one rotor and 256 W for both rotors.

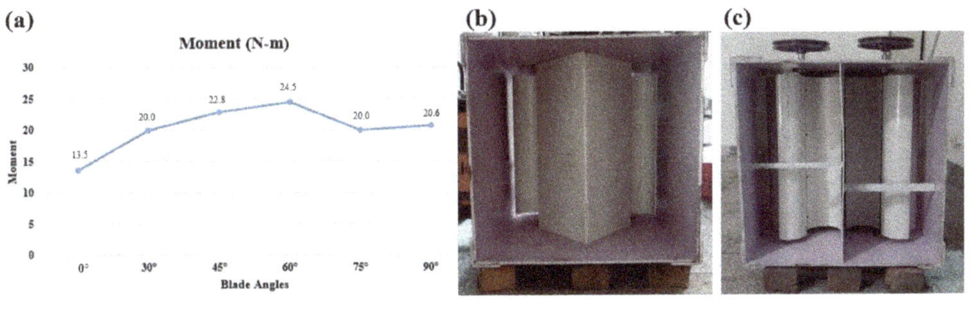

Figure 4. (**a**) Moment values measured through simulation for different blade angles, (**b**) Inlet of the prototype model, (**c**) Outlet of the prototype model.

3.2. Experimental Results

After completing the prototype model shown in Figure 4b,c, the location selected for experimentation was Soan River near Chakiyan stop, Islamabad. As mentioned earlier, the measured velocity of water was 0.5 m/s, which was used to calculate the theoretical RPM (=72) and the same velocity (0.5 m/s) was used for all simulations. However, the initial location (where the velocity was measured) was changed due to some of the constraints (like available depth and volumetric flow) of the stream. The velocity of the water was again measured at the new location (Chakiyan stop) and the noted average velocity was 0.67 m/s instead of 0.5 m/s. By using the same previous steps, the new value of theoretical RPM was 96 for a 0.67 m/s velocity. The prototype was tested under modified flow conditions and both of its rotors attained a sTable 60 RPM. Based on the experimental results, the RPM efficiency (experimental value of RPM ÷ theoretical value of RPM) of the modified design of the turbine is 62.5% providing twice the power as compared to the general cross-flow turbine due to the double rotors.

4. Conclusions

In the present study, the design of a cross-flow turbine for a low/zero-head water stream is modified and evaluated through ANSYS simulations. The speed of the water is increased and doubled by placing a divergence in the inlet channel. The two vertically positioned rotors after the divergence utilize the increased speed of the water and effectively transform it into rotational velocity. The rotor's blade angles are varied from 0 to 90° and the simulation results indicate that a 60° angle generates the maximum moment and power. The prototype model is prepared according to the optimum design parameters identified through simulations. The prototype is tested and shows stable performance with 62.5% RPM efficiency by providing twice the power as compared to the conventional cross-flow turbine.

Author Contributions: Conceptualization, methodology: G.A.; investigation, resources: A.A.; writing—original draft preparation: M.Z.A.; writing—review and editing: G.A.; visualization: M.F.; Supervision: G.A. All authors have read and agreed to the published version of the manuscript.

Funding: This research is funded by (NGIRI-2023-16887) National Gross-root ICT Research Initiative (NGIRI) 2022-23.

Institutional Review Board Statement: Not applicable.

Informed Consent Statement: Not applicable.

Data Availability Statement: The authors can provide relevant data on demand.

Conflicts of Interest: The authors declare no conflict of interest.

References

1. Desai, V.R.; Aziz, N.M. An Experimental Investigation of Cross-Flow Turbine Efficiency. *J. Fluids Eng.* **1994**, *116*, 545–550. [CrossRef]
2. MKhan, A.; Badshah, S. Research Article Design and Analysis of Cross Flow Turbine for Micro Hydro Power Application using Sewerage Water. *Res. J. Appl. Sci. Eng. Technol.* **2014**, *8*, 821–828.
3. Chattha, J.A.; Khan, M.S.; Wasif, S.T.; Ghani, O.A.; Zia, M.O.; Hamid, Z. Design of a cross flow turbine for a micro-hydro power application. *Power Conf.* **2010**, *49354*, 637–644.
4. Nasir, B.A. Design of high efficiency cross-flow turbine for hydro-power plant. *Int. J. Eng. Adv. Technol.* **2013**, *2*, 308–311.

engineering proceedings

MDPI

Proceeding Paper

A Novel Low-Cost Machine for Evaluating Helmet Performance in Bike Accident Scenarios †

Syed Mujahid Abbas [1], Haider Ali [2,*], Syed Aqib Ali Naqvi [2], Mohsin Iqbal [2], Hafiz Hamza Jabbar [2], Muhammad Azeem [2] and Manzar Masud [3]

1 Department of Mechanical Engineering, Yokohama National University, 79-1 Tokiwadai, Hodogaya Ward, Yokohama 240-0067, Kanagawa, Japan; syed.mujahid7589@gmail.com

2 Department of Mechanical Engineering, University of Wah, Rawalpindi 47040, Pakistan; syedaqibalinaqvi@gmail.com (S.A.A.N.); mohsiniqbalwec007@gmail.com (M.I.); hamzajabbar95@gmail.com (H.H.J.); m.azeem@wecuw.edu.pk (M.A.)

3 Department of Mechanical Engineering, Capital University of Science and Technology (CUST), Islamabad 44000, Pakistan; manzar.masud@cust.edu.pk

* Correspondence: haider.ali@wecuw.edu.pk

† Presented at the Third International Conference on Advances in Mechanical Engineering 2023 (ICAME-23), Islamabad, Pakistan, 24 August 2023.

Abstract: The increasing mortality rate in bike-related road accidents due to head injuries necessitates stronger helmets. This paper presents a cost-effective machine setup to evaluate helmet behavior during impacts. By employing unique methods to measure impact duration and post-impact energy absorption, this project eliminates the need for high-speed cameras. The results indicate that the developed machine enables scenario-based impact tests directly on helmets, providing valuable insights into collision effects. Validation confirms its effectiveness. The low-cost impact testing machine offers a practical solution for studying helmet performance and addressing the rising concerns surrounding bike accidents.

Keywords: pendulum-based impact testing; cost-effective; impact time; bike helmets; impact test

Citation: Abbas, S.M.; Ali, H.; Naqvi, S.A.A.; Iqbal, M.; Jabbar, H.H.; Azeem, M.; Masud, M. A Novel Low-Cost Machine for Evaluating Helmet Performance in Bike Accident Scenarios. *Eng. Proc.* **2023**, *45*, 39. https://doi.org/10.3390/engproc2023045039

Academic Editors: Mohammad Javed Hyder, Muhammad Mahabat Khan and Muhammad Irfan

Published: 13 September 2023

1. Introduction

Mechanical impact is the leading cause of brain injury, death, and disability in people aged under 45 in the USA, Europe, and developing countries [1]. Road accidents have contributed largely to this aspect. Although bikers are instructed to wear helmets while riding bikes, the quality of helmets still needs to be assessed before their commercialization [2]. Generally, there are three approaches to helmet testing: physical tests, analytical modeling, and numerical simulations [3]. Among them, physical testing has always been accepted worldwide because of its ability to produce real results, incorporating all the given mechanical conditions [4]. Several studies have investigated helmet performances using different impact environments [5]. However, expensive high-speed cameras for impact time measurements and post impact studies were used in these studies.

In countries such as Pakistan, where motorcycles comprise up to 75% of registered vehicles, local helmet manufacturing companies do not have the facilities to test the helmet material's strength and behavior during collision to make their product robust and reliable before bulk production. Existing global testing equipment is expensive and requires specialized expertise. In this study, a pendulum-based impact testing machine for helmets was designed, fabricated, and experimented directly on helmets. Limit switches were used for impact time measurement. The authors found that the results are approaching the standards and are fair to use. Comparing the cost and effectiveness of the machine, this could be a new commercial testing facility for helmets.

2. Materials and Methods

Table 1 shows the components and their respective materials. The height, length, and width of the machine were assumed to be 244 cm, 180 cm, and 120 cm, respectively.

Table 1. Components used in machine and the respective materials.

Component	Material
Column	Mild Steel (Both ends fixed)
Specimen (bike helmet)	Glass Fiber (E = 72 GPa)
Impactor (variable mass range of 5–20 kg)	Mild Steel (E = 200 GPa)
Instruments used	IR Sensor, Limit Switch, Timer
Guide Rails	Stainless steel (E = 190 GPa)

2.1. Design of Column

The different machine components were designed using the standard procedures of mechanical design. Columns were designed using the Euler formula given in Equation (1).

$$P_E = \frac{\pi^2 EI}{{L_e}^2} \tag{1}$$

where P_E is the critical load, E is the modulus of elasticity, L_e is the effective length, and I is the moment of inertia. The column was designed with a factor of safety equal to 2. The rectangular column had cross-sections of 30 mm × 60 mm × 2 mm. The height of the column was 244 cm.

2.2. Design of Impactor Arm

The machine should be as light as possible. Therefore, the impactor arm was made hollow. The safe diameter was calculated using bending criteria with a factor of safety 2. The pin was designed using both shearing and bending criteria where the bending gave a maximum diameter of 10.7 mm. Equation (2) gives the bending criteria.

$$d^3 = \frac{(B.M)(32)}{\pi(S_y)} \tag{2}$$

The head form was made by filling polyester foam in a head-shaped leather bag over a steel base. The manufactured pendulum-based testing machine is given in Figure 1.

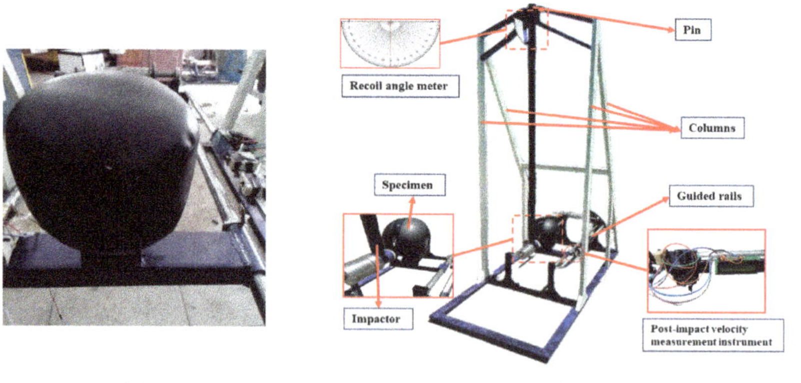

(a) (b)

Figure 1. (a) Head form for helmet rest, (b) pendulum-based impact testing machine.

2.3. Impact Duration Measurement

The duration of the impact was measured using the limit switch along with an Arduino controller. When the contact between the impactor and helmet was established, the current started flowing in the circuit, and it broke the circuit when the two were disconnected. The time span was calculated using the Arduino controller.

2.4. Measurement of Energy Absorbed by Helmet

The total energy of the impactor upon impact was converted into three energies, i.e., the energy absorbed by the helmet, the post-impact kinetic energy of the helmet, and the remaining energy in the impactor. Equation (3) is used for the calculation of the total energy absorbed in the helmet,

$$E_T = E_h + K{\cdot}E_h + E_{rec} \tag{3}$$

where E_T is the total energy, E_h is the energy absorbed by the specimen, $K{\cdot}E_h$ is the kinetic energy of the helmet after impact, and E_{rec} is the recoil/remaining energy of the impactor after the collision. The total energy of the impactor was calculated through the maximum kinetic energy achieved by the impactor just before impact. The potential energy of the impactor depended on the height achieved by the impactor measured by measuring the angle of inclination on the protractor attached. The kinetic energy of the helmet was measured through the total distance covered by the helmet on a steel rail, measured using a ruler. The frictional force was responsible for stopping the helmet. The higher the kinetic energy, the higher the distance traveled, as the friction force was assumed to be constant. The following Equations (4) and (5) were utilized to calculate the kinetic energy of the helmet.

$$K{\cdot}E_h = -F_f d \tag{4}$$

$$F_f = \mu_k N \tag{5}$$

3. Results and Discussion

The impact testing was performed on a standard available bike helmet, along with its catalogue in the local market. The experiments were conducted with varying angles of lift of the impactor for a constant mass of the impactor being taken as 10 kg. Table 2 gives the readings taken from the experiment.

Table 2. Experimental results obtained from the pendulum-based impact energy machine.

	Sr. No	1	2	3	4
Before Impact attributes of impactor	Angle of lift (degree)	30	45	60	75
	Height attained (m)	0.245	0.536	0.914	1.355
	Velocity just before impact (m/s)	2.19	3.24	4.23	5.15
	Total energy (J)	24.01	52.53	89.57	132.79
	Impact Duration (s)	0.13	0.23	0.35	0.4
Post Impact attributes of impactor	Impactor recoil angle (degree)	10	15	20	24
	Height attained (m)	0.028	0.062	0.11	0.16
	Recoil energy (J)	2.744	6.076	10.78	15.68
Post Impact attributes of helmet	Distance travelled by helmet (m)	0.18	0.27	0.3	0.33
	Kinetic energy of helmet (J)	3.70	5.56	6.17	6.79
	Energy absorbed by helmet (J)	17.56	40.89	72.62	110.32

The results highlight the relationship between the angle of lift of the impactor, the distance traveled by the helmet, and the energy absorbed by the helmet. The potential energy was directly proportional to the angle of lift of the impactor, thereby having a direct relation to both the distance traveled by the helmet and the energy absorbed by it. The apparatus in this study measured the total energy transferred to the helmet, along with the energy transferred to the neck. The limitations, however, include no measurement being taken of the peak acceleration values from the helmet to neck.

The maximum energy absorbed by the helmet was reported to be 110.32 J when the impactor was lifted at a 75° angle. The energy absorbed was 80.9 J for a corresponding impact energy of 101.13 J, which was ~80% of the impact energy, including the loss of energy in gaining momentum in the direction of the impact. The energy absorbed was ~88%, as per that given by Bhudolia et al. in 2021, not including the energy transferred to the neck [6].

4. Conclusions

In this research, a simple method for measuring the impact duration and post-impact energy absorption during helmet impact was presented. A pendulum-based impact energy machine was designed and fabricated, and then the experimentation was conducted with a varying angle of lift of the impactor for a constant mass of the impactor taken as 10 kg. The results concluded that a higher potential energy led to an increase in both the distance traveled by the helmet and the energy absorbed by it. The experimental results were then compared to the reference results and based on this comparison, it can be concluded that the novel machine used in the current study was economical and could be utilized for assessing helmet performance and enhancing the scrutiny of helmet designs.

Author Contributions: Conceptualization, S.M.A., H.A., M.A. and M.M.; methodology, S.M.A., H.A., S.A.A.N., M.I. and M.M.; software, S.M.A. and S.A.A.N.; validation, S.M.A. and H.H.J.; formal analysis, S.M.A., H.A., M.A. and M.M.; investigation, S.M.A., H.A., M.A. and H.H.J.; data curation, S.M.A., H.A., M.A. and M.M.; writing—original draft preparation, S.M.A., H.A., S.A.A.N., M.I. and H.H.J.; writing—review and editing, H.A., M.A. and M.M.; supervision, H.A.; project administration, H.A. All authors have read and agreed to the published version of the manuscript.

Funding: This research received no external funding.

Institutional Review Board Statement: Not applicable.

Informed Consent Statement: Not applicable.

Data Availability Statement: Not applicable.

Conflicts of Interest: The authors declare no conflict of interest.

References

1. Critchley, G.; Memon, A. Epidemiology of Head Injury. In *Head Injury: A Multidisciplinary Approach*; Cambridge University Press: Cambridge, UK, 2009; pp. 1–11. [CrossRef]
2. Kongwat, S.; Nueanim, T.; Hasegawa, H. FE Analysis of Motorcycle Helmet Performance under Severe Accidents. *Appl. Sci.* **2022**, *12*, 5676. [CrossRef]
3. Ali, H.; Mustansar, Z.; Talay, S.; Masood, A.; Zahoor, S.; Sherbaz, S. Time-Dependent Analysis and Classification of Deformation Mechanism in Head Impact Injury Using Image Based Finite Element Method (FEM). In Proceedings of the 2021 12th International Conference on Mechanical and Aerospace Engineering, ICMAE 2021, Athens, Greece, 16–19 July 2021; Institute of Electrical and Electronics Engineers Inc.: Piscataway, NJ, USA, 2021; pp. 576–581.
4. Whyte, T.; Stuart, C.A.; Mallory, A.; Ghajari, M.; Plant, D.J.; Siegmund, G.P.; Cripton, P.A. A Review of Impact Testing Methods for Headgear in Sports: Considerations for Improved Prevention of Head Injury Through Research and Standards. *J. Biomech. Eng.* **2019**, *141*, 070803. [CrossRef] [PubMed]
5. Mills, N.J.; Gilchrist, A. Oblique Impact Testing of Bicycle Helmets. *Int. J. Impact Eng.* **2008**, *35*, 1075–1086. [CrossRef]
6. Bhudolia, S.K.; Gohel, G.; Subramanyam, E.S.B.; Leong, K.F.; Gerard, P. Enhanced Impact Energy Absorption and Failure Characteristics of Novel Fully Thermoplastic and Hybrid Composite Bicycle Helmet Shells. *Mater. Des.* **2021**, *209*, 110003. [CrossRef]

engineering proceedings

MDPI

Proceeding Paper

Investigation of Progressive Delamination Growth Characterization in Composite Materials †

Nadeem Aslam [1], Muhammad Haroon [1,2,*], Muhammad Arsalan Munawar [1], Shummaila Rasheed [2] and Manzar Masud [2]

1 Department of Mechanical Engineering, International Islamic University, Islamabad 44000, Pakistan; nadeem.phdme49@iiu.edu.pk (N.A.); muhammad.phdme47@iiu.edu.pk (M.A.M.)
2 Department of Mechanical Engineering, Capital University of Science and Technology (CUST), Islamabad 44000, Pakistan; shummaila@cust.edu.pk (S.R.); manzar.masud@cust.edu.pk (M.M.)
* Correspondence: muhammad.haroon@cust.edu.pk

† Presented at the Third International Conference on Advances in Mechanical Engineering 2023 (ICAME-23), Islamabad, Pakistan, 24 August 2023.

Abstract: Composite-materials-based structures are extensively used in aerospace structures owing to their high strength-to-weight ratio and high specific modulus. There are different types of failures in a composite material subjected to multiple types of loading, but delamination is the most important one and occurs where the material fractures into layers. In this current research, a cohesive zone method approach is applied to investigate the fracture mechanics due to delamination. Finite element analysis was used for the delamination characterization in composite materials in which 2D and 3D models of double cantilever beams (DCBs) were used. ABAQUS Software was used for analysis completion. It is observed that the cohesive element's size must be 0.5 mm or less in order to forecast delamination precisely for double cantilever beams. It was also determined that the initial stiffness could not be less, or else the damage initiation cannot be forecast correctly. The value for initial stiffness used in this research was 10^6 kJ/m^2.

Keywords: delamination growth characterization; composite materials; double cantilever beam

Citation: Aslam, N.; Haroon, M.; Munawar, M.A.; Rasheed, S.; Masud, M. Investigation of Progressive Delamination Growth Characterization in Composite Materials. *Eng. Proc.* **2023**, *45*, 35. https://doi.org/10.3390/engproc2023045035

Academic Editors: Mohammad Javed Hyder, Muhammad Mahabat Khan and Muhammad Irfan

Published: 13 September 2023

1. Introduction

A composite is a type of material that comprises collective elements which are united at the macroscopic level and are insoluble in each-other. Delamination is a significant failure within the composite materials laminates due to weak strengthening through the thickness. The delamination among the laminae of composite materials is critical because it may lead to the debonding of the laminae and cause progressive damage to the entire structure.

In the literature, the cohesive zone method (CZM) is broadly used for investigating the crack progression. The cohesive zone method is founded on the idea of Barenblatt [1], who presented it for brittle materials. In CZM, the interlaminate separation around the cohesive zone is incurred. The fundamental idea of the CZM is based on the idea that the inelastic effects that occur at the crack vicinity can be lumped into a surface cohesive damage zone. In this regard, Elliott [2] considered nonlinear fabric failure and added an interatomic attracting pressure in line with unit area to research the fracture of a crystalline substance alongside a cleavage plane. Dugdale [3] employed an equal cohesive zone version to analyze yielding at a crack tip and size of the plastic region.

The objective of this research is to investigate the debonding of laminae and the progressive damage behavior of double cantilever beams manufactured by composite materials by using experimental and finite element techniques. For this purpose, a double cantilever beam was modelled in Abaqus to investigate the delamination behavior. The results of the 2D model of the double cantilever beam are prepared for two composite materials, i.e., glass epoxy and HTA-6376 carbon epoxy, and finally compared with experimental published data. Finally, the results of 2D and 3D models are compared for both materials and their delamination process is evaluated in detail.

2. Methodology

Modeling

Double cantilever beam (DCB) specimen was made up of 2 four-sided beams with a pre-cracked surface in the center plan of Figure 1. The dimensions of the double cantilever beam were 150 mm length, 25 mm width and 4.2 mm is thickness/depth. The initial crack length was 35 mm. A cohesive zone was also sandwiched between two rectangular beams. Cohesive zone thickness was 0.02 mm.

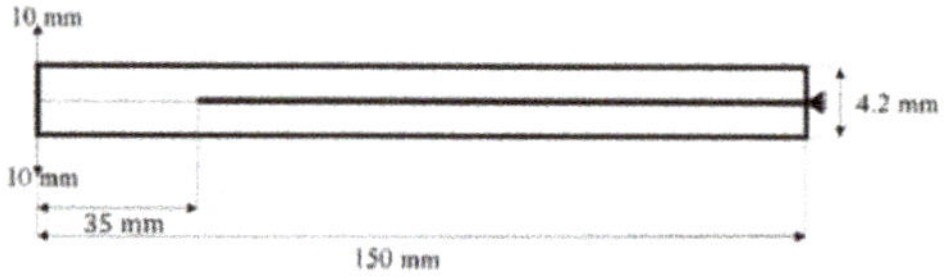

Figure 1. Double cantilever beam specimen.

The analysis was completed using displacement control boundary conditions, and the linear method was used. The magnitude of displacement was 10 mm, which was applied on the upper-left edge and the lower-left edge, while the other side was kept fixed. Throughout the analysis, the geometric properties of the double cantilever beam and applied displacement boundary conditions were kept constant. A detailed mesh-independent study was performed during the analysis. Several sets of simulations were performed with different mesh sizes from 0.1 to 0.35 million for glass epoxy, as shown in Figure 2. For glass epoxy, it was concluded that the results at 0.2 million mesh size were optimum, and were thus used for the rest of simulation process. Similarly, for carbon epoxy sample, 0.15 million mesh size was chosen.

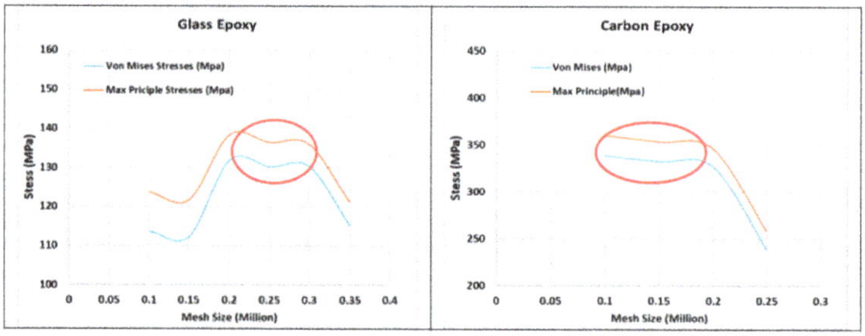

Figure 2. Mesh independence study.

3. Results and Discussion

3.1. Results for 2D Unidirectional Glass Epoxy and Carbon Epoxy

The DCB model's SDEG status in Abaqus is presented in Figure 3. The SDEG variable shows whether the criteria of damage initiation are fulfilled or not. If the value of "SDEG" is equivalent to one then the criteria of initiation is fulfilled. In other words, it shows the state of damage in the model. Similarly, the load vs. displacement curve for glass epoxy 2D, DCB and HTA-6376 carbon epoxy is shown in Figure 4. In this curve, the maximum load or peak load experienced by glass epoxy 2D DCB is 60 N. The load-displacement curve shows the elastic behavior or damage initiation phase at the start, and after reaching the peak load it starts to decline and show the damage progression phase. The minimum load experienced by glass epoxy 2D DCB is 38 N. The same behavior for load–displacement curve for the experimental result is experienced. The peak load for experimental result is 63 N. Similarly, the peak load experienced by HTA-6376 2D DCB is 142 N at 2 mm. The load–displacement curve shows a damage initiation phase at the start, and after reaching the peak load it starts to decline to show damage progression. The minimum load experienced by HTA-6376 carbon epoxy 2D DCB is 60 at the given displacement; the same load–displacement behavior is experienced for glass epoxy.

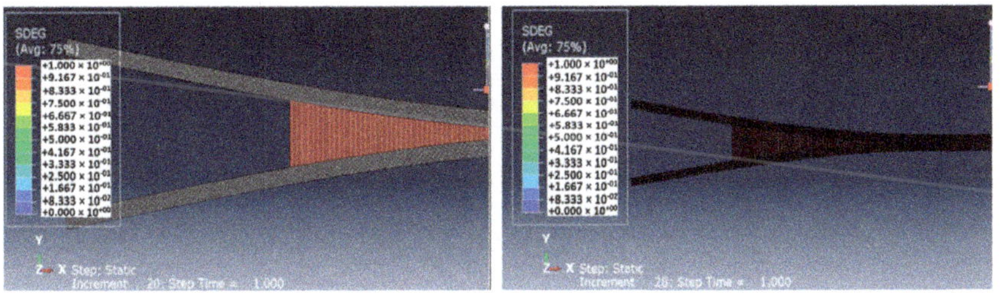

Figure 3. SDEG of carbon and glass epoxy.

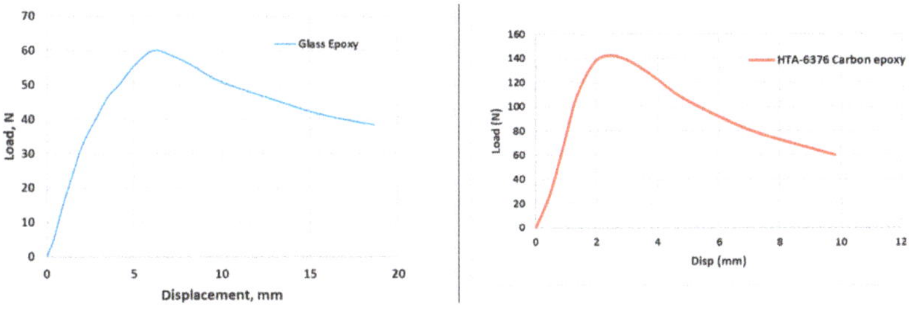

Figure 4. Comparison of peak load vs. displacement for both materials.

3.2. *Results for 3D Unidirectional Glass and Carbon Epoxy*

The contours of U Magnitude for 3D glass epoxy and HTA-6376 carbon epoxy are shown in Figure 5. The maximum value of strain energy is 306 J at time 1 s. As the crack increases in the DCB model, the strain energy increases, while the strain energy curve gradually rises with respect to time, and it reaches to the maximum value of 620 J for the HTA-6376 specimen. Figure 6 shows the strain energy comparison curve for both materials. Figure 7 encloses the near-matching results of 2D and 3D geometries for both specimens.

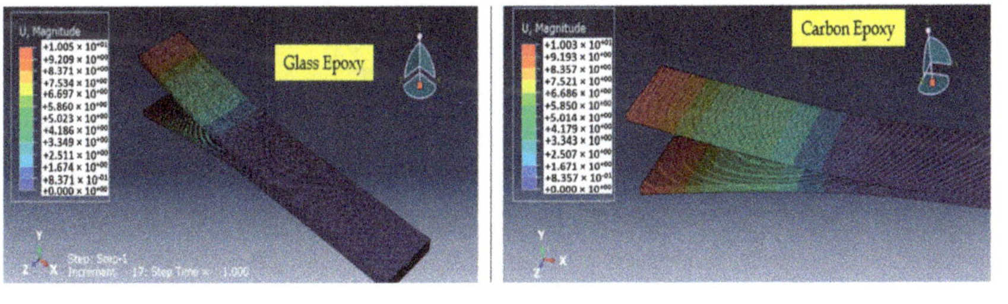

Figure 5. Comparison of U magnitudes for 3D specimens.

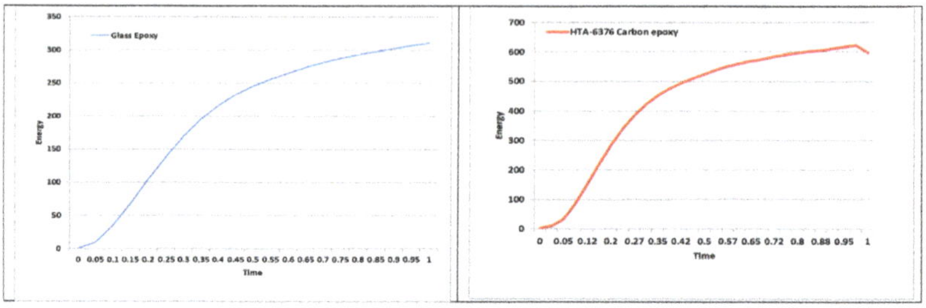

Figure 6. Strain energy comparisons for glass Epoxy and HTA-6376 carbon epoxy material.

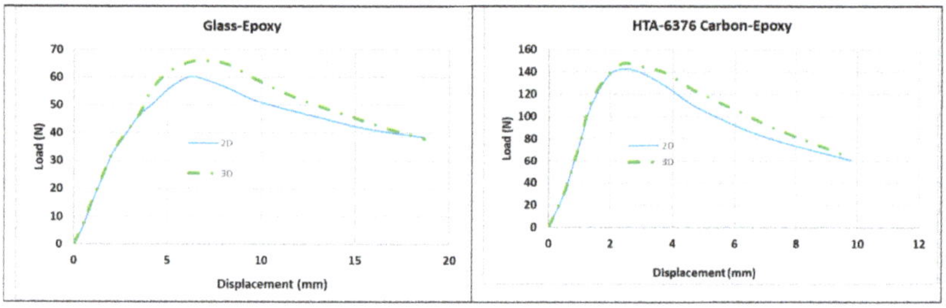

Figure 7. Both 2D and 3D load vs. displacement comparisons for both specimens.

4. Conclusions

In the present study, an investigation of progressive delamination growth characterization in composite materials was performed. It was noticed that the usage of higher fracture energy or lower fracture energy than the nominal value caused the over-assessment or under-assessment of delamination or damage initiation, respectively. Both the 2D and 3D finite element analysis results were compatible with the experimental results. The SDEG value was equivalent to 1, which means the initiation criterion is fulfilled.

Author Contributions: Conceptualization, N.A. and M.A.M.; methodology, S.R. and M.M.; formal analysis, M.H. and N.A.; investigation, M.H. and M.A.M.; data curation, S.R. and M.M.; writing—original draft preparation, N.A.; writing—review and editing, M.H. and M.M.; supervision, M.H. All authors have read and agreed to the published version of the manuscript.

Funding: This research received no external funding.

Institutional Review Board Statement: Not applicable.

Informed Consent Statement: Not applicable.

Data Availability Statement: Not applicable.

Conflicts of Interest: The authors declare no conflict of interest.

References

1. Barenblatt, G.I. The Mathematical Theory of Equilibrium Cracks in Brittle Fracture. In *Advances in Applied Mechanics*; Dryden, H.L., von Kármán, T., Kuerti, G., van den Dungen, F.H., Howarth, L., Eds.; Elsevier: Amsterdam, The Netherlands, 1962; Volume 7, pp. 55–129.
2. Elliott, H. Analysis of Condition for Rupture Due to Griffith Cracks. *Proc. Phys. Soc.* **2002**, *59*, 208. [CrossRef]
3. Dugdale, D.S. Yielding of Steel Sheets Containing Slits. *J. Mech. Phys. Solids* **1960**, *8*, 100–104. [CrossRef]

engineering proceedings

MDPI

Proceeding Paper

Tool Wear Parameter Optimization in Machining a Squeeze-Cast Metal Matrix Composite (Al6061-SiC) †

Asif Imran [1], Muhammad Waqas Hanif [1,*], Muhammad Sajid [1], Shahab Salim [1], Feroz Haider [1] and Muhammad Azeem [2]

1 Industrial Engineering Department, University of Engineering and Technology, Taxila 47050, Pakistan; asifimran.uet@gmail.com (A.I.); sajid.malik@uettaxila.edu.pk (M.S.); workingshahab@outlook.com (S.S.); feroz.haider@students.uettaxila.edu.pk (F.H.)

2 Mechanical Engineering Department, University of Wah, Wah Cantt 47040, Pakistan; m.azeem@wecuw.edu.pk

* Correspondence: waqas.hanif@students.uettaxila.edu.pk

† Presented at the Third International Conference on Advances in Mechanical Engineering 2023 (ICAME-23), Islamabad, Pakistan, 24 August 2023.

Abstract: In this research work, machining operations on an aluminum matrix composite (AMC) were optimized for improving the wear of high-speed steel tools. The squeeze casting method was used to manufacture the AMC, which had Al-6061 as matrix material and silicon carbide (wt. 15%) microparticles used as reinforcement. Feed rate (Fr), cutting speed (Cs), and depth of cut (Dc) were selected to optimize HSS tool wear rate. Using the Box–Behnken design, seventeen experiments were performed to analyze the single-factor effects and interaction effects of the process parameters on HSS tool wear rate. Experimental results show that optimal tool wear (0.964) was achieved at a Cs of 80 m/min, Fr of 0.2 rev/min, and Dc of 0.8 mm.

Keywords: aluminum matrix composite; tool wear; Al-6061; silicon carbide

Citation: Imran, A.; Hanif, M.W.; Sajid, M.; Salim, S.; Haider, F.; Azeem, M. Tool Wear Parameter Optimization in Machining a Squeeze-Cast Metal Matrix Composite (Al6061-SiC). *Eng. Proc.* **2023**, *45*, 1. https://doi.org/10.3390/engproc2023045001

Academic Editors: Mohammad Javed Hyder, Muhammad Mahabat Khan, Muhammad Irfan and Manzar Masud

Published: 7 September 2023

1. Introduction

Metal matrix composites (MMC) are a novel type of material and a swift substitute for conventional materials in manufacturing applications such as the aerospace and automobile industries [1]. Machining is one of the most common manufacturing processes for metal matrix composites to achieve the desired shapes. However, researchers face challenges in performing machining operations on MMC materials due to hard abrasive reinforcement particles, which are harder than cutting tools [2,3]. Therefore, it is important to find the optimal combination of process variables while performing turning operations on MMCs in order to achieve the desired shape and improve tool life. In this regard, Seeman et al. [2] found that Cs and Fr had a more significant effect as compared to machining time and Dc on flank wear and surface roughness. R. Suresh et al. [3] concluded that cutting speed was inversely proportional to surface roughness, which was caused by less contact between the tool and workpiece.

It has been revealed that there is still a research gap in optimizing turning operation process variables in order to reduce high-speed steel (HSS) tool wear, because machine tools account for roughly 70% of active machining production costs [4]. Therefore, the aim of this project was to optimize the variables of turning operations on Al6061-SiC composites, such as Cs, Fr, and Dc, in order to reduce HSS tool wear.

2. Materials and Methods

The aluminum matrix alloy used in this research work was a wrought 6061 aluminum alloy. A spectrometry test was performed to check the chemical composition of the 6061 Al alloy as shown in Table 1. Silicon carbide (SiC) particles were used as reinforcement in this study. Squeeze casting was used to incorporate the SiC particles in two distinct weight percentages, 7.5% and 15%, respectively. The sample was then constructed by cutting a billet of composite squeeze-cast aluminum to prepare a turned 20 × 150 mm specimen billet.

Table 1. Spectrometry results of aluminum 6061 alloy.

Elements	Al	Mg	Fe	Si	Cu	Ti	Mn	Cr	Zn
Weight Percent. (Al)	97.05	0.14	0.7	0.43	0.24	0.15	0.24	0.8	0.25

3. Experiment Design & Setup

The ranges of selected process parameters such as Cs (40 to 120 m/min), Fr (0.1 to 0.3 mm/rev) and Dc (0.4 to 1.2 mm) have been selected based on the literature review [2,3,5] and after performing the trial experiment. Three process parameters (k) and five central points (c) have been selected to design experiments using Equation (1) [6].

$$n = 2 \times k \times (k - 1) + c \quad (1)$$

A total of 17 runs for each billet were performed using the Box–Behnken design. The TX-75Y model turning center with a soluble oil-based coolant was used for the experimental process as shown in Figure 1b. Wear was measured using the tool–workpiece distance method. In this method, a micrometer, digital Vernier caliper, and electron microscope were used to measure the flank wear of the tool. Three experiments were performed under each experimental condition to ensure the accuracy of the output response. Table 2 displays the average of three responses.

Table 2. Experimentation using the Box–Behnken design.

Experiment No.	Input Variables			Output Response (Tool Wear)
	Cs	Fr	Dc	Average Value (mm)
1	40	0.1	0.8	1.884
2	120	0.1	0.8	1.956
3	40	0.3	0.8	1.436
4	120	0.3	0.8	1.448
5	40	0.2	0.4	1.4592
6	120	0.2	0.4	1.452
7	40	0.2	1.2	1.456
8	120	0.2	1.2	1.456
9	80	0.1	0.4	1.428
10	80	0.3	0.4	1.028
11	80	0.1	1.2	1.924
12	80	0.3	1.2	1.444
13	80	0.2	0.8	0.964
14	80	0.2	0.8	1.456
15	80	0.2	0.8	1.468
16	80	0.2	0.8	0.968
17	80	0.2	0.8	1.444

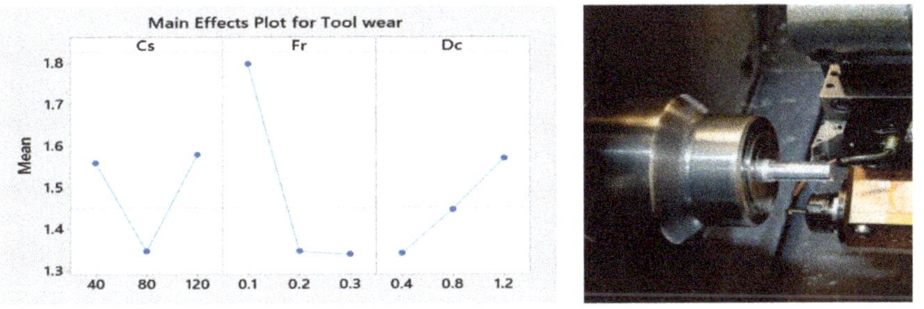

(**a**) (**b**)

Figure 1. (**a**) Single factor plot of mean ratios of tool wear; (**b**) experimental setup.

4. Results and Discussion

The experimental results showed that the minimum mean value of tool wear rate (0.964) was achieved at the Cs of 80 m/min, Fr of 0.2 rev/min, and Dc of 0.8 mm. Furthermore, standard deviation (0.019) and R-squared (0.9966) demonstrated that the variation of replicated mean values was lesser than the variation of the values anticipated or predicted in the design; therefore, the developed model was a good predictor [7,8]. At a 95% confidence level, an analysis of variance (ANOVA) table showed that the mathematical model was significant, and all the selected process parameters had a significant effect on tool wear because the *p* value was less than 0.05, as shown in Table 3.

Table 3. ANOVA table for wear rate.

Source	SS		Adj SS.	Adj MS.	F	*p*
Model	1.31	3	0.44	1267.99	<0.0001	significant
Cs	0.45	1	0.45	1302.4	<0.001	significant
Fr	0.47	1	0.47	1368.05	<0.0001	significant
Dc	0.39	1	0.39	1133.51	<0.0001	significant
Residual	4.85×10^{-03}	13	3.45×10^{-04}			
Lack of Fit	3.88×10^{-03}	9	4.31×10^{-04}	2.83	0.1178	not significant
Pure Error	6.08×10^{-03}	4	1.52×10^{-04}			
Cor Total	1.32	16				

Standard deviation (0.019); R square (0.9966); Adj. R square (0.9958); and Pred. R square (0.9933).

A single factor plot showed that tool wear was reduced when Cs increased from 40 to 120 m/min, as shown in Figure 1a. It was also revealed that tool wear was minimized when Fr was reduced from 0.1 to 0.3 rev/min. Similarly, tool wear linearly increased by increasing Dc from 0.4 to 1.2 mm. Furthermore, 3D Mesh plots created in our study analyzed the effects of two parameters at a time. The Cs vs. Fr plot showed that tool wear was linearly minimized by reducing Fr from 0.1 to 0.3 mm and increasing Cs from 40 to 120 m/min as shown in Figure 2a. The Fr vs. Dc plot showed that tool wear reduced by minimizing the Dc value from 0.4 mm to 1.2 mm and increasing Fr from 0.1 to 0.3 mm/rev as depicted in Figure 2b. Finally, the Cs vs. Dc plot showed that tool wear was gradually reduced by increasing the Dc from high level to low level and Cs from low level to high level as shown in Figure 2c. It is also evident from the studies of R. Suresh et al. [5] that tool wear is reduced when Cs is increased, Fr is reduced, and Dc is increased, because the combined impact of Fr and Cs in single point cutting tools makes them prone to flank wear.

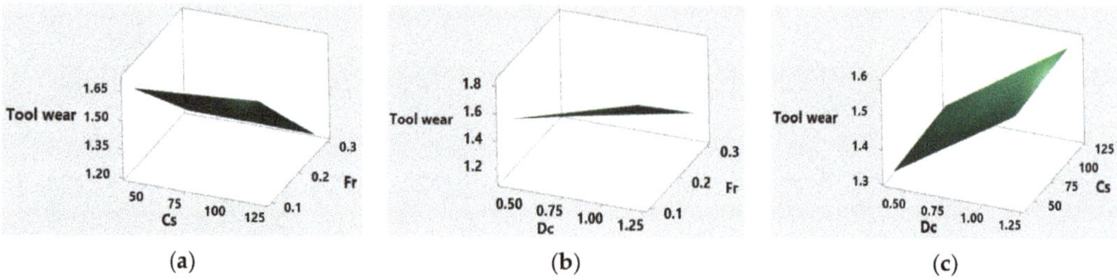

Figure 2. Impact of the following process variables on tool wear: (**a**) Cs vs. Fr; (**b**) Dc vs. Fr; and (**c**) Cs vs. Fr.

5. Validations

In order to validate the model achieved in the experiment, as shown in Equation (2), a 1.67 mm wear rate was observed under the first experimental validation condition, namely a Cs of 50 m/min, Fr of 0.15 rev/min, and Dc of 0.5 mm. A 1.47 mm wear rate was calculated using Equation (2). The percentage change in the actual and calculated values was computed as 11.88%, which shows the validity of the developed mathematical model of wear rate.

$$\text{Tool Wear} = 1.663 + 0.00024 \times (\text{Cs}) - 2.295 \times (\text{Fr}) + 0.285 \times (\text{Dc}) \tag{2}$$

6. Conclusions

It was concluded that the optimal value of HSS tool wear (0.964) during CNC turning operations of Al 6061 SiC composites was achieved using the following experimental settings: Cs of 80 m/min, Fr of 0.2 rev/min, and Dc of 0.8 mm. The ANOVA table showed that Fr had the most significant effect, compared to Cs and Dc. The low value of the percentage change in values validated the mathematical model.

Author Contributions: Conceptualization, A.I. and M.W.H.; methodology, M.W.H. and M.S.; software, F.H. and S.S.; validation, M.W.H. and M.S.; formal analysis, M.A. and A.I.; investigation, M.W.H. and F.H.; data curation, A.I.; writing—original draft preparation, M.W.H.; writing—review and editing, M.W.H. and M.S.; supervision, M.S.; project administration, M.W.H. All authors have read and agreed to the published version of the manuscript.

Funding: This research received no external funding.

Institutional Review Board Statement: Not applicable.

Informed Consent Statement: Not applicable.

Data Availability Statement: Not applicable.

Conflicts of Interest: The authors declare no conflict of interest.

References

1. Sarfraz, M.H.; Jahanzaib, M.; Ahmed, W.; Hussain, S. Multi-response parametric optimization of squeeze casting process for fabricating Al 6061-SiC composite. *Int. J. Adv. Manuf. Technol.* **2019**, *102*, 759–773. [CrossRef]
2. Seeman, M.; Ganesan, G.; Karthikeyan, R.; Velayudham, A. Study on tool wear and surface roughness in machining of particulate aluminum metal matrix composite-response surface methodology approach. *Int. J. Adv. Manuf. Technol.* **2010**, *48*, 613–624. [CrossRef]
3. Suresh, R.; Basavarajappa, S.; Gaitonde, V.; Samuel, G. Machinability investigations on hardened AISI 4340 steel using coated carbide insert. *Int. J. Refract. Met. Hard Mater.* **2012**, *33*, 75–86. [CrossRef]
4. Youssef, H.; EL-Hofy, H. Basic Elements and Mechanisms of Machine Tools. In *Traditional Machining Technology*, 2nd ed.; CRC Press: Boca Raton, FL, USA, 2020; pp. 11–57.
5. Baburaj, E.; Mohana Sundaram, K.; Senthil, P. Effect of high-speed turning operation on surface roughness of hybrid metal matrix (Al-SiC p-fly ash) composite. *J. Mech. Sci. Technol.* **2016**, *30*, 89–95. [CrossRef]

6. Hanif, M.W.; Wasim, A.; Sajid, M.; Hussain, S.; Jawad, M.; Jahanzaib, M. Evaluation of microstructure and mechanical properties of squeeze overcast Al7075−Cu composite joints. *China Foundry* **2023**, *20*, 29–39. [CrossRef]
7. Hanif, M.W.; Wasim, A.; Sajid, M. Evaluating the Effect of Process Parameters on the Mechanical Properties of an AA7075-Cu Overcast Joint Using the Taguchi Method. *Eng. Proc.* **2022**, *23*, 3.
8. Haider, F.; Jahanzaib, M.; Hanif, M.W. Optimizing the process parameters of Fiction Stir Welded dissimilar 2024Al-5754Al Joint using the Taguchi Method. In Proceedings of the 1st International Conference on Modern Technologies in Mechanical & Materials Engineering (MTME-2023), Topi, Pakistan, 6 May 2023; p. 02006.

engineering proceedings

MDPI

Proceeding Paper

Improving the Efficiency of Carbon-Based Perovskite Solar Cells with Passivation of Electron Transport Layer †

Sania Khan *, Adnan Daud Khan and Muhammad Noman

U.S.-Pakistan Center for Advanced Studies in Energy, University of Engineering and Technology, Peshawar 25000, Pakistan; adnan.daud@uetpehawar.edu.pk (A.D.K.); muhammad.noman@uetpeshawar.edu.pk (M.N.)

* Correspondence: 16pwche1136@uetpeshawar.edu.pk

† Presented at the Third International Conference on Advances in Mechanical Engineering 2023 (ICAME-23), Islamabad, Pakistan, 24 August 2023.

Abstract: Among perovskite solar cells (PSCs), carbon-based perovskite solar cells (C-PSCs) are regarded as one of the best advantageous designs centered on a number of desirable characteristics, such as outstanding scalability, long-term stability, and cost-effectiveness. In these C-PSCs, titanium oxide (TiO_2) has usually been utilized as the electron transport layer (ETL) because of its simplicity in preparation and low cost. In these hole transport layer-free C-PSCs, the quality of ETLs is essential for the high performance of PSCs. In this paper, we used $TiCl_4$ post-treatment for the passivation of the titania layer (TiO_2) to improve the quality of ETL. Consequently, after passivation, the charge recombination has been reduced, the efficiency increased from 3.15% to 4.16%, and resulted in a 32.06% improvement in power conversion efficiency (PCE).

Keywords: carbon-based perovskite solar cells; electron carrier TiO_2 layer; efficiency improvement; passivation

Citation: Khan, S.; Khan, A.D.; Noman, M. Improving the Efficiency of Carbon-Based Perovskite Solar Cells with Passivation of Electron Transport Layer. *Eng. Proc.* **2023**, *45*, 36. https://doi.org/10.3390/engproc2023045036

Academic Editors: Mohammad Javed Hyder, Muhammad Mahabat Khan, Muhammad Irfan and Manzar Masud

Published: 13 September 2023

1. Introduction

The interesting properties of perovskite materials have attracted the interest of numerous scholars in recent years, leading to the development of PSCs [1]. Among them, PSCs with carbon-based back electrodes have generated considerable concentration due to their exceptional durability, low cost, and resistance to a variety of environmental and operational circumstances [2]. Ku et al. initially demonstrated printable hole-transport layer (HTL)-free carbon-based perovskite solar cells (C-PSCs) in 2013 [3]. The quality of the ETL is critical in HTL-free C-PSCs, with TiO_2 being the most commonly used material. However, due to the presence of structural defects and surface inhomogeneity of TiO_2 ETL, charge recombination can occur at the TiO_2/perovskite contact quite easily, limiting the device's photovoltaic efficiency significantly [4]. The lack of expensive noble metals and expensive hole transport materials (HTMs) greatly lowers the cost of the device. In addition, the use of hydrophobic carbon electrodes in place of precious metals improves the device's overall stability while preventing moisture from penetrating the perovskite [5]. It has been discussed in many papers that this device's functional layers, the electron transporting layer, the insulating layer, and the carbon can be optimized to increase efficiency [6]. The fabricated device in this work, as shown in Figure 1, consists of a fluorine-doped tin dioxide (FTO) conductive glass substrate, a layer of compact titania, a mesoporous titania layer, a mesoporous insulating layer of ZrO_2, and a porous conductive carbon contact. By infiltrating liquid through the stack, the perovskite precursor solution is deposited [7].

In this work, we applied $TiCl_4$ post-treatment for TiO_2 ETL to increase C-PSCs performance. We dipped the samples in an aqueous $TiCl_4$ solution for 30 min, 40 min, and 60 min at 70 °C. The optimum results were obtained at 40 min of dipping time. The $TiCl_4$ post-treatment was first used in dye-sensitized solar cells (DSCs) and then carried over to PSCs [8]. TiO_2 with a small surface area frequently shows severe J-V characteristic hysteresis [9]. It has been shown that the $TiCl_4$ treatment can increase the surface area of TiO_2 while improving connectivity and passivating surface defect levels [10]. As a result, the surface roughness decreases with $TiCl_4$ passivation, and advancements in the photovoltaic characteristics of C-PSCs were seen.

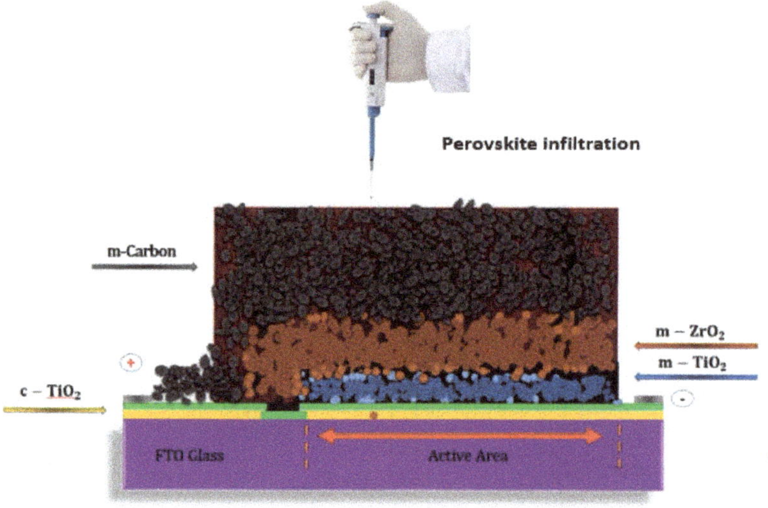

Figure 1. Schematic representation of standard carbon-based perovskite solar cells.

2. Experimental Section

Device Fabrication

First, for device fabrication, we take fluorine-doped tin oxide conductive glass substrate (FTO-coated). In an ultrasonic cleaner, distilled water, ethanol, and acetone were used to clean the FTO substrates. After cleaning, the compact titania layer was printed using the screen on the FTO glass substrates and the samples were annealed at 500 °C for 30 min. After that, we dipped some samples for passivation in $TiCl_4$ aqueous solution at 70 °C for 40 min; then, we rinsed them with distilled water followed by 30 min of annealing at 450 °C. A mesoporous titania layer was then applied to the samples, both passivated and non-passivated, and they were both annealed for 30 min at 500 °C. A mesoporous titania layer was then applied to the samples, both passivated and non-passivated, and they were both annealed for 30 min at 500 °C. Then, zirconia layer was deposited through screen printings and annealed at 500 °C for 30 min. Then, we deposited the carbon layer and hardened it at for 30 min at 400 °C using a hotplate. Next to carbon layer deposition, we infiltrated perovskite ($MAPbI_3$) into the stack through a micropipette, and waited for 40 min. After 40 min, we heated the samples at 70 °C for 20 min and waited for them to cool to room temperature [11]. Then, we checked the efficiencies of devices with and without passivation using a solar simulator as shown in Figure 2.

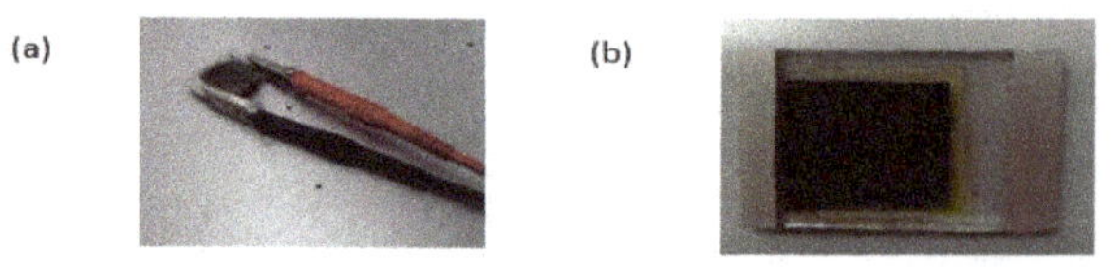

Figure 2. (**a**) The device under solar simulator. (**b**) Fabricated device.

3. Results and Discussion

The J-V curves were noted and placed into Table 1 to compare the performance of passivated and non-passivated devices in Figure 3. The non-passivated device exhibits low efficiency with an open-circuit voltage (Voc) of 756.9 mV, a short-circuit current density (Jsc) of 17.03 mA cm^{-2}, short circuit current (Isc) of 25.55 mA, a fill factor (FF) of 24.41%, and a power conversion efficiency (PCE) of 3.15%; which are improved to 802.0 mV, 19.38 mA cm^{-2}, 29.08 mA, 26.74% and a PCE of 4.16%, respectively, for the passivated device. This improvement in performance is due to the enhanced quality of TiO_2 ETL.

Table 1. Average values of photovoltaic parameters for passivated and non-passivated samples.

Device	V_{oc} (mV)	J_{sc} (mA/cm^2)	Isc (mA)	PCE (%)
Non-passivated	756.9	17.03	25.55	3.15
Passivated	802.0	19.38	29.08	4.16

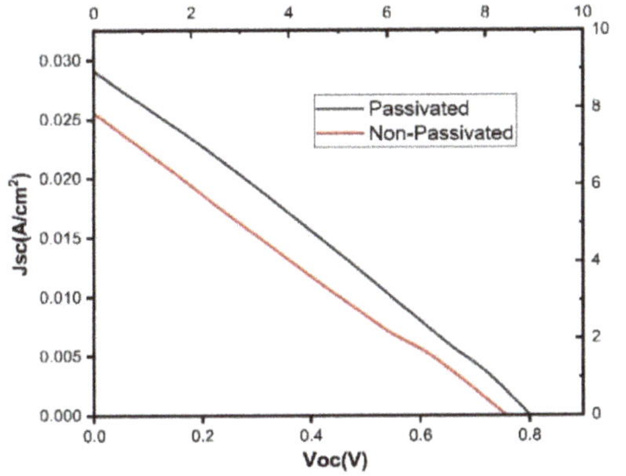

Figure 3. J-V curve of passivated and non-passivated sample.

4. Conclusions

In conclusion, it was possible to produce an efficient ETL to obtain proficient PSCs by modifying the compact TiO_2 surface with the $TiCl_4$ solution, which resulted in a 32.06% improvement in efficiency. The $TiCl_4$ treatment improves charge extraction and reduces charge recombination in the ETL layer, due to which efficiency increases. According to our findings, complex interface engineering offers a potentially effective means of obtaining high-performance PSCs.

Author Contributions: Conceptualization, M.N. and A.D.K.; methodology, M.N.; validation, M.N. and A.D.K.; formal analysis, M.N.; investigation, A.D.K.; resources, A.D.K.; data curation, M.N.; writing—original draft preparation, S.K.; writing—review and editing, S.K.; visualization, S.K.; supervision, M.N.; project administration, A.D.K.; funding acquisition, A.D.K. All authors have read and agreed to the published version of the manuscript.

Funding: This research received funding from the National Research Program for Universities (NRPU) Research Project No. 16011 "Third Generation Photovoltaics for Building Integration: A Smart and Sustainable Energy Solution", higher Education Commission (HEC), Pakistan.

Institutional Review Board Statement: Not applicable.

Data Availability Statement: Not applicable.

Acknowledgments: We really appreciate the financial assistance provided by the NRPU Research Project, Muhammad Noman, and Adnan Daud Khan for their invaluable guidance, support and mentorship throughout the course of this research.

Conflicts of Interest: The authors declare no conflict of interest.

References

1. Rong, Y.; Hu, Y.; Mei, A.; Tan, H.; Saidaminov, M.I.; Seok, S.I.; McGehee, M.D.; Sargent, E.H.; Han, H. Challenges for commercializing perovskite solar cells. *Science* **2018**, *361*, eaat8235. [CrossRef] [PubMed]
2. Bogachuk, D.; Saddedine, K.; Martineau, D.; Narbey, S.; Verma, A.; Gebhardt, P.; Herterich, J.P.; Glissmann, N.; Zouhair, S.; Markert, J. Perovskite Photovoltaic Devices with Carbon-Based Electrodes Withstanding Reverse-Bias Voltages up to −9 V and Surpassing IEC 61215: 2016 International Standard. *Sol. RRL* **2022**, *6*, 2100527. [CrossRef]
3. Ku, Z.; Rong, Y.; Xu, M.; Liu, T.; Han, H. Full Printable Processed Mesoscopic $CH_3NH_3PbI_3/TiO_2$ Heterojunction Solar Cells with Carbon Counter Electrode. *Sci. Rep.* **2013**, *3*, 3132. [CrossRef] [PubMed]
4. Bai, J.; Zhou, B. Titanium dioxide nanomaterials for sensor applications. *Chem. Rev.* **2014**, *114*, 10131–10176. [CrossRef] [PubMed]
5. Grancini, G.; Roldán-Carmona, C.; Zimmermann, I.; Mosconi, E.; Lee, X.; Martineau, D.; Narbey, S.; Oswald, F.; De Angelis, F.; Graetzel, M.; et al. One-Year stable perovskite solar cells by 2D/3D interface engineering. *Nat. Commun.* **2017**, *8*, 15684. [CrossRef] [PubMed]
6. Tian, C.; Mei, A.; Zhang, S.; Tian, H.; Liu, S.; Qin, F.; Xiong, Y.; Rong, Y.; Hu, Y.; Zhou, Y.; et al. Oxygen management in carbon electrode for high-performance printable perovskite solar cells. *Nano Energy* **2018**, *53*, 160–167. [CrossRef]
7. Mei, A.; Li, X.; Liu, L.; Ku, Z.; Liu, T.; Rong, Y.; Xu, M.; Hu, M.; Chen, J.; Yang, Y. A hole-conductor–free, fully printable mesoscopic perovskite solar cell with high stability. *Science* **2014**, *345*, 295–298. [CrossRef] [PubMed]
8. Sommeling, P.; O'Regan, B.C.; Haswell, R.; Smit, H.; Bakker, N.; Smits, J.; Kroon, J.M.; Van Roosmalen, J. Influence of a TiCl4 post-treatment on nanocrystalline TiO2 films in dye-sensitized solar cells. *J. Phys. Chem. B* **2006**, *110*, 19191–19197. [CrossRef] [PubMed]
9. Wang, Q.; Phung, N.; Di Girolamo, D.; Vivo, P.; Abate, A. Enhancement in lifespan of halide perovskite solar cells. *Energy Environ. Sci.* **2019**, *12*, 865–886. [CrossRef]
10. Roose, B.; Pathak, S.; Steiner, U. Doping of TiO_2 for sensitized solar cells. *Chem. Soc. Rev.* **2015**, *44*, 8326–8349. [CrossRef] [PubMed]
11. Adli, H.K.; Harada, T.; Nakanishi, S.; Ikeda, S. Effects of $TiCl_4$ treatment on the structural and electrochemical properties of a porous TiO_2 layer in $CH_3NH_3PbI_3$ perovskite solar cells. *Phys. Chem. Chem. Phys.* **2017**, *19*, 26898–26905. [CrossRef] [PubMed]

engineering proceedings

MDPI

Proceeding Paper

Fabrication of Carbon-Based Perovskite Solar Cell under Ambient Condition †

Abdul Basit *, Aimal Daud Khan and Adnan Daud Khan

US-Paksitan Center for Advanced Studies in Energy, University of Engineering and Technology, Peshawar 25000, Pakistan; aimaldaud.uspcase@uetpeshawar.edu.pk (A.D.K.); adnan.daud@uetpeshawar.edu.pk (A.D.K.)

* Correspondence: abdulbasit.ree@uetpeshawar.edu.pk; Tel.: +92-3151995853

† Presented at the Third International Conference on Advances in Mechanical Engineering 2023 (ICAME-23), Islamabad, Pakistan, 24 August 2023.

Abstract: In this study, the viability of the commercial production of ambient air perovskite solar cells is investigated. In order to address issues with overlapping and non-uniform layers and minimize recombination losses, this study suggests a novel device architecture. A 2 × 2cm^2 perovskite solar cell with a 4.2% power conversion efficiency (PCE) was successfully created utilizing a straightforward screen-printing technique. After humidity treatment, the short-circuit current (jsc) and open-circuit voltage (Voc) were stabilized at 5.36% due to layer optimization. These results show the potential for the widespread industrial production of more durable and efficient perovskite solar cells.

Keywords: perovskite solar cell; screen printing; ambient condition

Citation: Basit, A.; Khan, A.D.; Khan, A.D. Fabrication of Carbon-Based Perovskite Solar Cell under Ambient Condition. *Eng. Proc.* **2023**, *45*, 32. https://doi.org/10.3390/engproc2023045032

Academic Editors: Mohammad Javed Hyder, Muhammad Mahabat Khan, Muhammad Irfan and Manzar Masud

Published: 12 September 2023

1. Introduction

The perovskite solar cells (PSCs) have received a lot of interest in recent years from academics due to their unique characteristics [1–3]. Although the reported power conversion efficiencies (PCEs) of PSCs have improved significantly, the perovskite community is currently placing more attention on improving device stability [4]. The high durability of carbon-based back electrodes (CPSCs) across a variety of environmental and operating situations has been noted [5,6]. It is important to note that their PCEs still fall short of those of cutting-edge PSCs that use metal-based electrodes [7].

A hybrid organic–inorganic perovskite substance called $MAPbI_3$ is regarded as one of the most alluring options for inexpensive yet powerful absorber materials for solar energy applications [8,9]. The power conversion efficiency (PCE) of perovskite solar cells based on methylammonium lead iodide ($MAPbI_3$) has significantly improved recently, with reported PCEs rising from 3.8% to 20.01% [10]. In a recent study, $MAPbI_3$ was mechanically deposited onto the TiO_2 layer in order to create a completely printable mesoscopic solar cell utilizing TiO_2/ZrO_2/Carbon (TZC) sheets [11]. Because an organic hole-transporting material (HTM) is not present, TZC-based materials offer lower material costs and better stability.

During this research, we successfully fabricated a perovskite solar cell in the laboratory under full ambient conditions using a FTO/C-TiO_2/M-TiO_2/ZrO_2/carbon/$MAPbI_3$ architecture. By carefully optimizing the pattern of each layer to prevent the material from overlapping, we achieved a promising power conversion efficiency (PCE) of 4.2%. Moreover, the device's performance was significantly improved to a PCE of 5.36% through humidity treatment. These findings underscore the potential of ambient air perovskite solar cells as a cost-effective and efficient approach for commercial production. The demonstrated stability and performance enhancements lay the foundation for upscaling perovskite solar cell manufacturing on an industrial level, offering a hopeful path towards sustainable and renewable energy solutions.

2. Cell Fabrication by Using Screen Printing Methodology

The normal cleaning procedure is washing the samples with hand soap detergent, followed by 15 min of sonication in deionized (DI) water. The samples are then subjected to a 15 min sonication process with acetone, isopropyl alcohol (IPA), and acetone at a temperature of 40 °C. The samples are then dried with a flow of nitrogen gas after being sonicated with the abovementioned solvents, and they are then kept at room temperature for quick usage in the next phases. In the second phase, we created and deposited the perovskite solar architecture "FTO/C-TiO_2/M-TiO_2/ZrO_2/Carbon paste/$MAPbI_3$" using the screen printing technique, as shown in the Figure 1 [12]. To create a solar cell, several meshes are employed for various materials. The open mesh pores hold the viscous ink in place when it is applied to the screen by a squeegee. Compact TiO_2 is thinly placed over a glass surface that has been coated in fluorine-doped tin oxide (FTO) to prepare the substrate. Utilising ethanol and acetone, the compact TiO_2 layer's mesh structure is cleaned. The film is then given a 30 min annealing treatment at 500 °C to encourage crystallization. Mesoporous TiO_2 (m-TiO_2) is the fourth layer in the manufacturing process. This layer is essential to perovskite solar cells (PSCs) because it serves as a scaffold for perovskite crystal growth as well as being a selective contact for electron extraction. On top of the compact TiO_2 (C-TiO_2) layer that was previously deposited, the M-TiO_2 layer is applied. Following deposition, the substrate is subjected to a 30 min annealing process at 500 °C to encourage the crystallization and structural stability of the M-TiO_2 layer. Applying a ZrO_2 paste using a screen-printing process over the TiO_2 layer is the following step. After that, control cells are put through a 30 min sintering process at 400 °C. The carbon back electrode is placed onto the ZrO_2 layer after the prior procedures. The sample is subsequently heated to 40 °C for 30 min to guarantee adequate bonding and structural stability. The perovskite solar cell's performance and general efficiency are improved by this technique. Using the drop casting method, the last layer of perovskite is put onto the substrate. The perovskite film is then improved by annealing it for an hour at 70 °C to promote further crystallization [13].

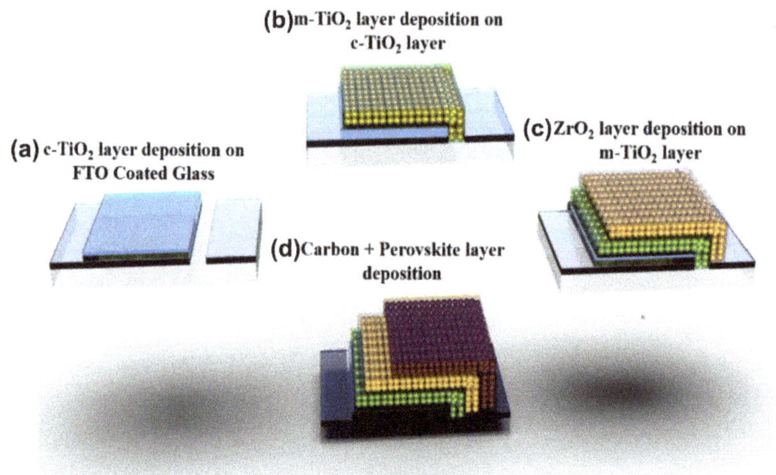

Figure 1. Fabrication steps of the perovskite solar cell. (**a**) c-TiO_2 layer deposition on FTO Coated Glass, (**b**) m-TiO_2 layer deposition on c-TiO_2 layer, (**c**) ZrO_2 layer deposition on m-TiO_2, (**d**) carbon + Perovskite layer deposition.

3. Result and Discussion

The fabricated perovskite solar cell comprised of the FTO/C-TIO_2/M-TIO_2/ZrO_2/carbon/$MAPbI_3$ architecture exhibited an initial efficiency of 4.2%. This efficient structure offered several advantages; notably, a well-defined interface which potentially improved the charge collection and light absorption capabilities, thus leading to increased efficiency.

Post fabrication, the solar cell underwent a humidity treatment, resulting in a noticeable enhancement in the overall efficiency to 5.36%. This could be attributed to the fact that a controlled humidity environment can facilitate the crystallization of perovskite, enhancing the perovskite layer's quality and, hence, the device performance. Moreover, the induced humidity could potentially have optimized the morphology of the $MAPbI_3$ layer, which can further contribute to efficiency enhancement. Figure 2a shows the J-V curve before and after humidity treatment of the perovskite solar cell.

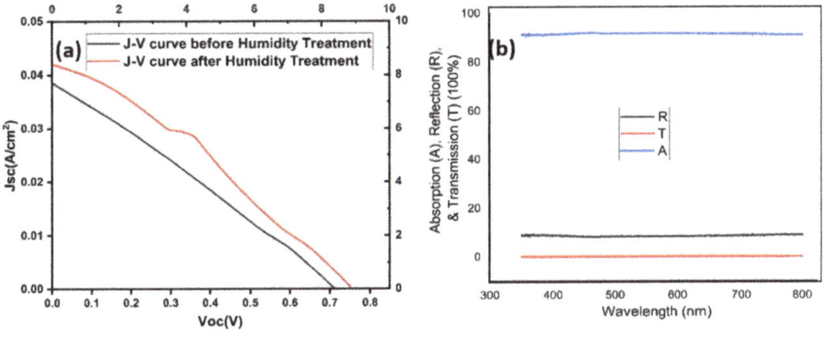

Figure 2. (**a**) J-V curve before and after humidity treatment of perovskite solar cell, and (**b**) absorption, reflection and transmission spectra of the perovskite solar cell.

Figure 2b shows the absorption, reflection, and transmission spectra of the perovskite solar cell, in which almost 90% of the absorption was achieved in our device and only a low amount of reflection has been observed of approximately 10%. Moreover, it can be seen that nearly 0% transmission has been observed.

4. Conclusions

A highly stable 5.36% efficient solar cell is reported which can be scaled up to fabricate large-area perovskite-based solar modules. This structure utilizes the mesoporous zirconia layer, which is used as an insulation layer, as an alternative for the hole transport layer, which is highly stable. Though this layer reduces the efficiency, it is cost effective and increases the stability of the structure, thus increasing its prospects for commercialization. A dual titania layer, i.e., thin compact titania along with a mesoporous titania, is used as an electron transport layer. The structure utilizes a carbon black blend along with graphite particles as the hole selector layer, as well as cathode. Due to its hydrophobic nature, this layer provides a protective cover to perovskite crystals, thus increasing the lifetime of this solar cell. The perovskite precursor solution 5-AVA-$MAPbI_3$ is infiltrated in the mesoporous layers. Due to the presence of aminovaleric acid along with organic cation, this perovskite is more stable compared to conventional methylammonium lead iodide crystals. Silver is used at both electrodes for charge collection due to its cost-effectiveness and efficient charge transportation. Thus, these solar cells are an excellent alternative for conventional silicon-based solar cells since these cells are cost effective and have simple fabrication processes. The screen printing technique is used for the fabrication of these solar cells, which can easily be used for layer thickness optimization and scaling up the module size. Moreover, these solar cells can be used as a basic model for the commercialization and initiation of perovskite solar manufacturing in Pakistan.

Author Contributions: Conceptualization, A.B.; methodology, A.B., A.D.K. (Aimal Daud Khan); fabrication, A.B., A.D.K. (Aimal Daud Khan); characterization, A.B., A.D.K. (Aimal Daud Khan); resources, A.D.K. (Adnan Daud Khan); writing—original draft preparation, A.B.; review and editing, A.D.K. (Aimal Daud Khan); supervision, A.D.K. (Adnan Daud Khan) All authors have read and agreed to the published version of the manuscript.

Funding: This research received no external funding.

Institutional Review Board Statement: Not applicable.

Informed Consent Statement: Not applicable.

Data Availability Statement: Not applicable.

Acknowledgments: All authors are grateful to US-Pakistan Center for Advanced Studies in Energy, University of Engineering & Technology, Peshawar for providing the facilitation to fabricate the perovskite solar cell.

Conflicts of Interest: The authors declare no conflict of interest.

References

1. Kim, J.Y.; Lee, J.-W.; Jung, H.S.; Shin, H.; Park, N.-G. High-efficiency perovskite solar cells. *Chem. Rev.* **2020**, *120*, 7867–7918. [CrossRef] [PubMed]
2. Correa-Baena, J.-P.; Saliba, M.; Buonassisi, T.; Grätzel, M.; Abate, A.; Tress, W.; Hagfeldt, A. Promises and challenges of perovskite solar cells. *Science* **2017**, *358*, 739–744. [CrossRef] [PubMed]
3. Rong, Y.; Hu, Y.; Mei, A.; Tan, H.; Saidaminov, M.I.; Seok, S.I.; McGehee, M.D.; Sargent, E.H.; Han, H. Challenges for commercializing perovskite solar cells. *Science* **2018**, *361*, eaat8235. [CrossRef] [PubMed]
4. Wang, Q.; Phung, N.; Di Girolamo, D.; Vivo, P.; Abate, A. Enhancement in lifespan of halide perovskite solar cells. *Energy Environ. Sci.* **2019**, *12*, 865–886. [CrossRef]
5. Mei, A.; Sheng, Y.; Ming, Y.; Hu, Y.; Rong, Y.; Zhang, W.; Luo, S.; Na, G.; Tian, C.; Hou, X. Stabilizing perovskite solar cells to IEC61215: 2016 standards with over 9,000-h operational tracking. *Joule* **2020**, *4*, 2646–2660. [CrossRef]
6. Bogachuk, D.; Saddedine, K.; Martineau, D.; Narbey, S.; Verma, A.; Gebhardt, P.; Herterich, J.P.; Glissmann, N.; Zouhair, S.; Markert, J. Perovskite Photovoltaic Devices with Carbon-Based Electrodes Withstanding Reverse-Bias Voltages up to–9 V and Surpassing IEC 61215: 2016 International Standard. *Solar Rrl* **2022**, *6*, 2100527. [CrossRef]
7. Bogachuk, D.; Zouhair, S.; Wojciechowski, K.; Yang, B.; Babu, V.; Wagner, L.; Xu, B.; Lim, J.; Mastroianni, S.; Pettersson, H. Low-temperature carbon-based electrodes in perovskite solar cells. *Energy Environ. Sci.* **2020**, *13*, 3880–3916. [CrossRef]
8. Idrissi, S.; Ziti, S.; Labrim, H.; Bahmad, L. Band gaps of the solar perovskites photovoltaic $CsXCl_3$ (X = Sn, Pb or Ge). *Mater. Sci. Semicond. Process.* **2021**, *122*, 105484. [CrossRef]
9. Lee, M.M.; Teuscher, J.; Miyasaka, T.; Murakami, T.N.; Snaith, H.J. Efficient hybrid solar cells based on meso-superstructured organometal halide perovskites. *Science* **2012**, *338*, 643–647. [CrossRef] [PubMed]
10. Yang, X.; Yu, H.; Guo, X.; Ding, Q.; Pullerits, T.; Wang, R.; Zhang, G.; Liang, W.; Sun, M. Materials Today Energy. *Mater. Today* **2017**, *5*, 72–78.
11. Ku, Z.; Rong, Y.; Xu, M.; Liu, T.; Han, H. Full printable processed mesoscopic $CH_3NH_3PbI_3/TiO_2$ heterojunction solar cells with carbon counter electrode. *Sci. Rep.* **2013**, *3*, 3132. [CrossRef] [PubMed]
12. Baker, J.; Hooper, K.; Meroni, S.; Pockett, A.; McGettrick, J.; Wei, Z.; Escalante, R.; Oskam, G.; Carnie, M.; Watson, T. High throughput fabrication of mesoporous carbon perovskite solar cells. *J. Mater. Chem. A* **2017**, *5*, 18643–18650. [CrossRef]
13. Duan, M.; Hu, Y.; Mei, A.; Rong, Y.; Han, H. Printable carbon-based hole-conductor-free mesoscopic perovskite solar cells: From lab to market. *Mater. Today Energy* **2018**, *7*, 221–231. [CrossRef]

engineering proceedings

MDPI

Proceeding Paper

Cost Estimation and Parametric Optimization for TIG Welding Joints in Dissimilar Metals Using Linear Regression Algorithm †

Ghulam Ameer Mukhtar [1,*], Sana Shehzadi [1,*], Abdullah Sajid [2], Jam Muhammad Talha Laar [1], Syed Ali Taqi [2], Rana Muhammad Usman [1] and Fakhar ul Hasnain [3]

1 Department of Industrial and Manufacturing Engineering, University of Engineering and Technology, Taxila 47050, Pakistan; 20-ie-7@students.uettaxila.edu.pk (J.M.T.L.); 20-ie-29@students.uettaxila.edu.pk (R.M.U.)
2 Department of Software Engineering, University of Engineering and Technology, Taxila 47050, Pakistan; 21-se-01@students.uettaxila.edu.pk (A.S.); 21-se59@students.uettaxila.edu.pk (S.A.T.)
3 Department of Mechanical Engineering, Capital University of Science and Technology, Islamabad 44000, Pakistan; fakhar.hasnain@cust.edu.pk
* Correspondence: 20-ie-43@students.uettaxila.edu.pk (G.A.M.); 20-ie-8@students.uettaxila.edu.pk (S.S.)
† Presented at the Third International Conference on Advances in Mechanical Engineering 2023 (ICAME-23), Islamabad, Pakistan, 24 August 2023.

Abstract: This study investigated the use of the linear regression algorithm (LRA) for estimating the cost of tungsten inert gas (TIG) welding of dissimilar metals, specifically, stainless steel 304 and aluminum 2024. Various cost analysis parameters, including weld time, power cost rate, labor cost, filler rod cost, and shielding gas cost, were considered. LRA was employed to develop a predictive model for welding costs based on these parameters. The model was then used to optimize welding parameters by identifying the combination that minimized overall welding costs. The results demonstrate that LRA effectively identified a more cost-effective parameter set compared to traditional methods. This highlights the potential of LRA in enhancing the cost-effectiveness of TIG welding processes. The findings have practical implications for field engineers and researchers, enabling the identification of optimal parameters for cost estimation and significant cost savings.

Keywords: linear regression algorithm (LRA); tungsten inert gas (TIG) welding; dissimilar metals; cost estimation of GTAW; linear programming technique (LPT); parametric optimization

Citation: Mukhtar, G.A.; Shehzadi, S.; Sajid, A.; Laar, J.M.T.; Taqi, S.A.; Usman, R.M.; Hasnain, F.u. Cost Estimation and Parametric Optimization for TIG Welding Joints in Dissimilar Metals Using Linear Regression Algorithm. *Eng. Proc.* **2023**, *45*, 50. https://doi.org/10.3390/engproc2023045050

Academic Editors: Mohammad Javed Hyder, Muhammad Mahabat Khan, Muhammad Irfan and Manzar Masud

Published: 19 September 2023

1. Introduction

This paper expounds the indispensable use of cost analysis for tungsten inert gas (TIG) welding for dissimilar and erratic metals. Incorporated factors are welding speed, cost of filler rod, labor charges, power cost [1]. Substantially, the usage of tungsten inert gas welding in the fabrication of structural frameworks as a perpetual fastening to generate single-pass, full-penetration welds and root passes of multi-pass welds commenced in the 1940s [2]. In fact, the effectiveness of cost estimation governs the transition of customer to a productive purchase, aids in setting the project budget, planning the required work, evaluating the feasibility of a project, managing new resources and one of the rudimentary imperative of any decision-making for enterprises like choosing material, manufacturing processes and morphological features of production and products [3]. Except for shredding or recycling the integral components of vehicles, this study addresses and guides cost estimation with a pertinent new way to reproduce vehicles assemblages that are comprised of stainless steel and aluminum. By means of TIG welding, using a unique filler metal can overcome this, while the cost estimation of this process is a practical means to determine and evaluate the overall prime expenditures of this course of action to optimize functions [3]. The parametric optimization of TIG welding using linear regression in Python is a significant aspect of this research. Case studies have addressed TIG welding on different materials, but this study specifically addresses the cost estimation of dissimilar metal welding [4].

2. Methodology

Nomenclature

t = time in hours, V = voltage, I = current, η = efficiency of machine, WT = weld time, PC = Power cost, P = power, W = filler metal weight (g), ρ = density (g/cm3), L = length (cm), E = deposition efficiency, R = root gap, T = thickness of base metal, F = root face, C.G = cost of shielding gas (S.G) cylinder, S = size of cylinder, F.G = flow rate of gas, V = volume of S.G, DLC= direct labor cost, AT = average time, LCOS = labor cost (1 Sec), NOS = number of seconds, LC = labor cost, S = speed, FRC = filler rod cost, SHC = shielded gas cost, U = unit cost.

2.1. Physical Experimentation

This experimental work design explains the quantitative analysis by performing experiments to join 100 × 100 mm^2 square plates of 3 mm thickness through butt joints of dissimilar metals using 17 samples and employing TIG welding in real time to facilitate the mathematical model and optimal parameters to consider multiple dependent and independent variables. As shown in Figure 1a, 17 experiments were performed in the lab while collecting data manually for cost estimation and parametric optimization. The results are as given in Table 1.

Table 1. Experimental readings against parameters.

Sr no.	I (Amp)	S mm/min	F.G L/min	V Volts	WT min	PC (PKR)	LC (PKR)	FRC (PKR)	SHC (PKR)
1	80	110	9	12.7	90	1.260587	540.90	96.96	259.2
2	70	110	8	12.2	135	1.601271	811.35	100.62	194.4
3	80	110	9	12.6	136	1.890137	817.36	94.57	172
4	80	100	10	13.1	90	1.300493	540.90	93.51	233
5	70	120	9	12	140	1.621819	841.40	95.64	167
6	80	110	9	12.1	76	1.014902	456.76	91.87	308
7	80	120	10	13.2	105	1.528905	631.05	93.05	200
8	70	110	10	13.3	135	1.732898	811.35	100.40	155
9	80	120	8	12	140	1.853209	841.40	92.19	187
10	90	100	9	11.9	90	1.328933	540.90	90.31	260
11	80	110	9	14.7	98	1.589388	588.98	91.37	238
12	70	100	9	10.7	180	1.858868	1081.8	93.55	129
13	90	110	8	16	140	2.781452	841.40	89.74	187
14	90	110	10	12	84	1.251117	504.84	89.47	250
15	80	110	9	15.8	130	2.26605	781.30	95.50	180
16	80	100	8	13.3	100	1.467082	601.00	95.85	263
17	90	120	9	16.3	110	2.224834	661.10	95.55	212
		Total				28.57194	11,893.79	1600.15	3594.6

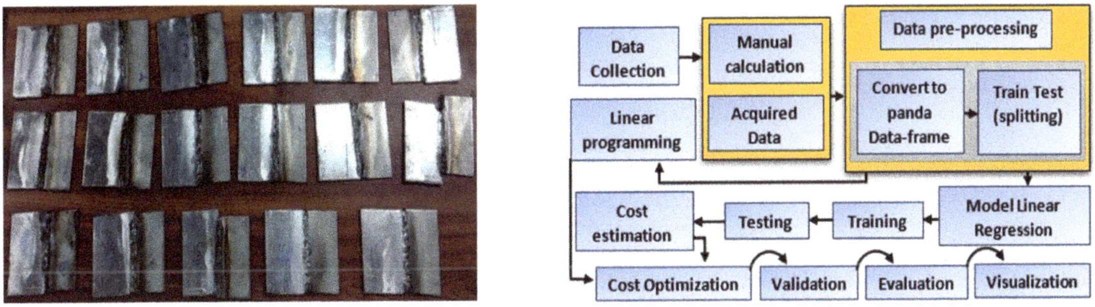

(**a**) (**b**)

Figure 1. (**a**) Physical experimentation; (**b**) schematic diagram.

The flow chart in Figure 1b shows the overall framework for performing experiments with collection of raw data and applying mathematical models manually to acquire lucrative data for further processing through a machine learning tool. Data pre-processing and modeling involve using linear regression and linear programming techniques. Acquired datasets are utilized for training with manual calculations. Cost estimation and optimization optimize resource allocation. The following equations show manual calculations for the following parameters:

$$\mathrm{PC} = \left\{ \mathrm{V} * \frac{\mathrm{I}}{\eta} \right\} * \mathrm{t} * \mathrm{U},\ \ \mathrm{W} = \mathrm{A} * \rho * \mathrm{L} * \frac{1}{\mathrm{E}},\ \mathrm{A} = (\mathrm{R} * \mathrm{T}) + (\mathrm{T} - \mathrm{F})2 + \tan\left(\frac{\theta}{2}\right)$$

$$SGC = VSG * CGUV,\ VSG = CSA\left(mm^2\right) * LJ * FRG,\ \ \mathrm{LCOS} = \frac{\mathrm{DLC}}{\mathrm{AT}},\ LC = LCOS * NOS$$

2.2. Experimentation Using Machine Learning

The TIG welding dataset, comprising 17 records with welding parameters and associated costs, underwent preprocessing to prepare it for analysis. To evaluate model performance, the dataset was split into training and testing subsets using the train_test_split.

The training subset contained 70% (11 records) of the data, while the testing subset comprised 30% (6 records), ensuring a representative distribution of data in both subsets. The actual setup and the schematic diagram are shown in Figure 2b.

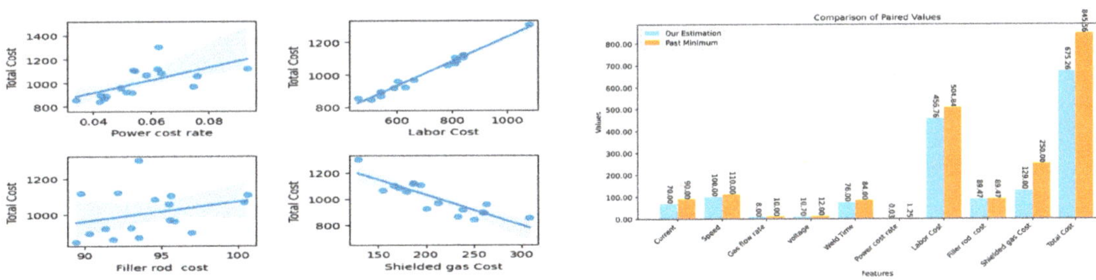

(**a**) (**b**)

Figure 2. (**a**) Total and parametric costs regression lines; (**b**) paired values comparison.

3. Discussion

To optimize the total cost of TIG welding, we utilized LP. Linear programming (LP) is a mathematical optimization method that helps identify the best values for decision variables while considering given constraints. The objective function of the LP problem consisted of the coefficients obtained from the linear regression model and the intercept. By minimizing this objective function (Equation (1) below), we aimed to find the optimal values for the welding parameters mentioned in the hope that it would result in the lowest total cost.

The results are shown the scatter plots of data generated from the model, as shown in Figure 2a. This analysis examines the correlations between cost factors and total cost using scatter plots with regression lines. A positive slope shows positive correlation and a negative slope shows a negative correlation.

Figure 2b compares the estimated optimized cost with the dataset's cost values, highlighting disparities in feature values using paired bars. The vertical height of each bar indicates the magnitude.

Figure 3 illustrates line plots showing the temporal evolution of cost factors. Each plot represents a specific cost factor, with the y-axis representing corresponding cost values and the x-axis representing chronological order.

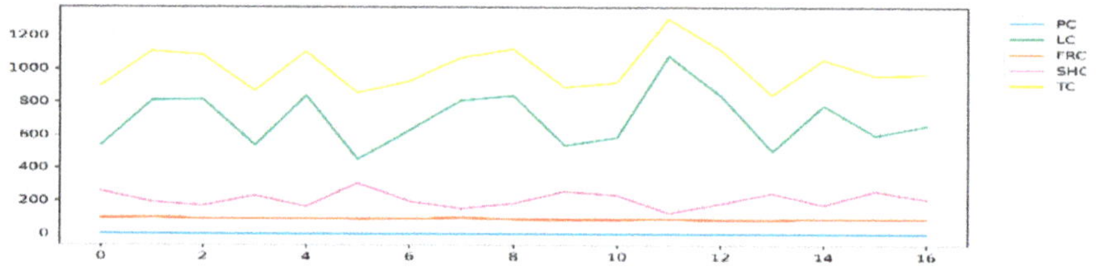

Figure 3. Parametric magnitudes vs. experiments.

The mentioned formula calculates cost as Y. The constants were calculated using Equation (1):

$$\begin{aligned} Y = \ & (-2.4395916331862\,84e-14).(I) + (-1.1102230246251565e-16) \\ & .(S) + (9.676959754345571e-14).(FG\_ + (-1.255003045930180 \\ & -13).(v) + (0.16190689141474435).(t) + (1.0000000000009737).(PC) \\ & +(0.9730604174018711).(DLC) + (1.0000000000000047).(FRC) \\ & +(1.0000000000000022).(SGC) + 1.2505552149377763e-12 \end{aligned} \quad (1)$$

4. Conclusions

This research utilized manual calculations and the Linear Regression Algorithm to estimate and optimize TIG welding costs for dissimilar metals. After analysis, the optimal values for welding parameters that minimize the total cost are current = 70 A, speed = 100.0 mm/min, gas flow rate = 8.0 L/min, voltage = 10.7 V, weld time = 76.0 min, power cost = 1.0149024 KPR, labor cost = 456.76 KPR, filler rod cost = 89.47 KPR, shielded gas cost = 129.0 KPR. These optimized values resulted in a total cost of 676.2449 KPR, reducing it by 169.31 KPR compared to the minimum cost in the original dataset (845.56 KPR).

Author Contributions: G.A.M.—conceptualization, technical writing, literature, data collection, manual calculations, draft preparation, reviewing, and editing; S.S.—formatting, vector designing, content writing, literature, reviewing, and editing; A.S.—applied ML tool; S.A.T.—applied ML tool; J.M.T.L.—manual calculations; R.M.U.—manual calculations; F.u.H.—final review and supervision. All authors have read and agreed to the published version of the manuscript.

Funding: This research received no external funding.

Institutional Review Board Statement: Not applicable.

Informed Consent Statement: Not applicable.

Data Availability Statement: Not applicable.

Conflicts of Interest: The authors declare no conflict of interest.

References

1. El-Kassas, A.M.; Sabry, I. A Comparison between FSW, MIG and TIG Based on Total Cost Estimation for Aluminum Pipes. *Eur. J. Adv. Eng. Technol.* **2017**, *4*, 158–163.
2. Vasudevan, M.; Bhaduri, A.K.; Raj, B.; Kalvala, P. Genetic-algorithm-based computational models for optimizing the process parameters of A-TIG welding to achieve target bead geometry in type 304 L(N) and 316 L(N) stainless steels. *Mater. Manuf. Process.* **2007**, *22*, 641–649. [CrossRef]
3. Patil, P.C.; Shelke, R.D. Review on Welding Parameter Effects on Tig Welding of Aluminium Alloy. *Int. J. Eng. Res. Gen. Sci.* **2015**, *3*, 1479–1486.
4. Dutta, P.; Pratihar, D.K. Modeling of TIG welding process using conventional regression analysis and neural network-based approaches. *J. Mater. Process. Technol.* **2007**, *186*, 56–68. [CrossRef]

engineering proceedings

Proceeding Paper

Investigating the Ability of Process Parameters to Reduce Defects during the Drilling of Carbon-Fiber-Reinforced Plastic (CFRP) and Aluminum Stack †

Muhammad Waseem *, Salman Hussain, Muhammad Waqas Hanif and Muhammad Jawad

Department of Industrial Engineering, University of Engineering and Technology Taxila, Punjab 47050, Pakistan; salman.hussain@uettaxila.edu.pk (S.H.); waqas.hanif@students.uettaxila.edu.pk (M.W.H.); engr.jawad@uettaxila.edu.pk (M.J.)

* Correspondence: waseemime@gmail.com

† Presented at the Third International Conference on Advances in Mechanical Engineering 2023 (ICAME-23), Islamabad, Pakistan, 24 August 2023.

Abstract: The effects of drilling process parameters, such as drill diameter and feed rate, on the delamination of holes has been examined while drilling Carbon-Fiber-Reinforced Plastics and aluminum (CFRP/Al2219-T6). The stack method was employed while drilling in order to reduce the defects in the drilling process and enhance the quality of the holes produced. A total of nine experiments were performed using the central composite design. An analysis of variance (ANOVA) revealed that feed rate and drill size have significant effects on the delamination of holes. The experimental results show that the minimum delamination value (1.018) was obtained at a feed rate of 26.5 mm/min and with a drill diameter of 6.5 mm.

Keywords: CFRP/Al2219-T6 stack; delamination; drill diameter; feed rate

Citation: Waseem, M.; Hussain, S.; Hanif, M.W.; Jawad, M. Investigating the Ability of Process Parameters to Reduce Defects during the Drilling of Carbon-Fiber-Reinforced Plastic (CFRP) and Aluminum Stack. *Eng. Proc.* **2023**, *45*, 53. https://doi.org/10.3390/engproc2023045053

Academic Editors: Mohammad Javed Hyder, Muhammad Mahabat Khan, Muhammad Irfan and Manzar Masud

Published: 19 September 2023

1. Introduction

Carbon-Fiber-Reinforced-Plastic (CFRP) is typically manufactured as "make-to-shape" parts through various molding techniques and utilized in the grouping of lightweight metal stacks such as titanium and aluminum alloys [1]. Once composite parts incorporating a combination of lightweight metal stacks, for example, aluminum alloys, have been manufactured, they need to be integrated into the main assembly or system using mechanical fasteners, such as rivets and screws, and this requires the drilling of holes. However, drilling these materials can be challenging, and defects such as delamination can occur due to differences in the properties of CFRP and stacked metals. Delamination is a type of failure. When the amount of thrust force exceeds a certain threshold, this can lead to the delamination of the layers in a multilayer material. This can cause a considerable reduction in mechanical durability and a decrease in interlaminar strength. Preventing delamination is of utmost importance when drilling CFRP and aluminum stack, particularly as they are increasingly utilized in the aerospace and automobile industries, where safety is a paramount concern.

Researchers have optimized different process parameters, such as feed rate, drill diameter, and spindle speed, to reduce the delamination of drilled holes. For instance, D'Orazio et al. [2] reported the effects of drill diameter and feed rate on delamination and thrust while drilling CFRP/AA 7075 stack using a nanocoated TiAlN drill bit. In this study, a total of 170 holes were drilled in the CFRP/Al stack, and the results show that the observed increment in delamination was due to increasing the number of holes. The delamination associated with nanocoated TiAlN was greater compared to that of a DLC-coated drill bit. The impact of spindle speed, feed rate, and drill diameter on the circularity, surface roughness, and torque has been studied by Redouane Zitoune et al. [3].

The authors concluded that feed rate has a greater impact on hole quality than spindle speed. In another research article, Alessio et al. [4] evaluated delamination and thrust during the taping of CFRP/Al7075/CFRP stack. With an increasing number of tapped holes, delamination and thrust forces increased due to the wear of the tool.

A literature review of [1–4] revealed that researchers have focused on optimizing cutting parameters to enhance the hole quality and trust force in the drilling process using high-speed steel tools. However, the effects of drilling process parameters on the stack of CFRP with Al2219-T6 must still be explored due to its application in the aerospace industry. Hence, the objective of this article is to evaluate the influence of drill diameter and feed rate on delamination while drilling CFRP/Al2219-T6 stack. Furthermore, a central composite design has been used in the experiment, and an ANOVA table has been used to analyze the significance of the process parameters.

2. Materials and Method

XRF spectrometer was used to perform a spectrometry test, and the result regarding the chemical composition of Al2219-T6 obtained from the spectrometry test is shown in Table 1. The stack method was used in the drilling process because Alessio et al. revealed that the stack method helped to reduce the defects in the drilling process [4]. In this method, Al2219-T6 plate was machined to achieve the required dimensions (425 × 225 × 3) in mm. Afterward, shot peening was applied to the surface of Al2219-T6, which would later be bonded with CFRP. A chemical agent (Trichloroethylene) and primer (EW-5000) were applied to polish the surface of the Al2219-T6. Finally, prepreg carbon fabric was placed in an autoclave at a pressure of 3 bar to increase the resulting material's bonding capability.

Table 1. Spectrometry results of Al2219-T6.

Composition (wt. %)	Cu	Fe	Al	V	Si	Mn	Ti	Mg
Al2219-T6	6.3	0.23	Bal.	0.10	0.18	0.35	0.06	0.02

After the preparation of the stack plate, drilling was performed using the HSS TiN Coated Twist Drill tool on the Taiwan-made universal milling machine center, as shown in Figure 1c. Delamination describes the damage inflicted on the surface of a composite workpiece either at the entry or exit point during the drilling process [2]. Therefore, the output response delamination was measured after performing experiments at each experimental setting. By dividing the damaged zone diameter by the drill bit diameter, the degree of delamination was calculated. The diameter of the damaged zone was measured using an optical microscope (VZM-200), and diameter of the drill bit was measured using a digital caliper.

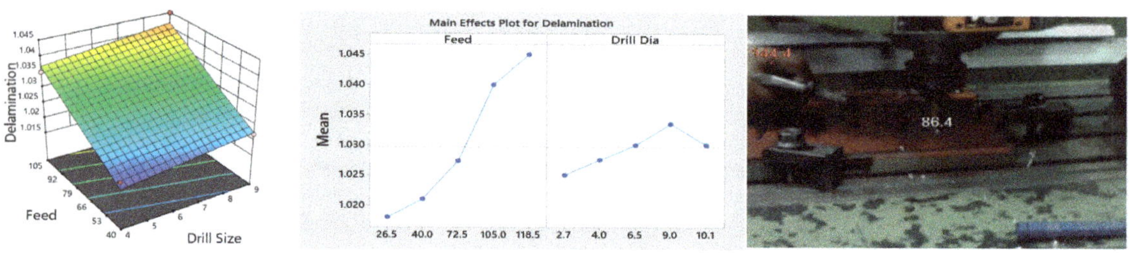

(a) (b) (c)

Figure 1. (**a**) 3D Mesh plot of feed and drill size; (**b**) single factor plots for delamination; (**c**) experimental setup.

3. Experiment Design

The lower and upper ranges of two dominant factors, namely, drill diameter and feed, in the drilling of CFRP and Al2219-T6 were selected based on trial runs and a literature review [2–4], as shown in Table 2. According to the central composite design, nine experiments were performed, and two process parameters and one center point were considered, as depicted in Table 3.

Table 2. Ranges of drilling process parameters.

Machining Parameters	Unit	Lower Level	Upper Level
Feed rate	mm/min	40	80
Drill diameter	mm	4	9

Table 3. Experiment design.

Experiment No.	Input Parameters		Output
	Feed Rate (mm/min)	Drill Diameter (mm)	Delamination
1	40	4	1.020
2	**26.5**	**6.5**	**1.018**
3	72.5	2.7	1.025
4	105	9	1.045
5	72.5	6.5	1.027
6	118.5	6.5	1.045
7	72.5	10.1	1.030
8	40	9	1.022
9	105	4	1.035

4. Results and Discussion

The experimental results show that the minimum value of delamination (1.018) was achieved in experimental run 2 (feed rate 26.5 mm/min and drill diameter 6.5 mm). The significance of the selected factors and their contribution to the output, i.e., delamination, were determined using analysis of variance (ANOVA) with 95% confidence intervals, as shown in Table 4. ANOVA indicated that the model of process parameters with the delamination is significant, as the p-value of model (0.0002) is lower than 0.05. p-values below 0.05 indicate that a model is significant[5]. The ANOVA table also shows that the reduction in delamination is significantly impacted by the drill diameter and feed rate, and the feed rate has a greater impact on delamination than the drill size, as evidenced by its lower p value (0.0001). Furthermore, the adjusted R-square [6] and predicted R-square [7] helped to determine the adequacy of the mathematical model. Therefore, R-square adjusted (0.9264) and R-square predicted (0.8751) depicted a high degree of model accuracy, as shown in Table 4.

Table 4. ANOVA results for selected factors.

Source	Sum of Squares	df	Mean Square	F-Value	p-Value	
Model	0.0008	2	0.0004	51.35	0.0002	Significant
A-Feed (f)	0.0007	1	0.0007	96.64	<0.0001	Significant
B-Drill Size	0	1	0	6.06	0.0491	Significant
Residual	0	6	7.51×10^{-6}			
Cor Total	0.0008	8				
R-sq. 94.48%, R-sq.(adjusted) 0.9264, R-sq.(predicted) 0.8751						

A three-dimensional mesh plot was created to analyze the combined effect of the drill diameter and feed rate, as shown in Figure 1a. It was clearly observed that delamination increases with an increasing drill diameter and feed rate. This is due to the increase in thrust force that accompanies an increase in the drill diameter and feed rate, which leads to the breaking of the inter-laminar layers of carbon fabric, which, in turn, leads to an increase in the delamination of holes [3]. Similarly, the single-factor plot showed that delamination gradually increases with an increasing feed rate up to a certain level, and then the increment of delamination slightly reduces when further increasing the value of the feed rate, as shown in Figure 1b. On the other hand, delamination increases when increasing the value of the drill diameter, but after a certain level, delamination starts to reduce with the increment in the drill diameter, as evident in Figure 1b. This happens due to the damage that starts to occur in the inter-laminar layers at higher values of the drill diameter [4].

5. Conclusions

It has been determined that the optimal value of delamination (1.018) was achieved under experimental conditions of a feed rate of 26.5 mm/min and a drill diameter of 6.5 mm. The ANOVA table concluded that feed rate and drill size have a significant impact on delamination during the drilling of CFRP/Al2219-T6. It was also concluded from the single factor plot and 3D Mesh plot that delamination increases with an increasing feed rate and drill size. Furthermore, it has been revealed that the feed rate's contribution to increasing delamination is significantly higher than that of drill size.

Author Contributions: Conceptualization, M.W. and S.H.; methodology, M.W.; software, M.W.; validation, M.W.; formal analysis, M.J. and M.W.H.; investigation, M.W.; resources, M.W.; data curation, M.W.; writing—original draft preparation, M.J. and M.W.H.; writing—review and editing, M.W.H. and S.H.; visualization, M.W.; supervision, S.H.; project administration, S.H.; funding acquisition, M.W. All authors have read and agreed to the published version of the manuscript.

Funding: This research received no external funding.

Institutional Review Board Statement: Not applicable.

Informed Consent Statement: Informed consent was obtained from all subjects involved in the study.

Data Availability Statement: Not applicable.

Conflicts of Interest: The authors declare no conflict of interest.

References

1. Gonczy, S.T. Federal Aviation Administration (FAA) airworthiness certification for ceramic matrix composite components in civil aircraft systems. In Proceedings of the Testing and Modeling Ceramic & Carbon Matrix Composites, Paris, France, 4–6 June 2014.
2. D'Orazio, A.; El Mehtedi, M.; Forcellese, A.; Nardinocchi, A.; Simoncini, M. Tool wear and hole quality in drilling of CFRP/AA7075 stacks with DLC and nanocomposite TiAlN coated tools. *J. Manuf. Process.* **2017**, *30*, 582–592. [CrossRef]
3. Zitoune, R.; Krishnaraj, V.; Collombet, F. Study of drilling of composite material and aluminium stack. *Compos. Struct.* **2010**, *92*, 1246–1255. [CrossRef]
4. D'Orazio, A.; Mehtedi, M.E.; Forcellese, A.; Nardinocchi, A.; Simoncini, M. Study of tapping process of carbon fiber reinforced plastic composites/AA7075 stacks. In Proceedings of the 21st International ESAFORM Conference on Material Forming (ESAFORM 2018), Palermo, Italy, 23–25 April 2018.
5. Hanif, M.W.; Wasim, A.; Sajid, M.; Hussain, S.; Jawad, M.; Jahanzaib, M. Evaluation of microstructure and mechanical properties of squeeze overcast Al7075– Cu composite joints. *China Foundry* **2023**, *20*, 29–39. [CrossRef]
6. Hanif, M.W.; Wasim, A.; Sajid, M. Evaluating the Effect of Process Parameters on the Mechanical Properties of an AA7075-Cu Overcast Joint Using the Taguchi Method. *Eng. Proc.* **2022**, *23*, 3.
7. Haider, F.; Jahanzaib, M.; Hanif, M.W. Optimizing the process parameters of Fiction Stir Welded dissimilar 2024Al-5754Al Joint using the Taguchi Method. In Proceedings of the 1st International Conference on Modern Technologies in Mechanical & Materials Engineering (MTME-2023), Topi, Pakistan, 6 May 2023; p. 02006.

engineering proceedings

MDPI

Proceeding Paper

Applying Ultrasonic-Assisted Incremental Sheet Forming to Al 5052 Aluminum Alloy †

Ky-Thanh Ho [1,*], Ngoc-Tuan La [2], Ngoc-Hung Chu [3], Nhu-Huynh Vu [4] and Tat-Loi Mai [2]

1 Faculty of Mechanical Engineering, Thai Nguyen University of Technology (TNUT), 250000 Thai Nguyen City, Vietnam
2 Faculty of Mechanical Engineering, Vinh University of Technology Education (VUTE), 430000 Vinh City, Vietnam; langoctuan.ktv@gmail.com (N.-T.L.); maitatloi7@gmail.com (T.-L.M.)
3 Faculty of Mechanical Engineering, TUETECH University, 250000 Thai Nguyen City, Vietnam; chungochung@vietbac.edu.vn
4 Faculty of Mechanical Engineering, Thai Nguyen Vocational Training College, 250000 Thai Nguyen City, Vietnam; huynhsen.tn@gmail.com
* Correspondence: hkythanh@tnut.edu.vn
† Presented at the Third International Conference on Advances in Mechanical Engineering 2023 (ICAME-23), Islamabad, Pakistan, 24 August 2023.

Abstract: In this paper, the influence of ultrasonic vibration on the forming forces and surface quality of products formed using an ultrasonic-assisted incremental sheet forming (UISF) method was investigated and compared with an incremental sheet forming (ISF) method. The elements and parameters used for research include a sheet of Al 5052 aluminum alloy with thickness of 1.0 mm, a lathe with a feed rate of 70–130–225 rpm, and step size of 1.0–1.5–2.0 mm. The results show that ultrasonic vibration significantly reduces the forming forces, of which the main forming force F_{zmax} is reduced by about 20%. Besides, the results also show that the surface quality of products formed via UISF is significantly improved compared to that formed by ISF.

Keywords: incremental forming; ultrasonic-assisted; force; forming

Citation: Ho, K.-T.; La, N.-T.; Chu, N.-H.; Vu, N.-H.; Mai, T.-L. Applying Ultrasonic-Assisted Incremental Sheet Forming to Al 5052 Aluminum Alloy. *Eng. Proc.* **2023**, *45*, 8. https://doi.org/10.3390/engproc2023045008

Academic Editors: Mohammad Javed Hyder, Muhammad Mahabat Khan, Muhammad Irfan and Manzar Masud

Published: 8 September 2023

1. Introduction

Known as a die-less method of manufacturing shell products, incremental sheet forming (ISF) has attracted global attention in the field of sheet forming processes due to its flexibility and elimination of costly dies [1,2]. In the forming process using ISF, as shown in Figure 1 [1], the forming tool only contacts the sheet workpiece in a very small area, so the forming force is much smaller than that of deep drawing. The potential applications of ISF are various: forming panels for cars, airplanes or home appliances [1,3]; rapid prototyping [1,3]; and applications in the biomedical field [1,4]. ISF is especially suitable for products with complex profiles and a small batch production size [1,2]. However, the ISF method is characterized by (1) low surface quality [3,5] and (2) low accuracy [1,3].

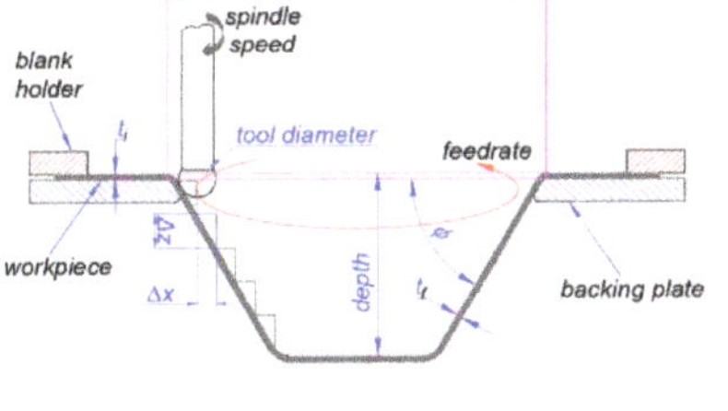

Figure 1. Basic principles of incremental sheet forming [1].

In order to improve forming accuracy while reducing the force component F_z, several research groups have recently implemented the application of ultrasonic-assisted ISF (called UISF or UVISF). When forming using the UISF method, the main force F_z decreases, and the shape accuracy of the product increases [6–8]. However, evaluations of surface quality have not been taken into account. This study aims to investigate the effect of ultrasonic vibration on the forming force and product surface quality formed using the UISF method.

2. Materials and Method

The experimental setup is depicted in Figure 2. The initial sheet is fixed on a jig which is clamped on the lathe chuck. The forming tool, made of hardened 90CrSi steel, and with a diameter of 12 mm and hardness of 61 HRC, is fixed on an ultrasonic transducer YP-5525-4Z. The force sensor (a Kistler three-component dynamometer, type 9257B) fixed to the carriage of the lathe is used to determine the forming forces. After the installation is complete, the system is scanned to determine the true resonant frequency, at 28.4 kHz.

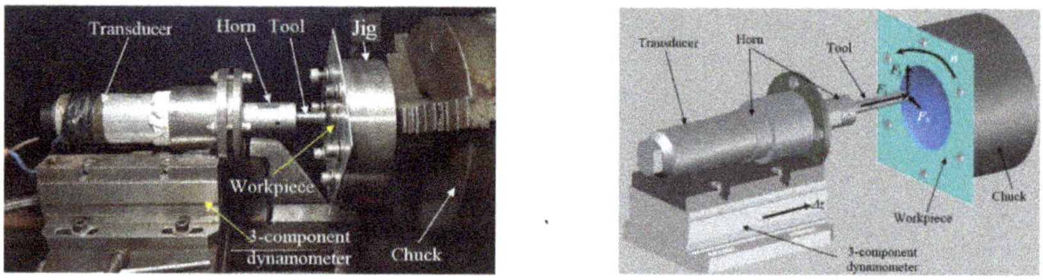

(a) (b)

Figure 2. Experimental setup: (**a**) photo and (**b**) model of the force components.

The initial workpiece is an Al 5052 aluminum alloy sheet in an annealed state with a thickness of 1.0 mm and a square edge of 120 mm × 120 mm in size. This is an alloy commonly used in the automotive and household industries [9]. In this study, the input parameters of the UISF and ISF methods include feed rate (n, chosen within a range of 70–130–225 rpm) and step size (Δz, within a range of 1.0–1.5–2.0 mm). Others parameters remained unchanged: the wall angle ϕ was equal to 45°, the depth of forming was 5 mm, and the diameter was 60 mm. All the parameters are illustrated in Figure 1. The experiment was designed according to the Taguchi method, with two parameters, three levels of values, and a total of 9 experiments, as shown in Table 1.

Table 1. Experiment results (maximum forming force F_{zmax}).

No	Input Parameters		F_{zmax} (N)		Reduction Ratio
	n (rpm)	Δz (mm)	UISF	ISF	$(F_{z(ISF)} - F_{z(UISF)})/F_{z(ISF)}$ (%)
1	70	1.0	359.1	466.4	23.01%
2	70	1.5	361.9	488.2	25.87%
3	70	2.0	387.6	512.5	24.37%
4	130	1.0	323.3	425.3	23.98%
5	130	1.5	347.8	457.5	23.98%
6	130	2.0	375.7	467.7	19.67%
7	225	1.0	297.6	301.2	1.20%
8	225	1.5	336.6	378.0	10.95%
9	225	2.0	353.5	421.4	16.11%

3. Results and Discussion

During the forming process in both ISF and UISF, the forming force F_z gradually increases until a maximum value is reached, as shown in Figure 3.

When the workpiece rotates for one full turn, due to the spring-back phenomenon, this force is still maintained, but tends to decrease gradually. The tangential force F_y and the radial force F_x also tend to be similar. The maximum forming forces (F_{zmax}) according to the input parameters are summarized in Table 1.

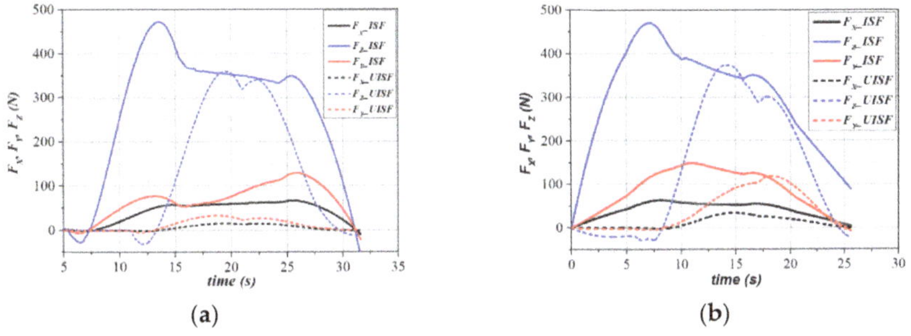

Figure 3. Force components (solid line—forces in ISF; dashed short line—forces in UISF) in various conditions: (**a**) n = 70 rpm, Δz = 1.0 mm; (**b**) n = 130 rpm, Δz = 2.0 mm.

From Figure 3 and Table 1, it is easy to see that under the same forming conditions, the forming force F_{zmax} in the UISF process is significantly reduced compared to that of the ISF process (about 20%). This may be due to the phenomenon of softening under the effect of ultrasonic vibration and the reduction in internal friction during the motion of dislocation when the material is deformed, so the force required for deformation in UISF is less than that in ISF [6–8]. Besides, ultrasonic vibration improves the contact conditions between the forming tool and the workpiece, thus reducing external friction [8]. Figure 4 shows images of the contact trace between the forming tool and the workpiece. It is easy to see that on the surface of the product formed using ISF, quite a lot of cracks appear (Figure 4a). Meanwhile, the product surface made using UISF is quite smooth (Figure 4b).

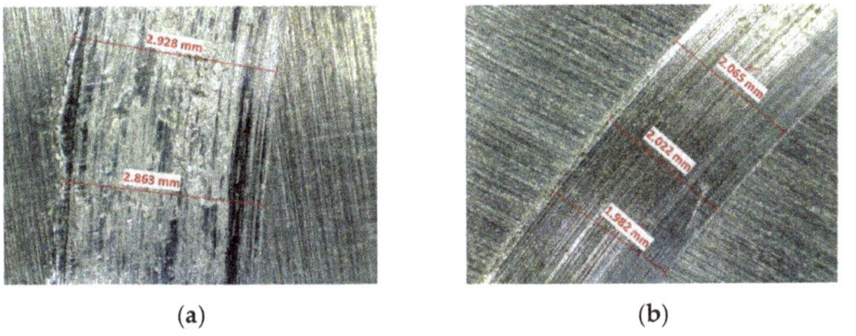

Figure 4. Photographs of (**a**) the area surface formed via ISF; and (**b**) the area surface formed via ISF.

The influence of the input parameters (including n and Δz) on F_{zmax} is also evaluated. As depicted in Figure 5, it can be seen that the main force F_{zmax} increases as Δz increases or/and n increases. When forming via the UISF method, the plot shows that the influence of Δz and n on the F_{zmax} is almost a straight line, with a rather low slope (see Figure 5a). This proves that these factors' influence on the main force F_{zmax} is not large. Meanwhile, when shaping with the ISF method, the effect of n on the main force F_{zmax} is more obvious, as shown by the line with a steeper slope (see Figure 5b).

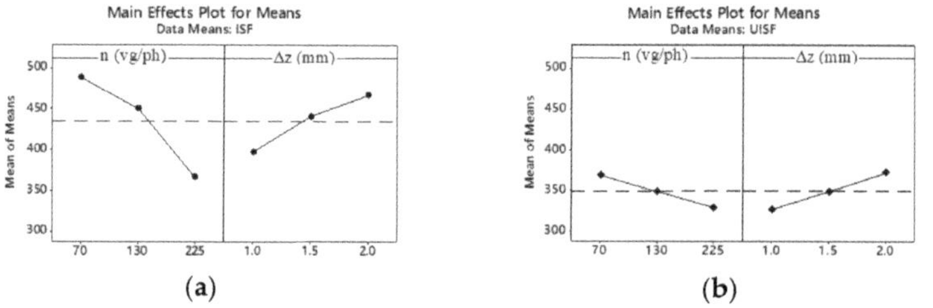

Figure 5. Plot of the influence of input parameters on the main force F_Z: (**a**) ISF; (**b**) UISF.

4. Conclusions

In this study, the influence of ultrasonic vibration on forming force and the surface quality of products formed via UISF was investigated and compared with ISF. The results show that the ultrasonic vibration significantly reduces the forming force, within which the main forming force F_{zmax} reduced by about 20%. The results also show that the surface quality formed using UISF is significantly improved compared to that formed using ISF.

Author Contributions: Conceptualization, K.-T.H.; methodology, N.-H.C. and N.-T.L.; validation, K.-T.H. and N.-H.C.; investigation, T.-L.M. and N.-H.V.; resources, K.-T.H.; writing—original draft preparation, K.-T.H.; writing—review and editing, K.-T.H.; visualization, N.-H.C.; project administration, N.-T.L.; funding acquisition, K.-T.H.; All authors have read and agreed to the published version of the manuscript.

Funding: This research was funded by the Vietnam Ministry of Education and Training, grant number B2022-TNA-26.

Institutional Review Board Statement: Not applicable.

Informed Consent Statement: Not applicable.

Data Availability Statement: Data are contained within the article.

Acknowledgments: The authors would like to express their thanks to Vietnam MOET, TNUT, VUTE, and TUETECH University for their support during this study.

Conflicts of Interest: The authors declare no conflict of interest.

References

1. Ajay; Mittal, R.K. *Incremental Sheet Forming Technologies: Principles, Merits, Limitations and Applications*, 1st ed.; CRC Press: Boca Raton, FL, USA, 2020. [CrossRef]
2. Uheida, E.H.; Oosthuizen, G.A.; Dimitrov, D.M. Toward understanding the process limits of incremental sheet forming of titanium alloys. In Proceedings of the Competitive Manufacturing, International Conference on Competitive Manufacturing (COMA'16), Stellenbosch, South Africa, 27–29 January 2016.
3. Pérez-Santiago, R.; Bagudanch, I.; Garcia-Romeu, M.L. Incremental Sheet Forming. In *Modern Manufacturing Processes;* Özel, M.K.a.T., Ed.; John Wiley & Sons, Inc.: Hoboken, NJ, USA, 2019; pp. 47–63. [CrossRef]
4. Bagudanch, I.; Centeno, G.; Vallellano, C.; Garcia-Romeu, M.L. Towards the Manufacturing of near Net Shape Medical Prostheses in Polymeric Sheet by Incremental Sheet Forming. In *Materials Forming, Machining and Tribology;* Springer: Cham, Switzerland, 2019; pp. 1–33. [CrossRef]
5. Hagan, E.; Jeswiet, J. Analysis of surface roughness for parts formed by computer numerical controlled incremental forming. *Proc. Inst. Mech. Eng. Part B J. Eng. Manuf.* **2004**, *218*, 1307–1312. [CrossRef]
6. Cheng, R.; Wiley, N.; Short, M.; Liu, X.; Taub, A. Applying ultrasonic vibration during single-point and two-point incremental sheet forming. *Procedia Manuf.* **2019**, *34*, 186–192. [CrossRef]
7. Bai, L.; Li, Y.; Yang, M.; Yao, Z.; Yao, Z. Influences of Process Parameters and Vibration Parameters on the Forming Force in the Ultrasonic-Assisted Incremental Forming Process. *Adv. Mater. Sci. Eng.* **2018**, *2018*, 5726845. [CrossRef]

8. Amini, S.; Hosseinpour Gollo, A.; Paktinat, H. An investigation of conventional and ultrasonic-assisted incremental forming of annealed AA1050 sheet. *Int. J. Adv. Manuf. Technol.* **2017**, *90*, 1569–1578. [CrossRef]
9. Ho, K.-T.; Nguyen, B.-H.; La, N.-T.; Le, T.-S.; Pham, V.-T. An Experimental Study on the Tensile Strength of Friction Stir Welded AA5052 Aluminum Alloy. In Proceedings of the 5th International Conference on Advances in Manufacturing and Materials Engineering, Kuala Lumpur, Malaysia, 9–10 August 2022; pp. 389–395.

engineering proceedings

Proceeding Paper

Analysis of Mechanical Strength of Indium-Doped SAC 105 Lead-Free Solder Alloy †

Muhammad Sohail Hameed [1,*], Aneela Wakeel [1], Riffat Asim Pasha [1], Barkat Ullah [2] and Umair Ali [3]

1 Department of Mechanical Engineering, Faculty of Mechanical & Aeronautical Engineering, University of Engineering and Technology, Taxila 74050, Pakistan; aneela.wakeel@uettaxila.edu.pk (A.W.); asim.pasha@uettaxila.edu.pk (R.A.P.)
2 Department of Mechanical Engineering, COMSATS University Islamabad, Wah Campus, Wah Cantt 47040, Pakistan; barkat@ciitwah.edu.pk
3 Department of Mechanical Engineering, CECOS University of Information Technology and Emerging Sciences, Peshawar 25000, Pakistan; umair@cecos.edu.pk
* Correspondence: miansohail910@gmail.com
† Presented at the Third International Conference on Advances in Mechanical Engineering 2023 (ICAME-23), Islamabad, Pakistan, 24 August 2023.

Abstract: The incorporation and doping of elements represent a widely used approach to enhance the solidity, integrity, and characteristics of pb-free solder joints. The present study summarizes the incorporation of indium and its impact on the mechanical aspects of the SAC105 pb-free solder alloy. To refine the mechanical impact of the solder alloy, the evaluation of samples were categorized into three groups: as-cast, low-thermal aged (at 125 °C), and high-thermal aged (at 180 °C). The tensile deformation data were obtained via the universal tensile machine (UTM). Investigational findings demonstrated the enhancement in mechanical characteristics, including ultimate tensile and yield strength of the solder alloy. The addition of 1 wt.% of indium to SAC105 led to a notable increase in ultimate tensile strength, rising from 29.6 MPa to 35.31 MPa, which corresponds to an approximate 19.30% increase over the initial value.

Keywords: lead-free soldering material; SAC105; UTM; ultimate tensile strength; indium-lead-tin-copper

Citation: Hameed, M.S.; Wakeel, A.; Pasha, R.A.; Ullah, B.; Ali, U. Analysis of Mechanical Strength of Indium-Doped SAC 105 Lead-Free Solder Alloy. *Eng. Proc.* **2023**, *45*, 18. https://doi.org/10.3390/engproc2023045018

Academic Editors: Mohammad Javed Hyder, Muhammad Mahabat Khan, Muhammad Irfan and Manzar Masud

Published: 11 September 2023

1. Introduction

Soldering is the process of joining metals by utilizing solder as a filler metal. For a long time, the electronics industry has used 63Sn-37Pb for component interconnection, owing to its outstanding and widely accepted qualities [1]. However, environmental and health considerations have resulted in limitations on the use of lead in the electronics industry [2]. Owing to the low recycling rate of electronics and the adverse effects of lead (Pb) on human health, its utilization in electronic components has been restricted [3]. Consequently, researchers have persistently worked to encourage the electronic industry to switch to lead-free soldering. The SAC family is the most effective and trustworthy alternative to conventional tin–lead soldering out of all lead-free solders [4]. Although, the SAC family consists of different doping compositions, which is to be considered as SAC305, SAC405, SAC105, SAC307, SAC396, and SA107 [5]. SAC305 is perceived as the most favorable choice among all these alternatives, while the concern of high cost is due to the high silver content [6]. The SAC105 is renowned for its suitability and attractiveness due to its cost-effectiveness and favorable thermal and mechanical properties [7].

Based on the existing literature, the primary focus of this investigation is to create a new, pb-free solder alloy (SAC105) infused with indium as a doping agent. Suchart Chantarmanee conducted a study on the mechanical characteristics of the (SAC305) pb-free solder [8]. Similarly, Sungkhaphaitoon et al. [9] investigated the impact of adding indium to SAC305 by the resulting hardness of the new alloy. In this study, the ultimate tensile and yield strength of the solder alloy were investigated by adding indium as the doping element to SAC105 under various thermal aging temperatures.

2. Experimental Procedure

Tensile specimens of Sn, Ag, and Cu with the addition of In were prepared using the casting process. Figure 1 illustrates the raw materials used for the fabrication of the tensile test specimen. Raw materials in powdered form were imported from China with almost a 99.89% purity level of each element. Subsequently, the powders were carefully weighed using a highly precise scale and then mixed at 98% tin, 1.0% Ag and 0.5% Cu by weight. A ball milling apparatus was used for mixing these elements for 45 min to acquire a uniform composition of alloys using the inclusion of propanol and isopropyl alcohol (IPA). The complete composition of the preparing samples at various percentages by weight are listed in Table 1. The elements were subsequently poured into an alumina crucible and then positioned within a muffle furnace at 1250 °C to attain the ultimate melting point of each element. Subsequently, the liquefied metal was poured in a specially prepared die to obtain the tensile samples, as shown in Figure 2. In a similar way, the casting process is discussed by Umair Ali et al. to prepare the tensile specimens [6].

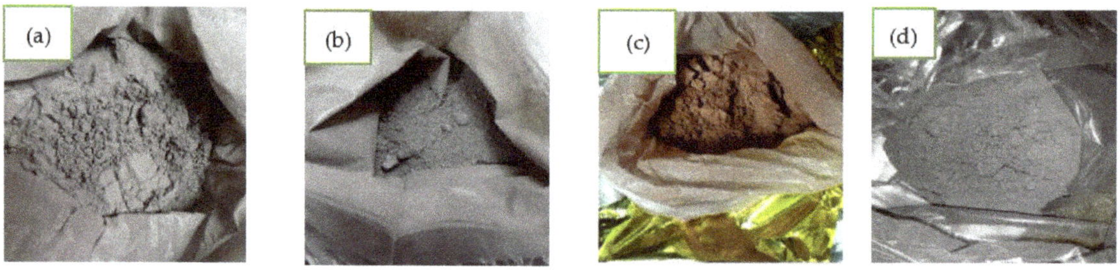

Figure 1. Raw materials were utilized in the preparation of samples: (**a**) tin, (**b**) silver, (**c**) copper, and (**d**) indium.

Table 1. The weight composition of doping elements.

Sr. No.	Alloy	Wt.%			
		Sn	**Ag**	**Cu**	**In**
1	SAC105	98.50	1.0	0.5	0.0
2	SAC105-1In	97.50	1.0	0.5	1.0

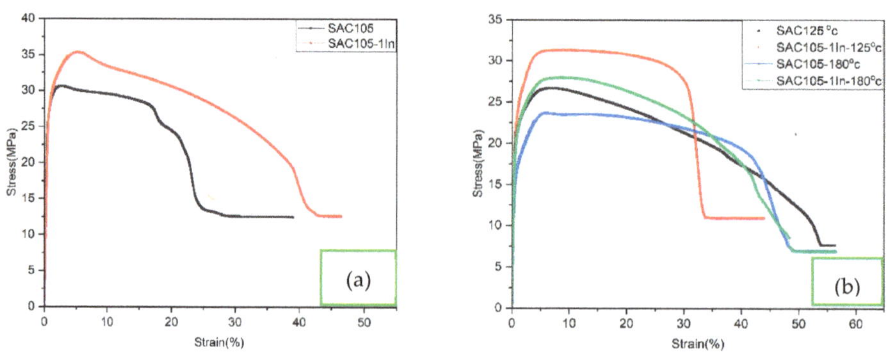

Figure 2. UTM curves (**a**) as a casted alloy and (**b**) thermally aged at 125 °C and 180 °C.

3. Results and Discussion

The reason for opting for a lead-free solder alloy was that the doping of an alloy to the pb-free soldering material has the capability to improve the characteristics of the SAC105 solder alloy. Figure 2 illustrates the stress–strain graphs observed during the tensile testing of SAC105 and SAC105+1In at a continual tensile rate of 0.5 mm/m at normal temperature. As a result, it has been observed that the addition of indium has increased the mechanical strength of SAC105. The highest strength of SAC105 was recorded with 1 wt.% of In. The yield strength (Y.S) was measured at 33 MPa, while the ultimate tensile strength (UTS) reached 35.31 MPa with 1 wt.% of indium.

Figure 2b demonstrates the deformation characteristics of specimens subjected to thermal exposure at 125 °C and 180 °C, respectively. The validation of the results and the comparison of the findings with the existing literature were undertaken. The corresponding outcomes were documented by M.H. Mahdavifard et al. [10], as well as H. Fallahi et al. reported the effect of indium upon the mechanical characteristics of the non-toxic solder alloy [11]. The study conclusion corelates and best matches with the previous findings and also makes a comparison of the indium-doped and without indium alloy. In this way, the findings were validated.

ASTM Standard

The ASTM Standard B32 especially deals with solder alloys having a melting zone less than 430 °C, which was followed in the preparation of the cast tensile specimen. The tensile test was performed by using a UTM at the tensile rate of 0.5 mm/m. Figure 3 depicts the geometry of a casted tensile specimen.

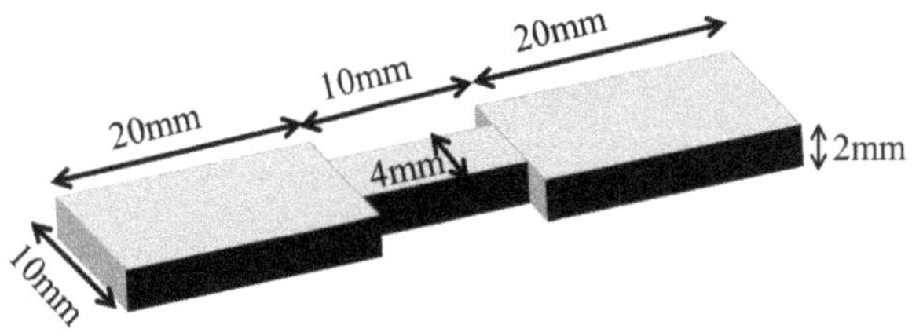

Figure 3. Geometry of casted tensile specimen.

4. Conclusions

The aim of this study was to examine the impact of indium on the mechanical characteristics of the SAC105 non-toxic solder alloy. Based on the empirical findings, the doping of indium could enhance the mechanical aspects of SAC105. In conclusion, it has been observed that the ultimate tensile strength of the indium-based SAC105 pb-free solder alloy exhibits the maximum UTS in comparison to all the other alloys that have been synthesized. Although the UTS decreases as the thermal aging increases, it is important to note that the mechanical properties of the indium-based alloy exhibited superior strength in comparison to both the undoped and aged SAC105.

Author Contributions: Conceptualization, M.S.H.; methodology, M.S.H. and U.A.; validation, M.S.H.; writing—original draft preparation, M.S.H.; writing—review and editing, M.S.H. and B.U.; supervision, A.W. and R.A.P. All authors have read and agreed to the published version of the manuscript.

Funding: This research received no external funding.

Institutional Review Board Statement: Not applicable.

Informed Consent Statement: Not applicable.

Data Availability Statement: Available on request.

Acknowledgments: The author wishes to express his heartful appreciation and gratitude to Muhammad Abid for the support and facilitation to perform the UTM testing at COMSATS University, Islamabad Sahiwal Campus.

Conflicts of Interest: The authors declare no conflict of interest.

References

1. Fazal, M.A.; Liyana, N.K.; Rubaiee, S.; Anas, A. A critical review on performance, microstructure and corrosion resistance of Pb-free solders. *Meas. J. Int. Meas. Confed.* **2019**, *134*, 897–907. [CrossRef]
2. Aamir, M.; Muhammad, R.; Tolouei-Rad, M.; Giasin, K.; Silberschmidt, V.V. A review: Microstructure and properties of tin-silver-copper lead-free solder series for the applications of electronics. *Solder. Surf. Mt. Technol.* **2020**, *32*, 115–126. [CrossRef]
3. Sun, L.; Zhang, L. Review Article: Properties and Microstructures of Sn-Ag-Cu-X Lead-Free Solder. *Adv. Mater. Sci. Eng.* **2015**, *2015*, 639028. [CrossRef]
4. El-Daly, A.A.; Hammad, A.E.; Fawzy, A.; Nasrallh, D.A. Microstructure, mechanical properties, and deformation behavior of Sn-1.0Ag-0.5Cu solder after Ni and Sb additions. *Mater. Des.* **2013**, *43*, 40–49. [CrossRef]
5. Ma, H.; Suhling, J.C. A review of mechanical properties of lead-free solders for electronic packaging. *J. Mater. Sci.* **2009**, *44*, 1141–1158. [CrossRef]
6. Ali, U.; Khan, H.; Aamir, M.; Giasin, K.; Habib, N.; Awan, M.O. Analysis of microstructure and mechanical properties of bismuth-doped SAC305 lead-free solder alloy at high temperature. *Metals* **2021**, *11*, 1077. [CrossRef]
7. Yang, T.; Chen, Y.; You, K.; Dong, Z.; Jia, Y.; Wang, G.; Peng, J.; Cai, S.; Luo, X.; Liu, C.; et al. Effect of Bi, Sb, and Ti on Microstructure and Mechanical Properties of SAC105 Alloys. *Materials* **2022**, *15*, 4727. [CrossRef] [PubMed]
8. Chantaramanee, S.; Sriwittayakul, W. Effects of Antimony and Indium Addition on Wettability and Interfacial Reaction of Sn-3.0Ag-0.5Cu Lead Free Solder on Copper Substrate. *Mater. Sci. Forum* **2018**, *928*, 188–193. [CrossRef]
9. Sungkhaphaitoon, P.; Chantaramanee, S. Effects of Indium Content on Microstructural, Mechanical Properties and Melting Temperature of SAC305 Solder Alloys. *Russ. J. Non-Ferrous Met.* **2018**, *59*, 385–392. [CrossRef]
10. Mahdavifard, M.H.; Sabri, M.F.M.; Said, S.M.; Rozali, S. High stability and aging resistance Sn-1Ag-0.5Cu solder alloy by Fe and Bi minor alloying. *Microelectron. Eng.* **2019**, *208*, 29–38. [CrossRef]
11. Fallahi, H.; Nurulakmal, M.S.; Fallahi, A.; Abdullah, J. Modifying the mechanical properties of lead-free solder by adding iron and indium and using a lap joint test. *J. Mater. Sci. Mater. Electron.* **2012**, *23*, 1739–1749. [CrossRef]

engineering proceedings

MDPI

Proceeding Paper

Preliminary Maturity Level Assessment of Industry 4.0 in the Context of Pakistani Industries †

Uzair Khan, Wasim Ahmad, Ahmad Sajjad * and Muhammad Jawad

Industrial Engineering Department, University of Engineering and Technology Taxila, Rawalpindi 47050, Pakistan; uzairkbaloch@gmail.com (U.K.); wasim.ahmed@uettaxila.edu.pk (W.A.); engr.jawad@uettaxila.edu.pk (M.J.)

* Correspondence: ahmad.sajjad@uettaxila.edu.pk

† Presented at the Third International Conference on Advances in Mechanical Engineering 2023 (ICAME-23), Islamabad, Pakistan, 24 August 2023.

Abstract: The primary objective of this study is to evaluate the adaptability and inclination of industrial sectors of Pakistan with respect to Industry 4.0. A questionnaire with nine questions was developed and disseminated to 20 sampled industries. To analyze the variability in responses, a one-way analysis of variance test was used. The statistical analysis revealed that there is an awareness of the basic concept behind Industry 4.0 in Pakistani industries, but there is a reluctance to adopt digitization and to shift from conventional production systems. This study will be helpful and will provide a guide for new and already existing enterprises for achieving Industry 4.0 requisite attributes precisely.

Keywords: Industry 4.0; readiness; maturity; cyber–physical production systems

Citation: Khan, U.; Ahmad, W.; Sajjad, A.; Jawad, M. Preliminary Maturity Level Assessment of Industry 4.0 in the Context of Pakistani Industries. *Eng. Proc.* **2023**, *45*, 29. https://doi.org/10.3390/engproc2023045029

Academic Editors: Mohammad Javed Hyder, Muhammad Mahabat Khan, Muhammad Irfan and Manzar Masud

Published: 12 September 2023

1. Introduction

Developments in manufacturing technologies by introducing systems based on cyber–physical concepts, Internet of Things (IOTs), and artificial intelligence (AI) are considered as Industry 4.0. The concept features two main aspects: integration and interoperability [1,2]. Linked with several applications and software, Industry 4.0 enables sustainability in production and service [3,4]. Developed countries are creating their own version of Industry 4.0-related strategies like "China 2025" in China [5], "Industry 4.1J" in Japan [6], and "Advanced Manufacturing Partnership (AMP 2.0)" in the USA [7]. However, this concept is still in its embryological stages in developing countries, specifically Pakistan. There is considerably less access to modern technologies and a collective reluctance in adopting its application, as indicated by its global ranking on various indices. Therefore, an evaluation of the maturity of developing countries in adopting Industry 4.0 poses a significant challenge. Pakistan, for instance, ranks 110th out of 141 countries in terms of global competitiveness. According to Khan [8], there is a significant decline in growth of the textile industries of Pakistan, which will impact the Key Index, i.e., the GDP of Pakistan, which was 4.24% [9], as the share of the industrial sector in terms of GDP is 12.4% of whole GDP. To increase the GDP, the manufacturing industry should shift toward Industry 4.0 in terms of large- and small-scale manufacturing systems [10]. Industries are also facing a shortage of skilled workers and knowledge sets [11]. The logistic network of the country is also deficient compared with other neighboring countries like India, China, and Bangladesh [12]. To tackle this issue, the use a framework involving evaluating the readiness level of individual companies using maturity models. The term "maturity" denotes a "state of being complete, ready or perfect" and implies development. The established seriousness models are generally used as tools to assess and measure the inclination of an organization or a process [13]. Readiness models are exclusively designed to secure a first perspective and to enable the start of the development route [14]. These models are established by renowned expert firms worldwide and have proven effective in assessing the manufacturing sector. By utilizing such models, countries can gain insights into their current state and take necessary measures to embrace Industry 4.0 technologies and practices. Several countries have developed unique models to assess their Industry 4.0 state, including the IMPULS model by the German Association of Mechanical Engineers (VDMA), the Singapore Smart Industry 4.0 Readiness Index, and the Smart Manufacturing Readiness Model formed by the National Institute of Standards and Technology (NIST) in the United States of America. These models have been tailored to suit the specific situations and circumstances of technologically advanced nations. However, a challenge arises when applying these models in developing countries, as they have not been extensively utilized in such contexts [10]. These developed models are comprehensive, detailed, and resource-intensive. Thus, it is necessary to assess the preliminary maturity stage of Industry 4.0 in the manufacturing sector of Pakistan using the limited resources and a cost-effective way to provide a preliminary understanding of the adoption of Industry 4.0. A questionnaire was developed to frame close-ended questions related to the Industry 4.0 concept and its awareness, willingness, and maturity level. It will provide a basic understanding and conceptualization of the readiness level to implement Industry 4.0 in Pakistan's industrial sector.

2. Methodology

The current report is exploratory and not based on any previously developed framework and primarily emphasizes an evaluation of the industrial sector of Pakistan, as well as the profiling of industrial concepts. The outcome of this study will be helpful and will provide guidelines for the development of strategies in Pakistani industries in the future. Close-ended questions are considered the most suitable way to conduct qualitative research. First, a comprehensive literature review was conducted to obtain a clear understanding of the Industry 4.0 paradigm. To find a suitable model for assessing Industry 4.0 readiness and maturity of manufacturing enterprises in Pakistan, the method in Figure 1 was adopted. It contains four steps to analyze the industrial sector of Pakistan [10]: the first step comprises the questionnaire development and its dissemination; the second step consists of data collection from industrial experts, top management, decision makers, and AI experts of the Pakistani industrial sector through a questionnaire; in the third step, the data are processed using statistical software; the last step consists of a data analysis, the findings, and the conclusions upon obtaining the answers and feedback.

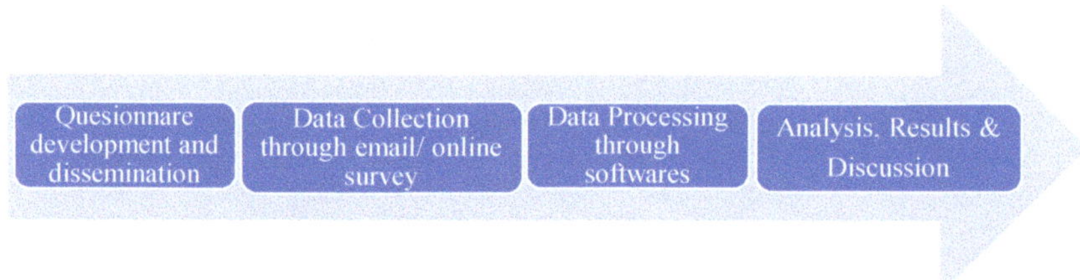

Figure 1. Steps adopted to analyze maturity and readiness.

2.1. *Questionaire Development*

The questions were in line with the major issues identified from the literature review and validated by the industrial and academic experts. The questionnaire consists of 9 questions, as given in Appendix A, with a complete set of data having the following three parts:

Part 1: Basic knowledge of Industry 4.0 (Questions 1–3);
Part 2: Readiness for Industry 4.0 (Questions 4–6);
Part 3: Maturity level (Questions 7–9).

2.2. Dissemination and Data Collection

In this step, an online survey was developed and disseminated through email, WhatsApp, and other social media platforms. The developed survey was filled out by industrial experts, production engineers, consultants, and top management from selected 20 industries. The sample size was calculated using random sampling (Equation (1)).

$$P = 1 - (1 - \left(\frac{1}{N}\right))n. \quad (1)$$

where n = sample size; N is the population size, which was 400; and P is the probability, which was 5% in this study.

The industries are categorized as the textile, automotive, and manufacturing industries, as tabulated in Table 1.

Table 1. List of companies.

Sr. no	Company Name	Sr. no	Company Name
1	Hattar Group of Industries, Haripur, Pakistan	11	Carriage Factory, Islamabad, Pakistan
2	HIT, Taxila, Pakistan	12	Macter International Limited, Karachi, Pakistan
3	Elektro Control Industries (Pvt.), Ltd. Islamabad, Pakistan	13	Rani & Company (Private) Limited, Karachi, Pakistan
4	Pothohar Industries Rolling Mills, Islamabad, Pakistan	14	International Polymer Industries (Pvt) Ltd., Islamabad, Pakistan
5	Dawn Electric Industries, Islamabad, Pakistan	15	The Indus Basin Company, Karachi, Pakistan
6	Faisalabad Textile Mills, Faisalabad, Pakistan	16	Heavy Mechanical complex, Taxila, Pakistan
7	Bestway Cement, Chakwal, Pakistan	17	Cherat Cement, Nowshera. Pakistan
8	Askari Fuels, Rawalpindi, Pakistan	18	POF Wah Cantt, Punjab, Pakitan
9	Auto Industry Chaklala, Rawalpindi, Pakistan	19	Wah Brass Mill, Wah Cantt, Pakistan
10	Poly Foils Pvt Ltd., Rawat, Pakistan	20	Coca Cola factories, Lahore, Pakistan

2.3. Data Processing

To process the data, Microsoft Excel was used to preprocess, compile, analyze, and visualize the responses. The collected data were analyzed to remove any blank entries and redundant data. Then, the data were compiled in such a way as to make it suitable for statistical analysis. The assembled data were modelled by amassing personal opinions about each variable into clusters to obtain a response rating distribution of these groups as a suitable means of reviewing the data.

2.4. Data Analysis

To investigate and assess deviations in the data, standard deviation (SD) is a useful tool in statistics. It is the distribution of a dataset compared with its normal value. Data points that are distant from the mean indicate a higher deviation within the dataset. As a result, the data become more dispersed, leading to an increased value for the standard deviation. An SD cannot be negative as it is calculated by squaring the parameter. Mean is the average of a certain set of observations. A one-way analysis of variance (ANOVA) was conducted to measure the inconsistency in the collected data. A p-value less than 0.05 depicts the significance of the data. A p-value of 0.003 was obtained in the comparison of data obtained directly from interviews and those obtained indirectly from mail.

3. Results and Analysis

In total, 150 responses were recorded and received from the 20 sampled industries. The average results of these responses are summarized in Table 2. The complete results regarding basic knowledge of, readiness/adaptability for, and level of understanding/maturity regarding Industry 4.0 were determined using a Microsoft Excel sheet and verified using SPSS 28.0.1; the overall average for the results is 1.9, as shown in Table 2. This means that the overall readiness/adaptability in the selected industries is below average. As far as awareness is concerned, the average score is 2.4, which means the bulk of industries in Pakistan are conscious of Industry 4.0 and only a few of them do not know about Industry 4.0. For example, cement manufacturing companies work manually via ordinary labor, as they work with ordinary methods and there are no digital manufacturing processes.

Table 2. Summary of averages for the results from the survey responses.

Question \ Customer	1.	2.	3.	4.	5.	6.	7.	8.	9.	10.	11.	12.	13.	14.	15.	16.	17.	18.	19.	20.	Average
Q 1	2	2	4	3	2	1	1	1	3	2	2	4	3	4	2	1	2	2	4	3	2.4
Q 2	1	1	2	1	2	1	1	2	2	1	1	2	1	2	3	2	1	1	2	1	1.5
Q 3	2	3	3	2	2	1	0	1	1	2	3	3	2	1	3	1	2	3	3	2	2
Q 4	3	1	2	2	1	2	1	2	1	3	1	2	2	3	2	1	3	1	2	2	1.85
Q 5	2	3	3	3	2	1	0	2	2	2	3	3	3	2	3	2	2	3	3	3	2.35
Q 6	1	2	2	4	2	1	0	1	6	1	2	2	4	2	2	1	1	2	2	4	2.1
Q 7	2	1	1	2	3	2	1	1	4	2	1	1	2	3	3	2	2	1	1	2	1.85
Q 8	1	1	2	1	2	2	0	2	3	1	1	2	1	1	4	4	1	1	2	1	1.65
Q 9	1	2	1	3	1	1	0	1	2	1	2	1	3	3	5	1	1	2	1	3	1.75
Average	1.6	1.7	2.2	2.3	1.8	1.3	0.4	1.4	2.6	1.6	1.7	2.2	2.3	2.33	3	1.6	1.6	1.7	2.2	2.3	1.9

Total number of questions = 9, number of respondents = 20, total score (rating) = 5, any data that were not known took a value of zero, 0 = least effective, 5 = most effective.

The commutative aggregate of the ratings of these three major categories are illustrated in Figures 2–4. It is depicted in Figure 1 that the number of peaks above average are more than below average, which means that most companies are aware of the basic idea of Industry 4.0, the suitability of Industry 4.0 technology for production, and that it is better than manual production systems. The rating for readiness for Industry 4.0 is shown in Figure 3. It can be noticed that the number of below-average peaks are greater than the number of above-average peaks, which clearly explains the lack of readiness of industries due to a preference for manual work rather than Industry 4.0 technology. Industries are assuming that conventional methods are better for manufacturing their products. Therefore, industries have less awareness about the efficiency of Industry 4.0 production processes, and they resist converting their conventional production systems into digital systems. The maturity levels of Industry 4.0 in selected industries are shown in Figure 4. It is evident that the number of below-average peaks are much greater than the number of above-average peaks, which depicts a lack of Industry 4.0 maturity for industries due to a greater adaptability to manual work than to Industry 4.0 technology. Industries assume that conventional methods are easier to use in manufacturing processes. Thus, industries are more adaptable to manual work, consider the quality of production through Industry 4.0 to be less, and believe that there are safety risks in the context of the modern paradigm.

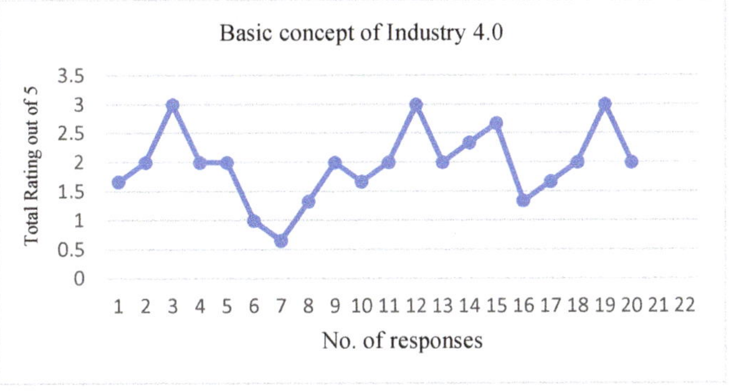

Figure 2. Basic concept of Industry 4.0 in Pakistan.

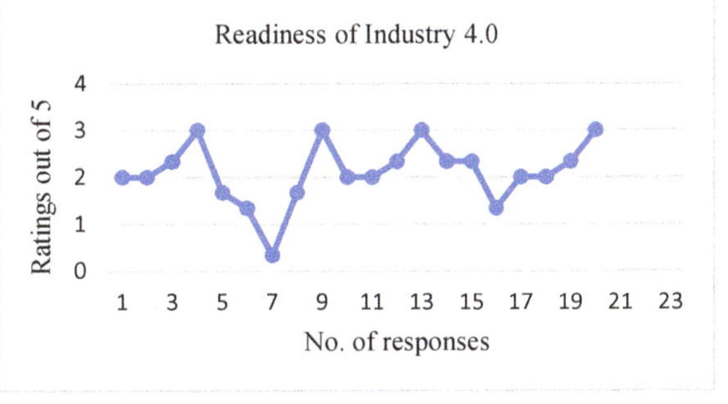

Figure 3. Readiness/adaptability.

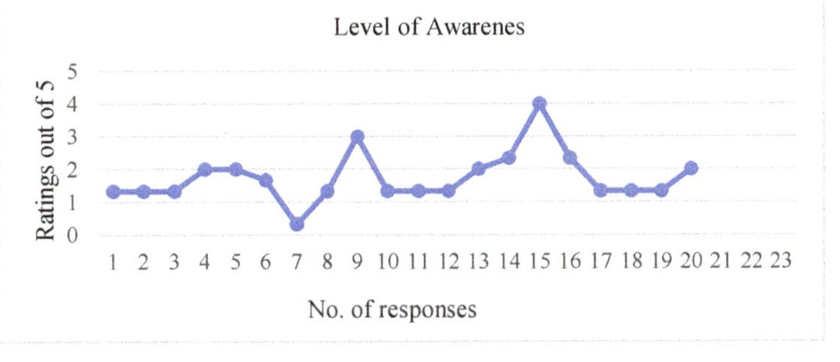

Figure 4. Maturity level of Industry 4.0.

A one-way analysis of variance (ANOVA) was carried out to assess the unevenness in data collected from industrial experts, production engineers, consultants, and top management. A p-value less than 0.05 depicts the significance of the data. The grouped data were modelled by amassing individual observations of a variable into groups, so that a response rating distribution of these groups was obtained. A variable was only significant when its f-stat was greater than 4 (>4) and its probability was less than 5% (<5%). Here, we can see that the variable under "within people", i.e., between treatments, is a major determinant in estimating Tukey's equal variances, having an f-stat of 20.4 and a probability of 0.001. Therefore, any variation in "between treatments" will cause a major variation in the dependent variable.

The result details are written in Table 3.

Table 3. ANOVA results.

	1	2	3	Total	
N	20	20	20	60	
$\sum X$	54	31.5	42	127.5	
Mean	2.4545	1.5	2	1.992	
$\sum X^2$	156	54.25	100	310.25	
Std.Dev.	1.0568	0.5916	0.8944	0.9449	
Source	**SS**		**df**	**f-Value**	***p*-Value**
Between treatments	9.7915		2	20.4	0.001
Within treatments	46.4545		61		
Total	56.2461		63		

4. Discussion

The results of this sectoral survey and its analysis reveal that the score for the first group in awareness of the basic concept is 1.96 out of 5 (40%), which is very low compared with the results of similar international industrial markets. The main reason behind this is the fact that management of main industrial sectors are significantly not related in terms of educational backgrounds and there is a loose arrangement in the job selection process. However, it also expresses that the respondents realize that Industry 4.0 technology is not suitable for their current production setups, as they feel more comfortable with manual work. These results correspond with those of with previous research [8]. The result for Group 2 is 2.1 out of 5 (42%). The readiness level is also below average. The main reason for this is that most industries prefer manual processing rather than digitization. The conventional methods are more suited for manufacturing the products. The result for maturity level (group 3) is 1.75 out of 5 (35%), which indicates that industries in Pakistan are not mature enough to embrace Industry 4.0, which corresponds with the outcomes of Hameed et al. [12]. Industries are adaptable to manual work, consider the quality of production through Industry 4.0 to be less, and believe that there are safety risks in the context of the modern paradigm. So, we can say that this maturity and readiness assessment research has critical value in the field of assessment of Industry 4.0 with regard to checking the fundamental knowledge, readiness, and maturity of Pakistani industries. Major stakeholders such as the Government of Pakistan and industrialists should take urgent measures to promote the use of this technology and to embrace the concept of Industry 4.0 in the manufacturing, hospital, and agriculture sectors as an increase in industry development means growth in the GDP of the country.

5. Conclusions

The motive behind this questionnaire-based survey was to assess the preliminary readiness and maturity level of Pakistan's industrial sectors with reference to Industry 4.0. It is concluded that the basic conceptualization of Industry 4.0 can be suitably checked using the designed questionnaire. Further, the distribution of responses in various industrial sectors such as the textile, automotive, and manufacturing industries also adds credibility. The questionnaire was validated using a statistical analysis. It was revealed that there is an awareness of the basic concept of Industry 4.0 across major Pakistani industries, but they are reluctant to adopt digitization and still prefer conventional production systems. A possible reason for this may be their better adaptability to conventional or manual technology and the low financial requirements. The current study provides a theoretical contribution to the subject in terms of assessments and has managerial implications as it can be used by top management involved in promoting the acceptance of Industry 4.0 processes. The findings of this research can be used for decision-making processes when converting conventional factories into smart manufacturing systems or when implementing Industry 4.0 processes.

Author Contributions: Conceptualization, U.K. and W.A.; methodology, U.K.; software, U.K.; validation, A.S. and M.J.; formal analysis, U.K.; data curation, U.K., M.J. and A.S.; writing—original draft preparation, U.K.; writing—review and editing, W.A. and A.S.; visualization, A.S. and M.J.; supervision, W.A. All authors have read and agreed to the published version of the manuscript.

Funding: This research received no external funding.

Institutional Review Board Statement: Not applicable.

Informed Consent Statement: Not applicable.

Data Availability Statement: Not applicable.

Conflicts of Interest: The authors declare no conflict of interest.

Appendix A. Questionnaire on Basic Knowledge, Readiness, and Maturity Level

Questions	Category	Detail	Rating				
			1	2	3	4	5
Q 1.	Concept of Industry 4.0	How much do you know about the fourth industrial revolution?					
Q 2.		How suitable is Industry 4.0 technology for production?					
Q 3.		Is one-command production suitable for production or manual work?					
Q 4.	Readiness of Industry 4.0 in Pakistan	Do you prefer to work manually or digitally?					
Q 5.		How appropriate are conventional methods of production for products?					
Q 6.		How much more efficient is Industry 4.0 than conventional methods?					
Q 7.	Maturity level of Industry 4.0	Which method makes work easier, digital or manual?					
Q 8.		What is the quality of production through Industry 4.0?					
Q 9.		Rate the safety risks in Industry 4.0.					

References

1. Romero, D.; Vernadat, F. Enterprise information systems state of the art: Past, present and future trends. *Comput. Ind.* **2016**, *79*, 3–13. [CrossRef]
2. Zhong, R.Y.; Xu, X.; Klotz, E.; Newman, S.T. Intelligent manufacturing in the context of industry 4.0: A review. *Engineering* **2017**, *3*, 616–630. [CrossRef]
3. Schiele, H.; Bos-Nehles, A.; Delke, V.; Stegmaier, P.; Torn, R.-J. Interpreting the industry 4.0 future: Technology, business, society and people. *J. Bus. Strategy* **2022**, *43*, 157–167. [CrossRef]
4. Ruggaber, R. Athena-Advanced technologies for Interoperability of heterogeneous enterprise networks and their applications. *Interoperability Enterp. Softw. Appl.* **2006**, *1*, 459–460.
5. Wübbeke, J.; Meissner, M.; Zenglein, M.J.; Ives, J.; Conrad, B. *Made in China 2025*; Papers on China; Mercator Institute for China Studies: Berlin, Germany, 2016; Volume 2, p. 4.
6. Lu, H.-P.; Weng, C.-I. Smart manufacturing technology, market maturity analysis and technology roadmap in the computer and electronic product manufacturing industry. *Technol. Forecast. Soc. Chang.* **2018**, *133*, 85–94. [CrossRef]
7. Dezhina, I.; Ponomarev, A. Advanced manufacturing: New emphasis in industrial development. *Foresight-Russia* **2014**, *8*, 16–29.
8. Khan, A.A.; Khan, M. Pakistan textile industry facing new challenges. *Res. J. Int. Stud.* **2010**, *14*, 21–29.
9. *Summary on Foreign Trade Statistics*; Pakistan Bureau of Statistics: Islamabad Pakistan, 2016.
10. Sajjad, A.; Ahmad, W.; Hussain, S.; Chuddher, B.A.; Sajid, M.; Jahanjaib, M.; Ali, M.K.; Jawad, M. Assessment by Lean Modified Manufacturing Maturity Model for Industry 4.0: A Case Study of Pakistan's Manufacturing Sector. *IEEE Trans. Eng. Manag.* 2023; *early access*.
11. Shamsi, M.I.; Syed, S.A. A study of the logistics capability factors for an e-commerce market. *FAST-NU Res. J. (FRJ)* **2015**, *1*, 143–149.
12. Hameed, W.-U.; Nadeem, S.; Azeem, M.; Aljumah, A.I.; Adeyemi, R.A. Determinants of e-logistic customer satisfaction: A mediating role of information and communication technology (ICT). *Int. J. Supply Chain. Manag. (IJSCM)* **2018**, *7*, 105–111.

13. Caiado, R.G.G.; Scavarda, L.F.; Gavião, L.O.; Ivson, P.; de Mattos Nascimento, D.L.; Garza-Reyes, J.A. A fuzzy rule-based industry 4.0 maturity model for operations and supply chain management. *Int. J. Prod. Econ.* **2021**, *231*, 107883. [CrossRef]
14. Dikhanbayeva, D.; Shaikholla, S.; Suleiman, Z.; Turkyilmaz, A. Assessment of industry 4.0 maturity models by design principles. *Sustainability* **2020**, *12*, 9927. [CrossRef]

engineering proceedings

MDPI

Proceeding Paper

Improved Spider Monkey Optimization Algorithm for Hybrid Flow Shop Scheduling Problem with Lot Streaming †

Jinhao Du, Jabir Mumtaz * and Jingyan Zhong

School of Mechanical and Electrical Engineering, Wenzhou University, Wenzhou 325035, China; jinhaoduwenda@163.com (J.D.); jingyzhong@163.com (J.Z.)

* Correspondence: jabirmumtaz@live.com

† Presented at the Third International Conference on Advances in Mechanical Engineering 2023 (ICAME-23), Islamabad, Pakistan, 24 August 2023.

Abstract: This paper investigates the hybrid flow shop scheduling problem with lot streaming, which integrates the order lot problem (OLP), order sequence problem (OSP), and lots assignment problem (LAP), with the objective of minimizing both the maximum completion time (C_{max}) and the total tardiness (TT) simultaneously. An improved spider monkey optimization (I-SMO) algorithm is proposed by combining the advantages of crossover and mutation operations of a genetic algorithm (GA) with the spider monkey optimization algorithm. The contribution value method is employed to select both global and local leaders. Experimental comparisons with classical optimization algorithms, including particle swarm optimization (PSO) and differential evolution (DE), were conducted to demonstrate the superiority of the proposed I-SMO algorithm.

Keywords: spider monkey optimization algorithm; multi-objective scheduling optimization; lot streaming; hybrid flow shop scheduling

Citation: Du, J.; Mumtaz, J.; Zhong, J. Improved Spider Monkey Optimization Algorithm for Hybrid Flow Shop Scheduling Problem with Lot Streaming. *Eng. Proc.* **2023**, *45*, 23. https://doi.org/10.3390/engproc2023045023

Academic Editors: Mohammad Javed Hyder, Muhammad Mahabat Khan, Muhammad Irfan and Manzar Masud

Published: 11 September 2023

1. Introduction

The Hybrid Flow Shop Scheduling Problem (HFSP), also known as the Flexible Flow Shop Scheduling Problem (FFSP), is an NP-hard problem. In the context of fast response requirements in the multi-variety small-batch customization production mode, the HFSP with Lot Streaming (LS) can effectively improve production efficiency, reduce production cycle, and enhance on-time delivery rate.

Zaky et al. address an integrated HFSP with LS for the objective of minimizing C_{max}. They formulate two mixed-integer nonlinear programming models to tackle this problem [1]. Danial and Fantahun proposed a two-stage GA for the HFSP with LS [2]. Beren and Ömer employed a GA to solve the HFSP with constraints on machine capabilities and limited waiting times [3]. However, there is limited research on HFSP with LS utilizing the Spider Monkey Optimization algorithm. In this paper, we propose an I-SMO algorithm for the HFSP with LS.

2. Problem Formulation

This study focuses on the scheduling problem of LS in a three-stage HFSP with different numbers of machines in each stage. Additionally, at least one of the stages has more than one machine. The objective is to minimize C_{max} and TT simultaneously. This problem mainly includes three sub-problems: OLP, OSP, and LAP. The corresponding decisions include for LSP, dividing the order into lots; for OSP, sorting of N^P; for LAP, determining each lot of machines that are processed in each process. In this case, the quantity of each order $c(c = 1, 2, \ldots, N^P)$, i.e., CN_c, delivery time d_c, number of lots in each order n_c^{lots}, and size of the lot $size_c^{lots}$.

The basic assumptions include the following: (1) All machines can be used at zero time; (2) A machine can only process one lot at the same time; (3) All lots can be processed at zero time, and any lot can enter the next process only after the previous stage is completed; (4) The mix-flow production is not considered; (5) There is a setup time before lots processing; (6) There is a buffer area between different stages, and the transportation time need to be considered; (7) The size of the lot is a decision variable, and the size of lot remains unchanged in all stages; (8) The sequence of lots of an order remains unchanged on all stages.

3. I-SMO Algorithm

3.1. Basic Flow of the I-SMO Algorithm

The SMO algorithm, introduced by Bansal et al. in 2014, is a novel swarm intelligence optimization algorithm that simulates the splitting and merging behavior of spider monkeys during their foraging process [4]. Mumtaz et al. proposed a new hybrid spider monkey optimization (HSMO) algorithm for PCB assembly scheduling problem [5]. In this study, the SMO algorithm is discretized using the crossover and mutation operations from the GA. Furthermore, the contribution value method is employed to select the global leader and local leaders. The I-SMO algorithm process is shown in Figure 1.

I-SMO

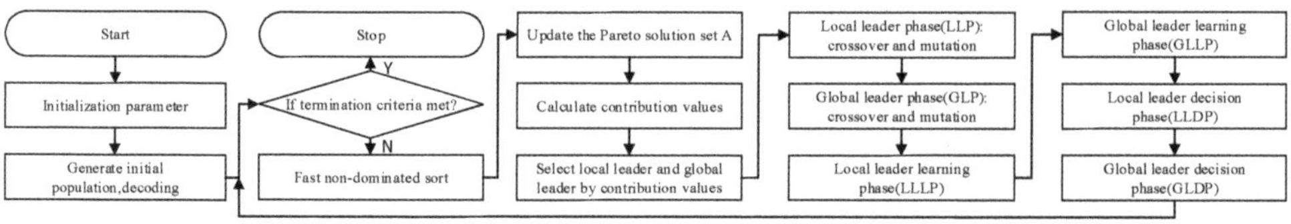

Figure 1. I-SMO algorithm process.

3.2. Solution Selection Based on Contribution Value

In this study, the hyper-volume performance indicator and the contribution value method proposed in the literature are used for selecting local and global leaders. This is because the greater the contribution values of the solution, the larger the area of the independent dominance of the solution, the more conducive to the distribution of the solution set. Specific steps are as follows: (1) Normalize the target value; (2) Pareto solution after the normalization is arranged in order of one of the goals; (3) According to Equation (1), the reference point $z^*(z_1^*, z_2^*)$, where the z_r^* indicates the value of the reference point on the target r, and σ is obtained by the test; (4) The dominance area of the calculation solution and its two adjacent solutions; (5) The area of the independent dominance of the calculation and solution is the contribution value of the solution. The contribution value is calculated based on Equation (2) at the edge or Pareto front end.

$$z_r^* = f_r^{max} + \sigma\left(f_r^{max} - f_r^{min}\right) \quad (1)$$

$$CV_A = \left(Z_1^* - f_1^A\right)\left(Z_2^* - f_2^A\right) - \left(Z_1^* - \max\left(f_1^A, f_1^B\right)\right)\left(Z_2^* - f_2^A\right) \quad (2)$$

3.3. Encoding

The hierarchical encoding scheme is utilized in this study. The first layer consists of a $2 \times N^P$ dimensional matrix, where the first row represents the lots quantity of N^P orders, and the second row represents the processing sequence of the orders. The second layer is a $4 \times N^{lot}$ dimensional matrix, where the first row represents the processing sequence of the lots, and the remaining layers represent the machine sequence to which the lots are allocated for each stage.

3.4. Local Leader Phase and Global Leader Phase (LLP and GLP)

During the LLP stage, members of each group will update their position by approaching their respective local leaders, while during the GLP stage, all members will update their position by approaching the global leader. Cross operation is divided into three layers: (1) Crossing the processing order by using the two-point crossing method. (2) According to the order sorting, the lots sorting is obtained; (3) For lots assignment, the partially matched crossover method, similar to the two-point cross method, is used to cross each process in order. An example of crossover is shown in Figure 2.

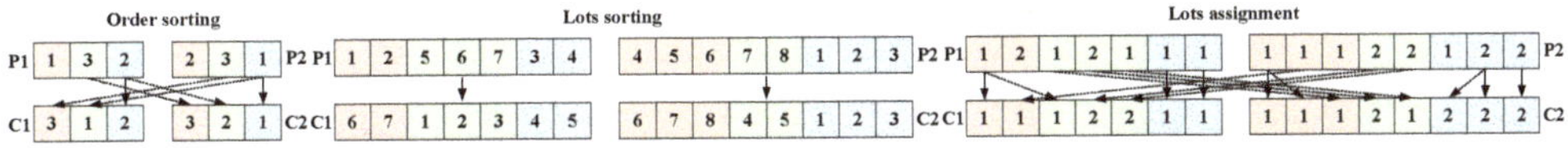

Figure 2. Crossover example.

Randomly select order c, and its lots coding sorting for reverse sequence processing. For the variation of the distribution and encoding of the process machine, the two batches of the order c randomly select 0 or 1 for each process. If it is 1, the corresponding gene is exchanged, otherwise it will not be exchanged. An example of mutation is shown in Figure 3.

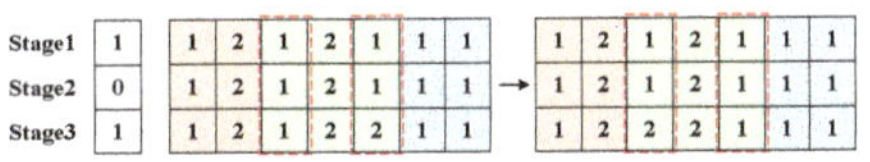

Figure 3. Mutation example.

4. Numerical Example and Analysis

4.1. Parameter Settings

The experimental data includes the number of orders CN_C, the number of orders N^P, the delivery date dc, the processing time of a sub-batch of orders on machine M in each stage P^{mc}_{stage}, number of machines in each stage n^{m}_{stage}, the setup time for the first processing of an order on machine M in each stage st^{moc}_{stage}, the setup time related to the sequence between different batches of orders $st^{mcc'}_{stage}$, and the transportation time during the order batch transition stage tt_{stage}. The range of batch quantities for orders is [1, 12]. The parameter ranges for different scale problems, based on production data from a specific company, are shown in Table 1. The parameter values for I-SMO and the comparative algorithms are obtained through Taguchi experiments, as shown in Table 2.

Table 1. Scale of experimental problems and parameter.

Parameter	Value			Parameter	Value	Parameter	Value	Parameter	Value
N^P	U [2, 3]	U [4, 6]	U [8, 12]	CN_C	100 × U [3, 10]	st^{moc}_{stage}	U [15, 20]	tt_{stage}	U [20, 30]
n^{m}_{stage}	U [1, 2]	U [1, 3]	U [2, 3]	P^{mc}_{stage}	U [50, 100]	$st^{mcc'}_{stage}$	U [10, 15]	d_c	U $[0,\ \rho \hat{C}_{max}]$

Table 2. Parameter values for different algorithms.

Parameter	Small			Medium			Large		
	PSO	DE	I-SMO	PSO	DE	I-SMO	PSO	DE	I-SMO
popsize	200	200	200	200	200	200	200	200	200
maxgen	200	100	200	200	150	200	200	200	200
p_c/p_m	/	/	0.85/0.1	/	/	0.85/0.25	/	/	0.85/0.25

4.2. Computational Experiments and Discussion

The computational experimental results for various algorithms are presented in Table 3. Each problem instance was independently run 10 times, where IGD stands for Inverted Generational Distance and NR represents Non-Dominated Rate.

Table 3. Performance results of different algorithms.

Size	PSO		DE		I-SMO	
	IGD	NR	IGD	NR	IGD	NR
Small	20,416.8006	0.0277	23,514.9404	0.0000	110.6899	0.9722
Medium	67,529.3626	0.0370	71,691.7599	0.0370	664.9987	0.9259
Large	218,301.1570	0.0000	244,254.7911	0.0000	0.0000	1.0000

IGD denotes the minimum Euclidean distance from the individual S^u to the optimal Pareto optimal solution, and the smaller this value is, the closer to the optimal solution the C_{max} and TT obtained by the algorithm are. In addition, NR denotes the proportion of solutions in the solution set of the current method that are also in the Pareto optimal frontier solution set; the larger this value is, the higher the proportion of the target solution obtained by the algorithm is in the Paret optimal frontier solution set. Therefore, I-SMO algorithm outperforms traditional population-based optimization algorithms in terms of IGD and NR.

5. Conclusions

This paper mainly includes the OLP, OSP, and LAP problems of HFSP with LS, and the experimental results show that the proposed I-SMO algorithm is superior to traditional algorithms and can obtain effective solutions. To study the HFSP with mixed-flow using the proposed ISMO should be further considered in the future.

Author Contributions: Conceptualization, J.M.; methodology, J.M., J.Z. and J.D.; software, J.M.; validation, J.M.; formal analysis, J.D.; investigation, J.D.; writing— original draft preparation, J.D.; writing—review and editing, J.M.; funding acquisition, J.M. All authors have read and agreed to the published version of the manuscript.

Funding: This research was funded by National Natural Science Foundation of China grant number [No. 51705370].

Institutional Review Board Statement: Not applicable.

Informed Consent Statement: Not applicable.

Data Availability Statement: The data that support the findings of this study are available from the corresponding author upon reasonable request.

Conflicts of Interest: The authors declare no conflict of interest.

References

1. Zaky, E.A.; Abdelmaguid, T.F.; Mohamed, T.A.; Mohamed, S.T. Lot streaming of hybrid flowshops with variable lot sizes and eligible machines. *J. Ind. Eng. Int.* **2023**, *43*, 238–264. [CrossRef]
2. Danial, R.; Fantahun, D. A Two-Stage Multi-Objective Genetic Algorithm for a Flexible Job Shop Scheduling Problem with Lot Streaming. *Algorithms* **2022**, *15*, 246.
3. Beren, G.Y.; Ömer, F.Y. Lot streaming in hybrid flowshop scheduling problem by considering equal and consistent sublots under machine capability and limited waiting time constraint. *Comput. Ind. Eng.* **2022**, *173*, 108745.
4. Bansal, J.C.; Sharma, H.; Jadon, S.S.; Clerc, M. Spider monkey optimization algorithm for numerical optimization. *Memet. Comput.* **2014**, *6*, 31–47. [CrossRef]
5. Mumtaz, J.; Guan, Z.L.; Yue, L.; Zhang, L.; He, C. Hybrid spider monkey optimisation algorithm for multi-level planning and scheduling problems of assembly lines. *Int. J. Prod. Res.* **2019**, *58*, 6252–6267. [CrossRef]

engineering proceedings

MDPI

Proceeding Paper

Neuro-Evolution of Augmenting Topologies for Dynamic Scheduling of Hybrid Flow Shop Problem †

Junjie Zhang, Yarong Chen *, Jabir Mumtaz * and Shengwei Zhou

School of Mechanical and Electrical Engineering, Wenzhou University, Wenzhou 325035, China; zhangjunjie6232@163.com (J.Z.); zhousw_zsw@126.com (S.Z.)
* Correspondence: yarongchen@126.com (Y.C.); jabirmumtaz@live.com (J.M.)
† Presented at the Third International Conference on Advances in Mechanical Engineering 2023 (ICAME-23), Islamabad, Pakistan, 24 August 2023.

Abstract: In this paper, the Neuro-Evolution of Augmenting Topologies (NEAT) algorithm is proposed to minimize the maximum completion time in a dynamic scheduling problem of hybrid flow shops. In hybrid flow shops, machines require flexible preventive maintenance and jobs arrive randomly with uncertain processing times. The NEAT-based approach is experimentally compared with the SPT and FIFO scheduling rules by designing problem instances. The results show that the NEAT-based scheduling method can obtain solutions with better convergence while responding quickly compared to the scheduling rules.

Keywords: hybrid flow shop; reinforcement learning; Neuro-Evolution of Augmenting Topologies; makespan; dynamic scheduling

Citation: Zhang, J.; Chen, Y.; Mumtaz, J.; Zhou, S. Neuro-Evolution of Augmenting Topologies for Dynamic Scheduling of Hybrid Flow Shop Problem. *Eng. Proc.* **2023**, *45*, 25. https://doi.org/10.3390/engproc2023045025

Academic Editors: Mohammad Javed Hyder, Muhammad Mahabat Khan, Muhammad Irfan and Manzar Masud

Published: 11 September 2023

1. Introduction

A hybrid flow shop is a kind of flow shop containing more than two stages and at least one stage with multiple parallel machines, also known as a flexible flow shop. The hybrid flow shop scheduling problem (HFSP) is of great theoretical significance and practical value, as it is widely applied in the chemical, textile, steel and semiconductor industries. In the context of intelligent manufacturing, dynamic scheduling based on reinforcement learning has become a research trend. Many research studies have been conducted in the literature. Han et al. first proposed a reinforcement learning method for HFSP [1]. Gil and Lee studied the use of the deep reinforcement learning approach to solve the material scheduling problem of many machines in a hybrid flow shop environment [2]. Cai et al. proposed a new shuffle frog-learning algorithm with Q-learning to solve a distributed assembly hybrid flow shop scheduling problem with fabrication, transportation and assembly [3]. Wang, J.J. and Wang, L. studied an energy-aware distributed hybrid flow shop scheduling method based on reinforcement learning [4]. Lang et al. presented a dynamic scheduling method based on the NEAT algorithm for a two-stage hybrid flow shop scheduling problem with family setup times [5].

The problem of dynamic scheduling in hybrid flow shops where machines require flexible preventive maintenance and where jobs arrive randomly and processing times are uncertain has not yet been identified, so this paper studies the HFSP scheduling problem with the objective of minimizing the makespan and designs a dynamic scheduling method based on NEAT reinforcement learning and compares it with scheduling rules.

2. Problem Description

The dynamic scheduling problem of the hybrid flow shop can be described as follows: n dynamically arriving jobs $J_i(i = 1, 2, \ldots, n)$ are processed continuously in $K(K \geq 2)$ stages, stage k has $m_k(m_k \geq 1; k = 1, 2, \ldots, K)$ machines, and there exists at least one stage with multiple machines. Flexible preventive maintenance of the machine is considered in accordance with production reality, i.e., the machine cannot be continuously used for more than the maintenance threshold UT, and the maintenance time is t^m. The scheduling objective is to minimize the makespan. The decision moment is triggered by the job being processed at a certain stage or by the arrival of a new job, which is then scheduled to the machine to optimize the objective value.

The problem studied in this paper considering the following assumptions: (1) Each job can be processed on any one of the machines at stage k. (2) Job processing time is the same for all machines at the same stage. (3) A machine can only process one job at a time. (4) All jobs are processed according to the same process sequence. (5) After determining the sequence of processing of the jobs in the first stage, the other stages are processed in this order.

3. Dynamic Scheduling Method for Hybrid Flow Shop Based on NEAT Algorithm

The interaction flow of the hybrid flow shop dynamic scheduling system based on reinforcement learning is shown in Figure 1.

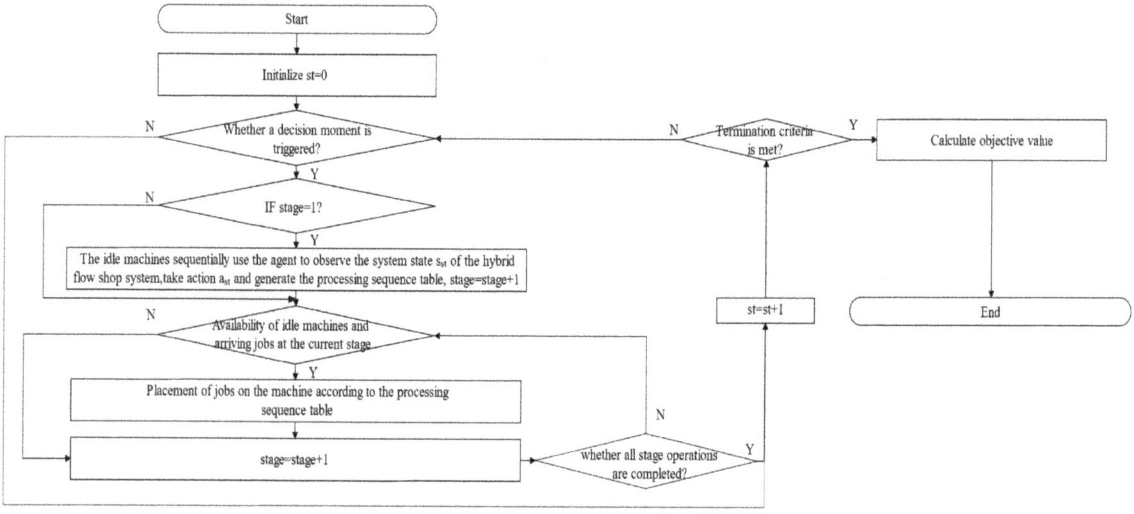

Figure 1. Interaction flow of hybrid flow shop system based on reinforcement learning.

3.1. State Space

Setting a suitable state space can reflect the change in the scheduling system in real time. For the characteristics of the dynamic scheduling problem of the hybrid flow shop, the environment is described in three dimensions (jobs, machines and stages) and the state space vector $S = \left[p_{ik}, T_i^r, T_{kc}, T_{kc}^a, U_{kc}^r\right]$ is defined. p_{ik} is the processing time corresponding to job J_i at stage k. T_i^r is the time interval between the current moment and the arrival time of job J_i. T_{kc} is the processing time of the job being processed by machine $M_{kc}(c = 1, 2, \ldots, m_k)$ at stage k, increasing the maintenance time t^m if maintenance is required. T_{kc}^a is the processing time of the job being processed by machine M_{kc} at stage k. U_{kc}^r is the maintenance threshold remaining at the current moment for machine M_{kc} at stage k.

3.2. Action Space

In the shop scheduling model, an action $a_t \in A_t$ is taken to determine a machine M_{kc} for processing job J_i at decision time t in state S_t.Based on the characteristics and objectives of the dynamic scheduling problem of a hybrid flow shop considering flexible preventive maintenance, the following three actions are designed.

Action 1: Selecting a job based on the SPT rule. The job J_i with the shortest processing time is selected from the waiting queue based on the SPT rule. One of the jobs J_i is selected at random if more than one job can be selected.

Action 2: Selecting a job based on FIFO rule. The job J_i with the earliest arrival time is selected from the waiting queue based on the FIFO rule. One of the jobs J_i is selected at random if more than one job can be selected.

Action 3: Waiting. No job is selected for machining.

The following constraints are set for several special cases: (1) When there are no jobs in the waiting queue, the agent can only select action 3. (2) When a new job arrives but there are no idle machines, action 3 is selected. (3) When there are jobs in the waiting queue and the machines are all idle, action 3 cannot be selected.

3.3. Fitness Function

The NEAT algorithm searches the optimal scheduling strategy through the optimization of the fitness function, so the fitness function of NEAT is designed to be $G = \frac{1}{C_{max}}$. Maximizing the fitness function is equal to minimizing the makespan. C_{max} can be described using Equation (1):

$$C_{max} = \max_i C_i \tag{1}$$

where C_i is the completion time of all stages of processing of job J_i.

4. Numerical Example and Analysis

4.1. Parameter Settings

According to the pre-experimentation, we set the evolutionary generations in the NEAT algorithm to 10 and the population size to 100. The job processing time p_{ik}, the arrival time r_i and the number of machines m_k at each stage were generated based on a uniform distribution. The problem was divided into three instances according to the number of stages K and the number of jobs n. Each type of instance was run three times for the experiments, and the parameters of the training set are shown in Table 1.

Table 1. The range of parameter values for the training set problem.

K	n	p_{ik}	r_i	UT	t^m	m_k
2	10	U (1,5)	U (1,5)	10	2	U (1,2)
3	20	U (1,5)	U (1,5)	10	2	U (1,2)
4	30	U (1,5)	U (1,5)	10	2	U (1,2)

4.2. Results Analysis

In this paper, the performance of the NEAT scheduling method and the two scheduling rules are evaluated based on the makespan C_{max}, running time T_t^R and response time T_s^R. The experimental results of the different algorithms are shown in Table 2.

Table 2. Experimental results of performance indicators for different scheduling methods.

Instance	NEAT			SPT			FIFO		
	C_{max}	T_t^R/s	T_s^R/s	C_{max}	T_t^R/ms	T_s^R/ms	C_{max}	T_t^R/ms	T_s^R/ms
$K = 2, n = 10$	25	1.06	0.05	25	6.37	0.32	26	6.68	0.33
	45	1.16	0.06	47	9.87	0.49	47	10.29	0.51
	51	1.21	0.06	51	10.72	0.54	51	10.30	0.51
$K = 4, n = 30$	128	11.66	0.10	129	59.60	0.50	131	62.31	0.52
	122	11.82	0.09	122	62.20	0.52	124	58.60	0.49
	98	10.21	0.08	100	48.57	0.40	105	58.40	0.49
$K = 6, n = 100$	392	177.16	0.29	398	530.40	0.88	406	601.94	1.00
	384	164.61	0.27	404	485.38	0.81	396	461.69	0.77
	416	190.16	0.32	434	493.33	0.82	432	478.78	0.80

It can be seen from Table 2 that the NEAT-based approach outperforms the scheduling rules for all three instances with the objective C_{max}. For the computation time, the SPT and FIFO rules have less time than the NEAT-based approach, because the NEAT reinforcement learning algorithm requires decisions to be made through interaction with the environment, while it is also able to respond the dynamic events quickly.

5. Conclusions

A NEAT-based approach is proposed for a dynamic scheduling problem of hybrid flow shops with machine preventive maintenance and the dynamic arrival of jobs to minimize maximum completion time. Our experimental results show that the NEAT algorithm is able to obtain better objective values while responding quickly. In future research, the proposed NEAT method can be used to solve dynamic scheduling problems considering some more realistic emergency events such as the insertion and withdrawal of orders.

Author Contributions: Conceptualization, J.Z. and Y.C.; methodology, J.Z. and S.Z.; software, J.Z. and J.M.; validation, J.Z., and J.M.; formal analysis, J.Z. and Y.C.; writing—review and editing, Y.C. and S.Z.; visualization, S.Z.; supervision, J.Z.; funding acquisition, Y.C. All authors have read and agreed to the published version of the manuscript.

Funding: This research was funded by the National Natural Science Foundation of China, grant number [No.51705370].

Institutional Review Board Statement: Not applicable.

Informed Consent Statement: Not applicable.

Data Availability Statement: The data that support the findings of this study are available from the corresponding author upon reasonable request.

Conflicts of Interest: The authors declare no conflict of interest.

References

1. Han, W.; Guo, F.; Su, X.C. A Reinforcement Learning Method for a Hybrid Flow-Shop Scheduling Problem. *Algorithms* **2019**, *12*, 222. [CrossRef]
2. Gil, C.B.; Lee, J.H. Deep Reinforcement Learning Approach for Material Scheduling Considering High-Dimensional Environment of Hybrid Flow-Shop Problem. *Appl. Sci.* **2022**, *12*, 9332. [CrossRef]
3. Cai, J.C.; Lei, D.M.; Wang, J.; Wang, L. A novel shuffled frog-leaping algorithm with reinforcement learning for distributed assembly hybrid flow shop scheduling. *Int. J. Prod. Res.* **2022**, *61*, 1233–1251. [CrossRef]
4. Wang, J.J.; Wang, L. A Cooperative Memetic Algorithm With Learning-Based Agent for Energy-Aware Distributed Hybrid Flow-Shop Scheduling. *IEEE Trans. Evol. Comput.* **2022**, *26*, 461–475. [CrossRef]
5. Lang, S.; Reggelin, T.; Schmidt, J.; Muller, M.; Nahhas, A. NeuroEvolution of augmenting topologies for solving a two-stage hybrid flow shop scheduling problem: A comparison of different solution strategies. *Expert Syst. Appl.* **2021**, *172*, 114666. [CrossRef]

engineering proceedings

Proceeding Paper

Intelligent RGV Scheduling Model Considering Advanced Movement under Repeated Single-Cycle Jobs †

Yinghao Meng, Zihao Xu, Faqun Qi * and Zhen Yin

School of Mechanical and Electrical Engineering, Wenzhou University, Wenzhou 325035, China; myh_goose@163.com (Y.M.); 15355919988@163.com (Z.X.); xy15258627399@163.com (Z.Y.)
* Correspondence: qiqizhezhe@126.com
† Presented at the Third International Conference on Advances in Mechanical Engineering 2023 (ICAME-23), Islamabad, Pakistan, 24 August 2023.

Abstract: At present, automated material processing systems have been adopted by numerous high-tech enterprises, which mainly include Rail Guided Vehicles (RGVs) and CNC Machines. However, in the actual operation process, RGVs are generally scheduled according to the preset procedures of the system, and the lack of intelligent models considering the advanced movement does not elevate the operation efficiency of the system. In this paper, we propose an intelligent scheduling model for RGVs with advanced movement, considering all the work of the RGVs and its matching CNCs in the system within the cycle time, and verify the accuracy and superiority of the model by taking an automated material processing system of a factory as an example.

Keywords: RGV; scheduling model; advanced movement; repetitive jobs; work efficiency

Citation: Meng, Y.; Xu, Z.; Qi, F.; Yin, Z. Intelligent RGV Scheduling Model Considering Advanced Movement under Repeated Single-Cycle Jobs. *Eng. Proc.* **2023**, *45*, 33. https://doi.org/10.3390/engproc2023045033

Academic Editors: Mohammad Javed Hyder, Muhammad Mahabat Khan, Muhammad Irfan and Manzar Masud

Published: 12 September 2023

1. Introduction

Automated material processing systems have been widely used in production plants with repetitive cycle work. As presented in Figure 1a, the system can be simplified to a number of CNCs, and RGVs. The CNCs are responsible for processing the raw material while the RGVs, with 2 grippers (as presented in Figure 1b), are responsible for loading the raw material and unloading the processed material. When one of the CNCs completes its work, a completion signal is sent and the RGV moves to the designated position for loading and unloading of the raw materials In actual production activities, the RGV may have to wait due to the different completion times of each CNC process, resulting in low productivity of the system.

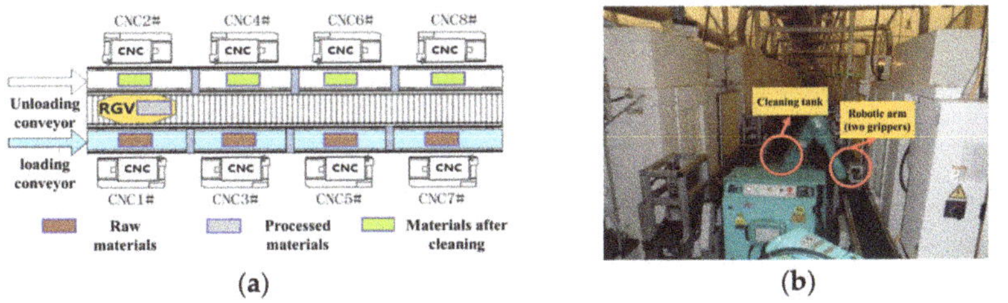

(a) (b)

Figure 1. (**a**) Schematic diagram of an automatic material processing system. (**b**) Structure of RGV.

Therefore, how to schedule the RGV and improve the efficiency of the system is an important issue to promote high-speed development in the manufacturing industry. Liu et al. considered the impact of equipment failures on production in parallel equipment scheduling and developed a mathematical model with the objective of minimizing the maximum completion time and task lead/delay time weighting [1]. Wu et al. proposed a group transport method for dynamic scheduling of RGVs under large material flow in which a genetic algorithm was used, and a coding method was proposed for specific problems [2]. Liu et al. proposed a stochastic fault dynamic scheduling model based on the optimal path. The dynamic scheduling strategy of RGVs and the system operation efficiency are solved by example analysis [3]. Most of these research studies are focused on modeling the scheduling strategy when a fault state occurs in CNCs; however, according to statistics, in the actual production, the probability of a CNC failure at work due to fixed maintenance is less than 1%, and there are fewer research studies on how to design a model to improve the efficiency of RGVs. Therefore, this paper aims to design a model that enables RGVs to move in advance to improve the efficiency of the system.

2. Intelligent RGV Scheduling Model Considering Advanced Movement

2.1. Hypothesis

- Due to regular preventive maintenance, the probability of failure of CNCs and RGVs during the production process is extremely low;
- Only one production process is completed for each material.

2.2. Notations

The notations involved in the model and their definitions are listed in Table 1.

Table 1. Notations and their definitions within the model.

Notations	Definitions	Unit
$x_{ij} \equiv \begin{cases} 1 \\ 0 \end{cases}$	Whether the *ith* CNC is processing the *jth* material; 1 if yes, 0 otherwise.	
$w^{lr} = \begin{cases} 1 \\ 0 \end{cases}$	Whether RGV completes its left and right CNC loading and unloading work; 0 if yes, 1 otherwise.	
P	Work efficiency of the system in a single cycle.	
T_0	Processing time for each raw material.	s
λ	The time for RGV to service the entire CNC.	s
N_0	The maximum number of finished products that can be produced in the time period.	
y_i^{ud}	Loading and unloading time of the *ith* CNC single job.	s
T	Total time of the system working at once.	s
$x_i^w(j)$	Waiting time of the *ith* CNC before processing the *jth* material.	s
t_0	RGV cleaning time for a processed material.	s
t_i	RGV moves *i* units of distance on the track ($i = 1.2.3$).	s
C_T	The number of CNC.	s
y_j^m	The moving time for the RGV when the CNC is processing the *jth* material.	s
y_j^w	The waiting time for the RGV when the CNC is processing the *jth* material.	s
$x_i^u(j)$	The loading start time of the *jth* material produced by the *ith* CNC.	s
$x_i^d(j)$	The unloading start time of the *jth* material produced by the *ith* CNC.	s
τ_1	Time required for one CNC loading and unloading for odd numbered CNCs.	s
τ_2	Time required for one CNC loading and unloading for even numbered CNCs.	s

2.3. Model Description

The RGV scheduling model considering advanced movement is shown below.

$$p = \max \frac{\sum\limits_{i=1}^{C_T} \sum\limits_{j=1}^{N_0} x_{ij}}{N_0}$$

$$
s.t.\begin{cases}
N_0 = \sum_{i=1}^{C_T}\left[\frac{T}{T_0+y_i^{ud}}\right] \\
\sum_{j=1}^{N_0}\left[T_0 + y_i^{ud} + w^{lr}x_i^w(j)\right] \le T \\
\sum_{i=1}^{C_T}\sum_{j=1}^{N_0} x_{ij}\left[y_j^m + y_j^w + y_i^{ud} + t_0\right] \le T \\
\sum_{i=1}^{C_T}\sum_{j=1}^{N_0} x_{ij}y_j^m \ge (n_1t_1 + n_2t_2 + n_3t_3)\frac{N_0}{C_T} \\
\lambda = n_1t_1 + n_2t_2 + n_3t_3 + \frac{C_T}{2}\tau_1 + \frac{C_T}{2}\tau_2 \\
x_i^u(k+1) = x_i^u(k) + \lambda + w^{lr}x_i^w(k) + (C_T-1)t_0\left(k = 1,2,\ldots,\left[\frac{N_0}{T}\right]\right) \\
x_i^d(k) = x_i^u(k+1) \\
x_i^d(j) - x_i^u(j) = x_{ij}\left[T_0 + y_i^{ud} + w^{lr}x_i^w(j)\right]
\end{cases} \quad (1)
$$

The efficiency can be expressed as (Actual quantity of finished products) in the time period dividing by the maximum number of finished products that can be produced, which is $\frac{\sum_{i=1}^{C_T}\sum_{j=1}^{N_0} x_{ij}}{N_0}$, where N_0 is denoted as $\sum_{i=1}^{C_T}\left[\frac{T}{T_0+y_i^{ud}}\right]$. As the working contents of CNCs and RGVs are different, the working time of CNCs includes the manufacturing process, loading, unloading and waiting time, while the working time of RGVs only includes the loading, unloading, moving, waiting time; therefore, the working time of CNCs and RGVs needs to be constrained separately according to the actual system working time. These two constraints are reflected in $\sum_{j=1}^{N_0}\left[T_0 + y_i^{ud} + w^{lr}x_i^w(j)\right] \le T$ and $\sum_{i=1}^{C_r}\sum_{j=1}^{N_0} x_{ij}\left[y_i^m + y_j^w + y_i^{ud} + t_0\right] \le T$. In a single-cycle multi-cycle working time, the time used in each cycle can be found and then the relationship between the unloading time of each CNC and the loading time of the next CNC can be established to dynamically represent the process of change; they are $x_i^d(k) = x_i^u(k+1)$ and $x_i^d(j) - x_i^u(j) = x_{ij}\left[T_0 + y_i^{ud} + w^{lr}x_i^w(j)\right]$. When the RGV completes the existing work, it will evaluate the received signal strength (the default judgement factors change the greater the signal, the greater the distance, the stronger the signal), and it will choose the CNC with the stronger signal for service, which saves the cost of waiting time for the RGV, and reduces the proportion of the system's moving time for the RGV. Using the unique constraints of the above model, the dynamics of the system over a cycle can be modelled using MATLAB.

3. Numerical Examples

In order to test the improved model, three sets of operational parameters of an automated material processing system of a high-tech enterprise were collected in this paper, as presented in Table 2. The existing model without considering advanced movement [4] was compared; the comparison results are presented below.

Table 2. Three sets of data for the operating parameters of the intelligent processing system (Unit: s).

System Operation Parameters	Group 1	Group 2	Group 3
The time for an RGV to move 1 unit (t_1)	20	23	18
The time for an RGV to move 2 units (t_2)	33	41	32
The time for an RGV to move 3 units (t_3)	46	59	46
The time for a CNC to complete a one-step material (T_0)	560	580	545
Time required for one CNC loading and unloading for odd numbered CNCs (T_1)	400	280	455
Time required for one CNC loading and unloading for even numbered CNCs (T_2)	378	500	182
The time for an RGV to complete the cleaning operation of a material (t_0)	28	30	27

Note: The data were obtained from the literature [4].

The parameters are brought in, and the model is solved to obtain the optimal movement route of the RGV in a single cycle: 1,2→3,4→7,8→5,6→1,2. The comparison of the working process and efficiency of the system before and after optimization is shown in Figure 2a,b.

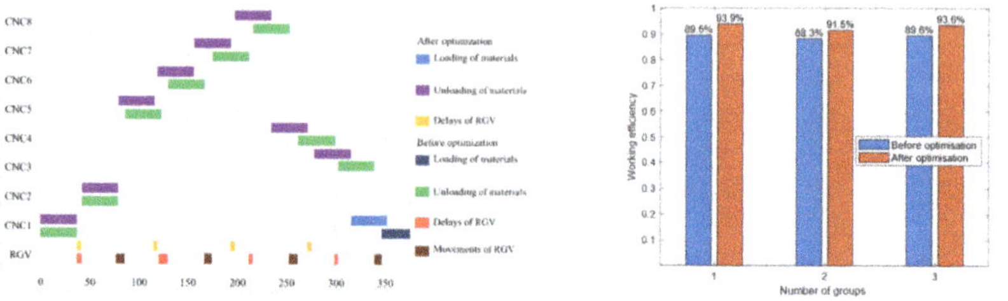

(a) **(b)**

Figure 2. (**a**) Gantt chart of the system working before and after optimization. (**b**) The total efficiency of the system before and after optimization in terms of working time.

As can be clearly seen in Figure 2, the movement time of the RGV becomes less after optimization, leaving only the unoptimizable waiting time (due to the rotation of the mechanical gripper of the RGV), At the same time, the three sets of data collected throughout the working time are brought into the model and the calculated productivity increases are all above 90%. These 5% increases in efficiency are formally due to the reduction in wait times and excess movement time of the RGV.

4. Conclusions

In response to the low production efficiency in the RGV scheduling process, an intelligent RGV scheduling model considering advanced movement is designed. The feasibility of the model is verified based on data provided in an actual factory production, and it is found that the production cycle time is shortened, and efficiency is improved.

Author Contributions: Conceptualization, F.Q.; methodology, Y.M.; software, Y.M.; validation, Y.M.; investigation, Z.Y.; resources, Y.M.; writing, Y.M. and F.Q.; visualization, Z.X.; supervision, F.Q., project administration, F.Q.; funding acquisition: Y.M. All authors have read and agreed to the published version of the manuscript.

Funding: This research received no external funding.

Institutional Review Board Statement: Not applicable.

Informed Consent Statement: Written informed consent has been obtained from the patient(s) to publish this paper.

Data Availability Statement: Not applicable.

Conflicts of Interest: The authors declare no conflict of interest.

References

1. Liu, G.B.; Zhang, J. A dynamic scheduling method based on improved rolling time-domain optimization strategy. *J. Mech. Eng.* **2013**, *49*, 182–190. [CrossRef]
2. Wu, Y.; Liu, Y.; Zhang, D. Research on dynamic scheduling of RGV based on genetic algorithm. *Lift. Transp. Mach.* **2012**, *6*, 20–23.

3. Liu, F.; Xing, K.; Zhang, Y. Dynamic scheduling strategy for random fault intelligent RGV based on optimal path. *Electron. Technol. Softw. Eng.* **2018**, *23*, 160.
4. Han, C.G.; Mei, Z.Y. Mathematical modelling of dynamic scheduling strategy problem for intelligent RGVs. *Math. Model. Its Appl.* **2019**, *8*, 53–65+83.

engineering proceedings

Proceeding Paper

Differential Evolution Algorithm to Solve the Parallel Batch Processing Machine Scheduling Problem with Multiple Jobs †

Xue Zhao, Yarong Chen *, Mudassar Rauf * and Chen Wang

School of Mechanical and Electrical Engineering, Wenzhou University, Wenzhou 325035, China; zx2710750937@163.com (X.Z.); wangchen32130@163.com (C.W.)
* Correspondence: yarongchen@126.com (Y.C.); mudassarrauf16@gmail.com (M.R.)
† Presented at the Third International Conference on Advances in Mechanical Engineering 2023 (ICAME-23), Islamabad, Pakistan, 24 August 2023.

Abstract: We conducted this study with the aim of resolving the scheduling problem of parallel batch processing machines (PBPM) with different capacity constraints and different energy consumption per unit of time, as well as jobs with different processing times, arrival times, delivery dates and sizes, with the goal of simultaneously minimizing the maximum completion time, ET and total energy consumption. The IUDRLM rule is used to batch and sort jobs, and a decomposition-based multi-objective differential evolution algorithm MODE/D is proposed. Simulation experiments are performed to compare the performance of the proposed algorithm to those of existing algorithms. The proposed MODE/D algorithm outperformed NSGA-III in terms of NR value (0.96) and IGD (6.6) measures.

Keywords: parallel batch processor; multi-objective scheduling; DE algorithm; energy consumption

Citation: Zhao, X.; Chen, Y.; Rauf, M.; Wang, C. Differential Evolution Algorithm to Solve the Parallel Batch Processing Machine Scheduling Problem with Multiple Jobs. *Eng. Proc.* **2023**, *45*, 22. https://doi.org/10.3390/engproc2023045022

Academic Editors: Mohammad Javed Hyder, Muhammad Mahabat Khan, Muhammad Irfan and Manzar Masud

Published: 11 September 2023

1. Introduction

The parallel batch processing machine scheduling problem (PBPMSP) is essential to production scheduling problems. Therefore, studying the PBPMSP is of great significance in the context of intelligent manufacturing.

Extensive literature can be found on PBPMSP. However, the literature on PBPMSP, such as its different capacity constraints and energy consumption, is relatively scarce. Majumder et al. [1] minimized the maximum completion time PBPMSP and designed a cuckoo algorithm based on the discrete Levy Flights strategy to improve local search ability and enhance solution diversity. Wang et al. [2] considered the deterioration effect constraint of the machine and proposed an improved algorithm based on the drosophila algorithm to solve the problem. Zhou et al. [3] studied the scheduling problem affecting parallel batch processors using dynamic arrival and usage time pricing schemes. They designed a multi-objective DE algorithm to solve large-scale problems. Li et al. [4] generated individuals through heuristic rules and adopted an angle-based environmental selection strategy to select individuals. They designed an improved algorithm based on a genetic algorithm to minimize both delay and total pollution emission costs.

It can be surmised from the above literature that research on PBPMSPs mainly focuses on minimizing the completion time, total process time and other single objectives under the same capacity constraints of batch processors. Therefore, this article studies the multi-objective PBPMSP by placing different capacity constraints on batch processors and varying energy consumption per unit of time, and it proposes a multi-objective differential evolution (MODE/D) algorithm.

2. Materials and Methodology

2.1. Problem Description

The problem studied in this paper is described as follows: n jobs needed to be processed on m PBPM. The job had the different processing time p_j, arrival time r_j, job size s_j and delivery time d_j. The capacity constraints Q_i of PBPM were different, and the energy consumption per unit of time varied due to machine updates. At the same time, three optimization objectives involved in minimizing total lead time/lag time ET, maximum completion time C_{max} and total energy consumption TEC were considered.

The scheduling problem was denoted as $P_m \left| Q_i, p_j, r_j, d_j, s_j \right| C_{max}, ET, TEC$; the main decisions included determining the batch mode of the job assigned to the machine and the processing sequence of the batches. The basic assumptions to be met were as follows: (1) The sum of all the job sizes in the batch B_b could not exceed the capacity constraint of the batch machine. (2) The batch processor could not be interrupted while processing. (3) The batch arrival time was determined based on the latest arrival time of the job in the batch; the batch processing time was equal to the maximum processing time of the job in the batch.

2.2. Multi-Objective Differential Evolution Algorithm Based on Decomposition

The DE algorithm is a simple, rapid and efficient global optimization algorithm that has been successfully applied in communication and scheduling. This article combined the characteristics of the PBPM multi-objective scheduling problem, decomposed multi-objective problems into single-objective subproblems and introduced the DE algorithm to design the MODE/D algorithm.

2.2.1. Initial Solution Generation

In this paper, the scheduling solution of the 2 × n array representation problem was generated via double-layer coding. The first layer represented the random number generated via each job gene in the interval [−1, 1], which constituted the job-ordering layer. The second layer represented the machines assigned to the job and randomly generated an integer value in the interval [1, m] for each job, with m representing the number of machines.

2.2.2. Decoding and Target Value Calculation

Zhou et al. [5] designed three heuristic batch rules of first row start (FRS), minimum distance start (MDS) and updated distance (UD). In this paper, UD batch rules were selected to enable improvement according to the characteristics of PBPMSP, and the improved UD-right-light moving (IUDRLM) decoding rules were designed.

Direct use of the UD rules could not optimize target values other than C_{max}. For PBPM multi-objective scheduling problems considering artifacts, if two jobs p_j, r_j and d_j were very close, and the sum of s_j of the two jobs was less than or equal to Q_i, the two jobs were assigned to the same batch. The IUD rule distance designed in this paper was calculated via Formula (1).

$$d(j, w) = \begin{cases} \sqrt{\alpha\left(p_j - p_w\right)^2 + \beta\left(r_j - r_w\right)^2 + \gamma\left(d_j - d_w\right)^2} & s_j + s_w \le Q_i \text{ and } j < w \\ \infty & s_j + s_w \le Q_i \text{ or } \quad j \ge w \end{cases} \quad (1)$$

We used the IUDRLM rule to sort all jobs in batches, before using randomly generated 0–1 variables to sequentially decide whether to perform left and right movement operations on each processing batch/block and, finally, obtain the scheduling solution.

2.2.3. Differential Mutation

The job-ordering layer used the formula $V_i^t = X_a^t + F \times (X_b^t - X_c^t)$ to perform differential mutation operations, where F was the shrinkage factor, and the value of F could be determined through pre-experiments. The job distribution machine layer adopted the method of two-point variation and carried out the variation operation on the job distribution machine layer of individual X_a^t, randomly generated two-position numbers, and it exchanged the machine information of the corresponding gene position of the job distribution machine layer to obtain the job distribution machine layer after the mutation. Assuming that F = 0.5, the mutation operation shown in Figure 1 was used.

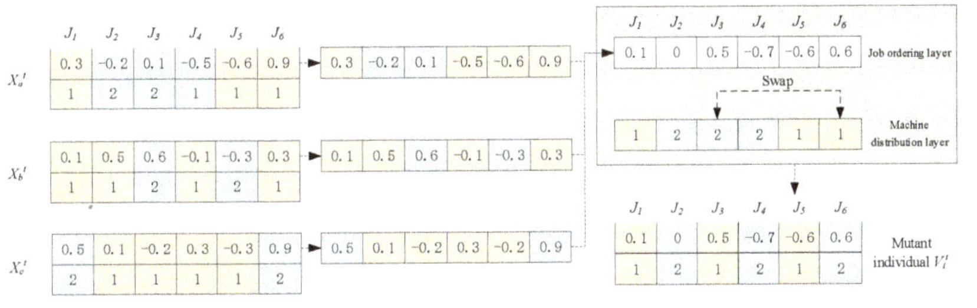

Figure 1. Differential mutation operation.

2.2.4. Crossover

In the interval [0, 1], n random numbers were randomly generated to form a 1 × n-dimensional array *rand*. When the value contained in each position in the *rand* array was less than or equal to the cross probability CR, the gene at the position of the test individual U_i^t directly inherited the corresponding position gene of the mutant individual V_i^t. Otherwise, the gene at the position of the test individual U_i^t directly inherited the gene at the corresponding position of the target individual X_i^t.

2.2.5. Multi-Objective Solution Set Update Based on Decomposition

After the difference variation and cross-operation, the test individual U_i^t was obtained, and the three target values of the test individual U_i^t were calculated. The aggregation function decomposition method was used to update the neighborhood and accelerate the convergence rate of the population.

2.3. *Data Generation*

The experimental problem associated with data generation is shown in Table 1.

Table 1. The scale and parameter range of the experimental problem.

Instance	n	m	p_j	r_j	d_j	s_j
Small	10, 20	2			$d_j = p_j + U[d_{min}, d_{min} + \rho \times P]/2$,	
Medium	40, 60	2	U[1, 100]	U[0, 50]	$d_{min} = P \times (\tau - \rho/2)$,	U[1, 10]
Large	80, 100	3			$P = \sum_{j=1}^{n} p_j, \tau = 0.5, \rho = 0.05$	

Six sets of experiments were generated for different artifacts and machine combinations, with each set generating 10 test data for use in experiments, resulting in a total of 60 test data.

3. Analysis of Results

The NSGA-III algorithm was selected as the comparison algorithm, and the algorithm's performance was evaluated using the Nondomination rate (NR) index and the inverted generation distance (IGD) index. The larger the NR value, the smaller the IGD value, as well as the better the comprehensive performance of the algorithm. This study's experimental results are shown in Table 2.

Table 2. NR and IGD values of MODE/D and NSGA-III algorithms.

Instance	n	m	NR		IGD	
			MODE/D	NSGA-III	MODE/D	NSGA-III
Small	10	2	**0.82**	0.47	**6.3**	19.2
	20	2	**1.00**	0.00	**0.0**	161.7
Medium	40	2	**1.00**	0.00	**0.0**	671.5
	60	2	**1.00**	0.00	**0.0**	1547.7
Large	80	3	**1.00**	0.00	**0.0**	978.1
	100	3	**0.96**	0.04	**33.3**	1026.2

Table 2 shows that for six instances of three scales, the NR values of the MODE/D algorithm are greater than those of the NSGA-III algorithm, while the IGD values are all smaller than those of the NSGA-III algorithm. This result indicates that the performance of the MODE/D algorithm is superior to that of the NSGA-III algorithm. As the MODE/D algorithm can decompose the multi-objective optimization problem into a single objective subproblem, it can achieve a better distribution and convergence effect on the Pareto frontier.

4. Conclusions

In this paper, the PBPM multi-objective optimization problem was studied, which simultaneously minimizes the three objectives of C_{max}, ET and TEC. The MODE/D algorithm was designed by decomposing multi-objective subproblems into single-objective subproblems and introducing the DE algorithm. The comprehensive performance of the algorithm was evaluated through simulation experiments, and the final results showed that the designed MODE/D algorithm was better than the comparison algorithm NSGA-III. In future research, we will consider the impact of dynamic events and other dynamic scheduling methods to better meet the actual production needs of enterprises.

Author Contributions: Conceptualization, X.Z. and Y.C.; methodology, X.Z. and M.R.; software, X.Z. and C.W.; validation, X.Z. and C.W.; formal analysis, X.Z. and Y.C.; writing—review and editing, Y.C. and M.R.; visualization, M.R.; supervision, X.Z.; funding acquisition, Y.C. All authors have read and agreed to the published version of the manuscript.

Funding: The National Natural Science Foundation of China, grant number [No.51705370].

Institutional Review Board Statement: Not applicable.

Informed Consent Statement: Not applicable.

Data Availability Statement: The data are available from the corresponding author upon request.

Conflicts of Interest: The authors declare no conflict of interest.

References

1. Majumder, A.; Laha, D.; Suganthan, P.N. A hybrid cuckoo search algorithm in parallel batch processing machines with unequal job ready times. *Comput. Ind. Eng.* **2018**, *124*, 65–76. [CrossRef]
2. Wang, R.; Jia, Z.; Li, K. Scheduling parallel-batching processing machines problem with learning and deterioration effect in fuzzy environment. *J. Intell. Fuzzy Syst.* **2021**, *40*, 12111–12124. [CrossRef]
3. Zhou, S.; Li, X.; Du, N.; Pang, Y.; Chen, H. A multi-objective differential evolution algorithm for parallel batch processing machine scheduling considering electricity consumption cost. *Comput. Oper. Res.* **2018**, *96*, 55–68. [CrossRef]
4. Li, K.; Zhang, H.; Chu, C.; Jia, Z.; Chen, J. A bi-objective evolutionary algorithm scheduled on uniform parallel batch processing machines. *Expert Syst. Appl.* **2022**, *204*, 117487. [CrossRef]
5. Zhou, S.; Chen, H.; Xu, R.; Li, X. Minimising Makespan on a single batch processing machine with dynamic job arrivals and non-identical job sizes. *Int. J. Prod. Res.* **2014**, *52*, 2258–2274. [CrossRef]

engineering proceedings

MDPI

Proceeding Paper

A Study of Mixed-Flow Human–Machine Collaborative Disassembly Line Balancing Problem Based on Improved Artificial Fish Swarm Algorithm †

Gaofei Wang, Yarong Chen *, Jabir Mumtaz and Lixia Zhu *

School of Mechanical and Electrical Engineering, Wenzhou University, Wenzhou 325035, China; highfly1998@163.com (G.W.); jabirmumtaz@live.com (J.M.)

* Correspondence: yarongchen@126.com (Y.C.); m18200289139@163.com (L.Z.)

† Presented at the Third International Conference on Advances in Mechanical Engineering 2023 (ICAME-23), Islamabad, Pakistan, 24 August 2023.

Abstract: A mixed-flow human–machine collaborative disassembly line balancing problem is introduced, considering the various recycling methods for waste products and the relationship between the attributes of each product part and the corresponding disassembly operator. The problem aims to optimize the number of workstations, balance the idle time, and minimize the disassembly cost. To address this, an Improved Artificial Fish Swarming Algorithm (IAFSA) was designed based on the combination of the problem characteristics, and the IAFSA algorithm was applied to a mixed-flow television (TV) disassembly example and compared with two different algorithms. The solution shows that the proposed algorithm optimizes the proposed algorithm by 14.3%, 52.3%, and 9.8%, respectively, on the three objectives. Finally, the performance of the three algorithms is compared using Non-dominant rate (NR) and Generation distance (GD) metrics.

Keywords: mixed-flow disassembly; human–machine collaboration; disassembly line balancing; IAFSA algorithm

Citation: Wang, G.; Chen, Y.; Mumtaz, J.; Zhu, L. A Study of Mixed-Flow Human–Machine Collaborative Disassembly Line Balancing Problem Based on Improved Artificial Fish Swarm Algorithm. *Eng. Proc.* **2023**, *45*, 40. https://doi.org/10.3390/engproc2023045040

Academic Editors: Mohammad Javed Hyder, Muhammad Mahabat Khan, Muhammad Irfan and Manzar Masud

Published: 14 September 2023

1. Introduction

Disassembly is the key to the recycling of waste products, and efficient disassembly methods are of great significance to achieving sustainable economic development. With the advancement of science and technology, the human–machine collaborative disassembly method gradually replaces the traditional manual disassembly method, which has greater production potential.

In a related literature study, Ci et al. [1] demonstrated that robots have great potential to be applied to disassembly lines by proposing a heuristic algorithm based on ant colony optimization to solve the single-operator robot disassembly line balancing problem. Yin et al. [2] proposed an incomplete disassembly line balancing problem for multi-product multi-robot disassembly. Huang [3] proposed a new method of human–robot collaboration for disassembling stamped parts based on the active flexibility of collaborative robots, and the feasibility of the method was proven through practical cases. Liu et al. [4] studied the task classification and task assignment of human–robot disassembly, and considered the safety strategy between the operator and the robot to ensure the safety of human–robot disassembly in a workstation. Xu et al. [5] studied the human–machine collaborative disassembly line balancing problem, considering the safety of workers' operations, constructed a multi-objective mathematical model for this problem, and solved this multi-objective optimization problem using an improved discrete bee colony algorithm.

The above-mentioned literature mainly focuses on improving the disassembly efficiency and safety of disassembly workers, with less consideration given to human–machine collaborative disassembly of hybrid products and the assignment of disassembly operators based on the characteristic properties of parts. Thus, this paper proposes a mixed-flow human–machine collaborative disassembly line balancing problem (MHMC_DLBP) with the minimization of the number of workstations, the idle time balance index and the disassembly costs as the optimization objectives, and designs an improved artificial fish swarm algorithm to solve it.

2. Problem Description

The mixed-flow human–machine collaborative disassembly line balancing problem studied in this paper involves the construction of a mixed-product disassembly information model as well as the categorization and assignment of human/machine task operators with different attributes of parts, in addition to the need to satisfy general DLBP constraints. The disassembled parts are classified into three categories according to their characteristic properties: hazardous parts, complex parts and common parts, where hazardous parts correspond to machines, complex parts correspond to humans and common parts are without restrictions. Respectively, where the machine and human disassembly times are different for the same disassembly task, only one type of operator can be assigned to each workstation.

3. IAFSA Algorithm

This paper proposes an improved artificial fish swarming algorithm for the studied problem, which includes foraging behavior, clustering behavior and tailgating behavior.

3.1. Foraging Behavior

To guide the artificial fish to converge in the optimal direction, the crossover operator of the genetic algorithm is introduced, and the crossover operation is performed on it to guide the artificial fish to forage. First, generate 2 random crossover points, P_1 and P_2, on X_1, and the sequence before and after the crossover point satisfies the disassembly priority relationship. Then, a reference fish, X_2, is randomly generated, and the sequence between P_1 and P_2 is obtained by mapping X_2 with the remaining sequence of X_1 to form a new artificial fish,X_{new}.

3.2. Clustering Behavior

With artificial fish, $X_1 = (a_1, a_2, \ldots a_n)$,$X_2 = (b_1, b_2, \ldots b_n)$, the distance between two artificial fish is defined by combining the DLBP problem features as

$$D(X_1, X_2) = \sum_{i}^{n} \operatorname{sgn}|X_1 - X_2| \quad (1)$$

where sgn is a 0–1 variable indicating the similarity and difference between vectors a_i and b_i in X_1 and X_2.

If the artificial fish horizon is V, the crowding degree is $delta$ and the population size is $fish\_num$. When $D(X_1, X_2) \leq V$, it is determined that X_2 is an artificial fish searched by X_1 in its field of view. Find all artificial fish $fish_{all}$ in the fields of view, judge whether the conditions are satisfied by the crowding degree calculation formula $delta = fish_{all}/fish\_num$, choose NSGA-II crowding distance mechanism to sort the artificial fish that satisfy the conditions and select the one with the largest crowding degree as the central fish X_{center} obtained from the clustering behavior.

3.3. Tailgating Behavior

Explore the number of partners N_f within X_1 field of view V, determine whether the tail-chasing condition is satisfied and then place the current artificial fish in the tail-chasing behavior bulletin board Q_t for non-dominated sorting to filter out the non-inferior solutions in the bulletin board. The tail-chasing behavior accelerates the movement of the artificial fish towards a more optimal state.

4. Example and Analysis

The proposed algorithm is applied to the human–machine collaborative disassembly example of a hybrid TV product to verify the solution performance. The algorithm parameters are $CT = 100$ s, $fish\_num = 50$, $Gen = 500$, $V = 30$, $try\_number = 10$, $delta = 0.8$. The Genetic Simulated Annealing Algorithm (GASA) [6], and Teaching Optimization (HTLBO) algorithm [7] were introduced for comparison, and the parameters of the comparison algorithms can be referred to in the literature.

From Figure 1, it can be seen that for F_1, the value obtained via the IAFSA algorithm is 6, the GASA algorithm is 7 and the HTLBO algorthim is 8; for objective F_2, the three algorithms converge at the 219th, the 385th and the 485th generations, respectively; for F_3, the IAFSA algorithm converges to 78.69 at the 250th generation and the HTLBO converges to 99.08 at the 390th generation; the GASA algorithm converges to 87.19 at the 489th generation. Thus, the convergence speed and the convergence results of each optimization objective for the proposed IAFSA algorithm are all better than the other two comparison algorithms. Figure 2 shows the Gantt chart of the disassembly task sequence for a solution in the Pareto solution set solved by the IAFSA algorithm.

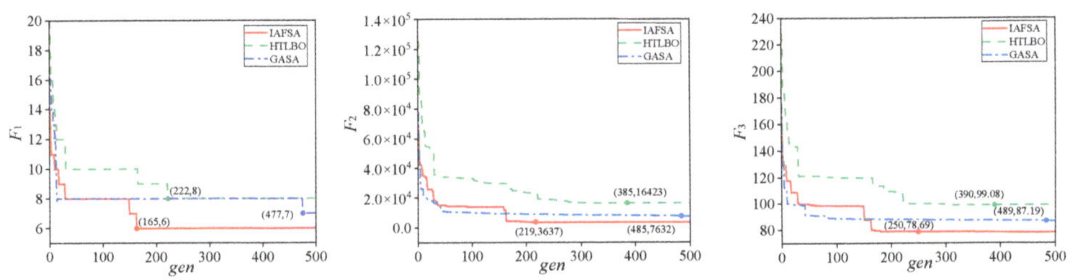

Figure 1. Single-objective convergence curves of the three algorithms at 500 iterations.

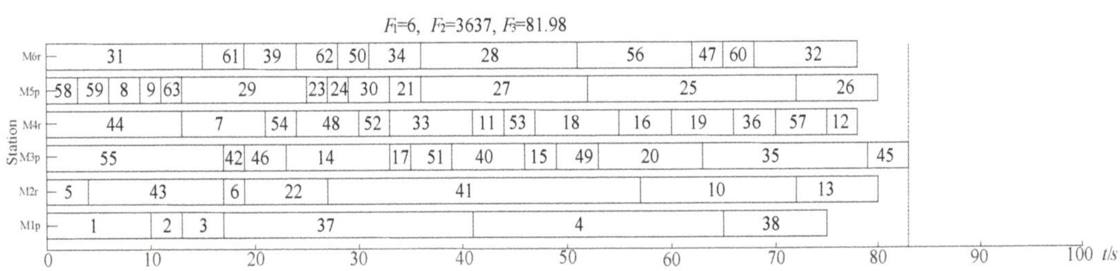

Figure 2. Gantt chart of disassembly task sequence assignment for hybrid TV products.

The three algorithms were executed five times for an equal duration to evaluate the effectiveness of the algorithms and assess the convergence and distribution of the Pareto solutions. This allowed us to determine the non-dominance rate (NR) and generation distance (GD) metrics. As given in Table 1, in the five runs, the NR metrics of the IAFSA were larger than those of the HTLBO and GASA algorithms, and the GD metrics were smaller than those of the other two comparison algorithms. In summary, the NR and GD metrics verify that the IAFSA algorithm is superior in terms of solution diversity and convergence.

Table 1. Multi-objective evaluation values via three algorithms for the same time.

Alg	IAFSA		HTLBO		GASA			IAFSA		HTLBO		GASA	
	NR	GD	NR	GD	NR	GD	3	1	0	0	1010.06	0	230.11
1	1	0	0	1547.03	0	770.6	4	0.59	0	0	682.60	0.41	35.16
2	0.53	19.24	0.47	34.51	0	229.26	5	1	0	0	2187.57	0	2035.77

5. Conclusions

A mixed-flow human–machine collaborative disassembly line balancing problem was studied in this paper, and an improved IAFSA algorithm was designed according to the problem characteristics, in order to solve the disassembly instances of two kinds of TVs mixing, and the effectiveness and the superiority of the IAFSA algorithm was verified by comparing it with the GASA and HTLBO algorithms from a convergence performance perspective as well as by the NR and GD two metrics. Subsequently, research can also be carried out on the layout form of the human–machine disassembly line, and multi-operator joint disassembly.

Author Contributions: Conceptualization, G.W. and Y.C.; methodology, J.M. and L.Z.; software, G.W. and J.M.; validation, G.W. and J.M.; formal analysis, G.W. and Y.C.; writing—review and editing, Y.C. and L.Z.; visualization, L.Z.; supervision, G.W.; funding acquisition, Y.C. All authors have read and agreed to the published version of the manuscript.

Funding: This research was funded by National Natural Science Foundation of China, grant number [No. 51705370].

Institutional Review Board Statement: Not applicable.

Informed Consent Statement: Not applicable.

Data Availability Statement: The data that support the findings of this study are available from the corresponding author upon reasonable request.

Conflicts of Interest: The authors declare no conflict of interest.

References

1. Çil, Z.A.; Mete, S.; Serin, F. Robotic Disassembly Line Balancing Problem: A Mathematical Model and Ant Colony Optimization Approach. *Appl. Math. Model.* **2020**, *86*, 335–348. [CrossRef]
2. Yin, T.; Zhang, Z.; Zhang, Y.; Wu, T.; Liang, W. Mixed-integer programming model and hybrid driving algorithm for multi-product partial disassembly line balancing problem with multi-robot workstations. *Robot. Comput-Integr. Manuf.* **2022**, *73*, 102251. [CrossRef]
3. Huang, J.; Pham, D.T.; Wang, Y.; Qu, M.; Ji, C.; Su, S.; Xu, W.; Liu, Q.; Zhou, Z. A case study in human–robot collaboration in the disassembly of press-fitted components. *Proc. Inst. Mech. Eng. Part B-J. Eng. Manuf.* **2019**, *234*, 654–664. [CrossRef]
4. Liu, B.; Xu, W.; Liu, J. Human-Robot collaboration for disassembly line balancing problem in remanufacturing. In Proceedings of the ASME 2019 14th International Manufacturing Science and Engineering Conference, Erie, PA, USA, 10–14 June 2019.
5. Xu, W.; Cui, J.; Liu, B.; Liu, J.; Yao, B.; Zhou, Z. Human-robot collaborative disassembly line balancing considering the safe strategy in remanufacturing. *J. Clean Prod.* **2021**, *324*, 129158. [CrossRef]
6. Wang, K.; Zhang, Z.; Zhu, L.; Zhou, S. Pareto genetic simulated annealing algorithm for multi-objective disassembly line balancing problem. *Robot. Comput-Integr. Manuf.* **2017**, *23*, 9.
7. Xu, P.; Zhang, Z.; Guan, C. Modeling of human-machine co-located parallel disassembly line balance problem and hybrid teaching optimization algorithm. *Robot. Comput-Integr. Manuf.* **2023**, *29*, 2175–2190. (In Chinese)

engineering proceedings

Proceeding Paper

Ant Colony Optimization Algorithm for the Hybrid Flow Shop Scheduling Problem with Integrated Consideration of Fixture Resources †

Longlong Xu, Jian Guo * and Lixia Zhu

School of Mechanical and Electrical Engineering, Wenzhou University, Wenzhou 325035, China; 13395771275@163.com (L.X.); m18200289139@163.com (L.Z.)

* Correspondence: 20210701@wzu.edu.cn

† Presented at the Third International Conference on Advances in Mechanical Engineering 2023 (ICAME-23), Islamabad, Pakistan, 24 August 2023.

Abstract: A solution method based on the ant colony optimization (*ACO*) algorithm is established for the hybrid flow shop scheduling problem with the objective of minimizing the maximum completion time and the number of fixtures. By using the linear weighting method, the dual-objective optimization problem can be effectively converted into a single-objective optimization. Taking the scheduling problem of an automated production line of precision parts as an example, the feasibility of the proposed algorithm is verified, and the integrated optimization of fixture resource allocation and production scheduling is realized.

Keywords: automatic production line; matlab; ant colony algorithm; fixture resources

Citation: Xu, L.; Guo, J.; Zhu, L. Ant Colony Optimization Algorithm for the Hybrid Flow Shop Scheduling Problem with Integrated Consideration of Fixture Resources. *Eng. Proc.* **2023**, *45*, 37. https://doi.org/10.3390/engproc2023045037

Academic Editors: Mohammad Javed Hyder, Muhammad Mahabat Khan, Muhammad Irfan and Manzar Masud

Published: 13 September 2023

1. Introduction

The hybrid flow-shop scheduling problem (*HFSP*) is a classical problem in the field of production scheduling. In automated production lines, the cost of fixture resources is high due to the fact that each fixture system can only hold one job at a time. Therefore, in this paper, an ant colony algorithm is used to simultaneously consider minimizing the maximum completion time and the number of fixtures for hybrid flow shop scheduling.

Some classic studies on ACO are as follows. Qin et al. studied the dynamic hybrid flow shop scheduling problem with uncertain processing time and proposed a rescheduling method based on the ant colony algorithm [1]. Zhao analyzed the objective function and boundary conditions of the emergency distribution routes based on the ant colony algorithm and solved the optimal routes based on the pheromone content of each route [2]. Jiang et al. proposed to guide the optimization approach by improving the pheromone update rule based on the experience of the global optimum [3]. Li et al. improved the pheromone updating method by improving the path selection strategy and path transfer probability function and utilizing a max–min ant colony system [4].

2. Problem Formulation

The problem can be described as n jobs $J_i(i = 1, 2, \ldots, n)$ are to be processed in an automated production line with $m(m \geq 2)$ processes and the same processing sequence for each job. $K_j(K_j \geq 1; j = 1, 2, \cdots, m)$ is the set of parallel machines on each process. Process m can be processed on any K_j. Scheduling decisions are made to determine the sequencing of all jobs for each process and the selection of machines for each process. The goal of scheduling is to minimize the maximum completion time and reduce the number of fixtures.

The basic assumptions include: (1) jobs on automated production lines must be processed using fixtures, each fixture follows a job until the end of processing, and the number of fixtures is limited; (2) only one job can be processed per machine at any one time without interruption; (3) if the machine required for the processing of the job is in operation, the job waits at the caching station; (4) negligible movement of the robot arm as well as disassembly and clamping time of jobs.

3. ACO Algorithm with Consideration of Fixture Resources

3.1. Flow of the ACO Algorithm

The ACO algorithm is a heuristic bionic algorithm based on positive feedback of information [5]. By transforming the scheduling process into the search path of an ant colony, a feasible and effective ant colony algorithm model is constructed to solve the complex automatic production line shop scheduling problem. The ACO algorithm flow is shown in Figure 1. C_{max} represents maximum completion time. *item* represents the current number of iterations. $item_{max}$ represents the maximum number of iterations.

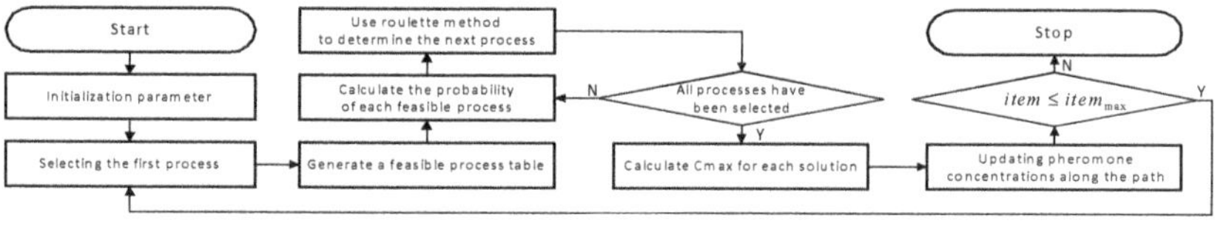

Figure 1. ACO algorithm flow.

3.2. Feasible Process Table for the ACO Algorithm

$N(i)$ represents the serial number of the process that J_i is processing $0 < N(i) < m$ indicates that the job started processing but not finished processing and will occupy a fixture until the last process is completed. The number of jobs that meet $0 < N(i) < m$ is equal to G_{now}. G_{now} represents the number of fixtures currently occupied. G_{max} represents the maximum number of fixtures on a production line.

If $G_{now} < G_{max}$, the table of feasible process includes the first process of a raw job and the tightening process of the job has begun. If $G_{now} = G_{max}$, the table of feasible process includes the tightening process of the job has begun.

3.3. Optional Process Probabilities for the ACO Algorithm

The probability of optional processes is calculated as shown in Equation (1). y is one of the optional processes, u is any of the optional processes, *Allowed* is the set of optional processes. ρ_{xy} represents the probability that an ant's current process is x and the next process is chosen as y, τ_{xy} represents its pheromone concentration, μ_{xy} represents the value of its heuristic function value.

$$\rho_{xy}^{p} = \begin{cases} \dfrac{(\tau_{xy})^{\alpha}\cdot(\mu_{xy})^{\beta}}{\sum\limits_{u\in Allowed}\left[(\tau_{xu})^{\alpha}\cdot(\mu_{xu})^{\beta}\right]} & , y \in Allowed \\ 0 & , y \notin Allowed \end{cases} \quad (1)$$

3.4. Feasible Solution Generation for the ACO Algorithm

Use an array of $2 * w$ to represent a feasible scheduling solution to the problem, w is the total number of processes. The first layer is the total process gene and the second layer is the machine gene. Randomly generate an integer in the range 1 to n as the first process chosen by an ant. After generating the corresponding feasible process table, all optional processes and their selection probabilities are determined. Update the feasible process table and recalculate the selection probabilities for all available processes after determining the next process by means of the roulette method, until all processes have been selected. This is the step by step for the ACO algorithm to generate feasible solutions. A feasible solution for case $n = 3$, $m = 3$ is shown in Figure 2.

	1	2	3	4	5	6	7	8	9
Total process gene	1	4	5	6	7	2	8	3	9
Machine gene	1	1	2	4	1	2	3	4	4

Figure 2. Feasible solution example.

3.5. Pheromone Update for the ACO Algorithm

Two adjacent processes in the feasible solution are considered as a process path. The increment of pheromones on the path is inversely proportional to the maximum completion time, as shown in Equation (2). Q represents the pheromone constant factor. C^p_{max} represents the maximum completion time of ant p. $\Delta\tau^p$ represents the pheromone increment of each ant on its own path through the process.

Renewal of pheromones includes addition and volatilization as shown in Equation (3). ρ represents pheromone volatile factor. $\sum_{p=1}^{q}\Delta\tau^p_{xy}$ represents the pheromone increment of all ants on the process path x to y.

$$\Delta\tau^p = Q/C^p_{max} \tag{2}$$

$$\tau^{new}_{xy} = (1-\rho)\tau^{old}_{xy} + \sum_{p=1}^{q}\Delta\tau^p_{xy} \tag{3}$$

4. Numerical Example and Analysis

4.1. Parameter Settings

Based on the automated production line for precision parts, the experimental parameter obtained through Taguchi's experiments are shown in Table 1.

Table 1. Experimental parameter.

q	α	β	ρ	Q	$item_{max}$
50	1	5	0.1	1	200

4.2. Computational Experiments and Discussion

Normalization is performed by Equation (4) to determine the fitness of each solution. f_r^{min} represents the minimum value of the target r. f_r^{max} represents the minimum value of the target r.

$$f_r = \frac{f_r - f_r^{min}}{f_r^{max} - f_r^{min}} \tag{4}$$

A linear weighting method is used to convert the optimization objective into a single-objective problem, as shown in Equation (5). Weighting factors ω_i determined by multiple qualified experts with extensive experience and expertise. The weights of C_{max} and the weights of fixtures number are set to $\omega_1 = 0.7$ and $\omega_2 = 0.3$.

$$minf(x) = \sum_{i=1}^{2}\omega_i minf_i(x) \tag{5}$$

According to Table 2, $minf(x)$ achieves the minimum value when the G_{max} = 3, and the corresponding production line scheduling Gantt chart is shown in Figure 3. Compared with $G_{max} = 10$, the number of fixtures is reduced by 7, and C_{max} is increased by 53. Compared to heuristic rules, the ACO algorithm gives better results.

Table 2. Correspondence table between G_{max} and C_{max}.

G_{max}	1	2	3	4	5	6	7	8	9	10
f_1	0	0.11	0.22	0.34	0.44	0.56	0.67	0.78	0.89	1
C_{max}	1071	616	466	441	432	431	431	421	413	413
f_2	1	0.309	0.081	0.043	0.029	0.027	0.027	0.013	0	0
$minf(x)$	0.7	0.2493	0.1227	0.1321	0.1523	0.1869	0.2199	0.2431	0.267	0.3

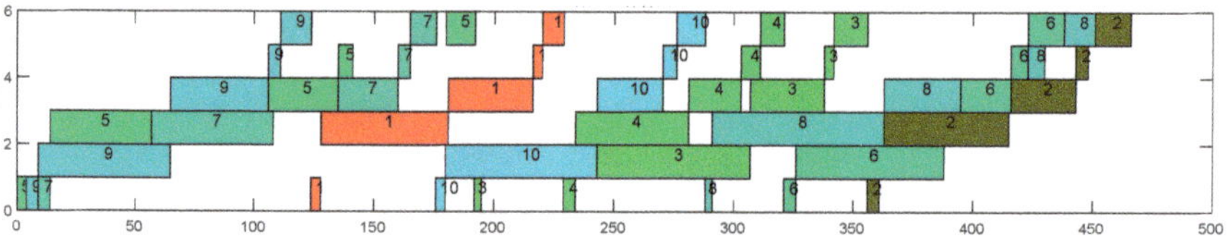

Figure 3. Gantt chart of automated production line scheduling for $G_{max} = 3$.

5. Conclusions

In this paper, the ant colony algorithm is applied to the hybrid flow-shop scheduling problem with integrated consideration of fixture resources, and the experimental results show that it is feasible. More complex cases for HFSP are to be further studied and considered.

Author Contributions: Conceptualization, J.G.; methodology, J.G., L.X., and L.Z.; software, L.X.; validation, L.X.; formal analysis, L.Z.; investigation, L.X.; writing—original draft preparation, L.X.; writing—review and editing, J.G.; funding acquisition, J.G. All authors have read and agreed to the published version of the manuscript.

Funding: The National Natural Science Foundation of China, grant number [No. 51705370].

Institutional Review Board Statement: Not applicable.

Informed Consent Statement: Not applicable.

Data Availability Statement: The data are available from the corresponding author upon request.

Conflicts of Interest: The authors declare no conflict of interest.

References

1. Qin, W.; Zhang, J.; Song, D. An improved ant colony algorithm for dynamic hybrid flow shop scheduling with uncertain processing time. *J. Intell. Manuf.* **2018**, *29*, 891–904. [CrossRef]
2. Zhao, L. Research on Emergency Distribution Route Planning Based on the Ant Colony Algorithm. In Proceedings of the ISMHI 2019, Paris, France, 21–23 November 2019; Francis Academic Press: London, UK, 2019; p. 5.
3. Jiang, X.; Tao, Q. Multi-objective flexible job shop schedule based on ant colony algorithm. In Proceedings of the DCABES 2015, Guiyang, China, 18–24 August 2015; p. 4.
4. Li, X.; Li, Q.; Zhang, J. Research on global path planning of unmanned vehicles based on improved ant colony algorithm in the complex road environment. *Meas. Control* **2022**, *55*, 945–959. [CrossRef]
5. Qin, W.; Zhuang, Z.; Liu, Y.; Tang, O. A two-stage ant colony algorithm for hybrid flow shop scheduling with lot sizing and calendar constraints in printed circuit board assembly. *Comput. Ind. Eng.* **2019**, *138*, 106115. [CrossRef]

engineering proceedings

MDPI

Proceeding Paper

Unrelated Parallel Batch Machine Scheduling Using a Modified ABC Algorithm †

Ke Ke, Yarong Chen, Jabir Mumtaz and Shenquan Huang *

School of Mechanical and Electrical Engineering, Wenzhou University, Wenzhou 325035, China; kekewzu@163.com (K.K.); yarongchen@126.com (Y.C.); jabirmumtaz@live.com (J.M.)

* Correspondence: hshenquan@163.com

† Presented at the Third International Conference on Advances in Mechanical Engineering 2023 (ICAME-23), Islamabad, Pakistan, 24 August 2023.

Abstract: This paper introduces a multi-objective discrete artificial bee colony (MODABC) algorithm, which aims to simultaneously minimize the makespan, total earliness and tardiness (ET), and total energy consumption (TEC) by efficiently scheduling a variety of jobs on unrelated parallel batch machines. Machines have different capacities and consume varying amounts of processing energy, whereas the jobs differ in sizes, release times, and due dates. In the employed bee and follower bee phase, three neighborhood structures are employed to generate feasible solutions, improving the population's quality. In the scout bee phase, three multi-objective local search strategies are used to fully search the solution space. The experimental results show that the MODABC algorithm is superior to the NSGA-III algorithm in terms of convergence and diversity.

Keywords: artificial bee colony algorithm; multi-objective optimization; unrelated parallel batch machine scheduling

Citation: Ke, K.; Chen, Y.; Mumtaz, J.; Huang, S. Unrelated Parallel Batch Machine Scheduling Using a Modified ABC Algorithm. *Eng. Proc.* **2023**, *45*, 19. http://doi.org/10.3390/engproc2023045019

Academic Editors: Mohammad Javed Hyder, Muhammad Mahabat Khan, Muhammad Irfan and Manzar Masud

Published: 11 September 2023

1. Introduction

It is of great practical significance to study the scheduling problem of parallel batch processing machine (PBPM) in intelligent manufacturing, such as chemical processing, pharmaceutical, semiconductor manufacturing, and other industries.

At present, the research on the PBPM scheduling problem is very extensive. For example, Wang and Chou studied the identical parallel batch machine scheduling problem with different sizes and arrival times of jobs and proposed a hybrid genetic algorithm to minimize the maximum completion time [1]. Beldar et al. studied the equivalent parallel batch machine scheduling problem considering machine variable maintenance, job dynamic arrival, and due date and proposed two meta-heuristic methods to minimize the total tardiness time [2]. Zhou et al. proposed a multi-objective differential evolution algorithm to simultaneously minimize the maximum completion time and total power cost, considering the uniform parallel batch machine scheduling problem with the dynamic arrival of jobs [3]. Li et al. studied the uniform parallel batch machine scheduling problem of jobs with different processing times, sizes, and due dates considering dynamic arrivals and proposed a differential evolution algorithm to minimize the maximum delay and total pollution emission cost [4].

In contrast, few studies exist on unrelated parallel batch machine scheduling methods based on artificial bee colony algorithms. In this paper, the MODABC algorithm is proposed for the multi-objective optimization problem of unrelated parallel batch machine scheduling and compared with the classical NSGA-III algorithm.

2. Problem Formulation

There are n independent jobs to be processed on m unrelated parallel batch machines. The processing time of the jobs J_j on the machine M_i is $p_{i,j}$, the release time is r_j, the job size is s_j, and the due date is d_j. The capacity Q_i of different batch machines are different, and the energy consumption l_i per unit time is also different. The optimization objective is to minimize makespan, ET, and TEC simultaneously. The scheduling problem is denoted by $R_m|Q_i,r_j,d_j,s_j|C_{max},ET,TEC$. The main decisions include: 1. How the n jobs $J=\{J_1,J_2,\cdots,J_n\}$ are assigned to m machines $M=\{M_1,M_2,\cdots,M_m\}$; 2. How to group batches of n_i jobs assigned to machine M_i; 3. The processing sequence of batches on the machine M_i.

3. MODABC Algorithm

3.1. Basic Flow of the MODABC Algorithm

ABC is a global optimization algorithm, including three search phases: employed bee, follower bee, and scout bee. The process of MODABC algorithm is shown in Figure 1.

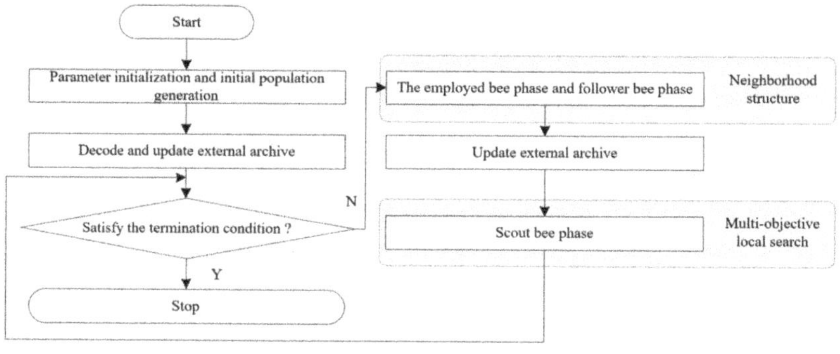

Figure 1. Flow chart of the proposed MODABC.

3.2. The Employed Bee Phase and Follower Bee Phase

To improve the quality of the solution and improve the convergence of the algorithm, in this study, three neighborhood structures are designed based on the characteristics of the problem, which are used to perform a local search on the food sources in the population at the phase of employed bee and follower bee. In the employed bee, each food source performs a local search operation in order. In the following bee phase, the population uses the roulette rule to select the food source according to the HV index value of the food source (the volume of the cuboid whose diagonal is formed by the position of the food source and the reference point) and performs a local search operation in turn. An example of neighborhood structure can be seen in Figure 2, i represents the machine, and b represents the batch on machine i.

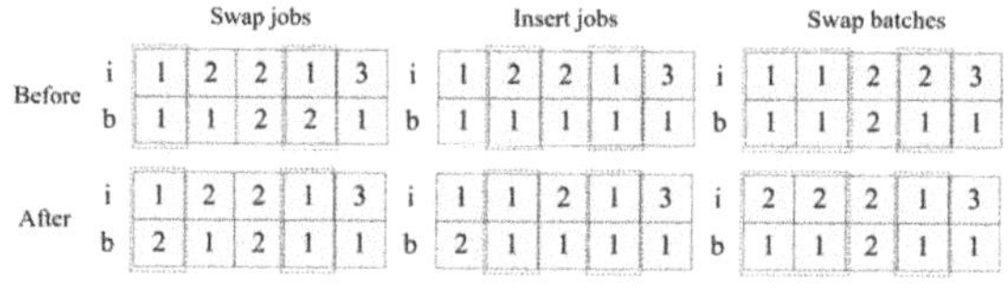

Figure 2. Example of neighborhood structure.

Among them, job swapping is to select any job in any two batches on any two machines for swapping; job insertion is to select any job in any batch to be inserted into another batch; and batch exchange is to select two batches on any two machines for exchange.

3.3. Scout Bee Phase

To enhance the diversity of the algorithm and make the obtained Pareto front as close to the real Pareto front as possible and evenly distributed, we propose three multi-objective local search strategies for different target values at this phase. These strategies aim to perform a local search on non-dominated solutions in external archive sets and food sources, obtain three non-dominated solution sets, and merge them with the current population. Next, according to the non-dominated sorting method, we select individuals to enter the next generation. The following will introduce these three multi-objective local search strategies, respectively.

(1) Minimizing Makespan Local Search: The machines with the largest and smallest makespans are defined as CM_{max} and CM_{min}, respectively. Each batch on CM_{max} is exchanged with all possible batches in CM_{min}.

(2) Total energy consumption local search: The machines with the maximum and minimum processing energy consumption are defined as TM_{max} and TM_{min}, respectively. Each batch on TM_{max} is exchanged with all possible batches in TM_{min}.

(3) Total earliness/tardiness local search: The batch with the largest ET is selected and exchanged with any batch other than this in all machines.

4. Numerical Example and Analysis

4.1. Parameters Setting

Based on the actual enterprise production data, the test data generation method of reference [4] is used to generate the $p_{i,j}$, r_j, d_j, s_j, Q_i, and l_i. A total of 6 sets of test data; the number of small-scale jobs n = 10, 20, the number of machines m = 2; the number of scale jobs n = 40, 80; the number of machines m = 3; the number of large-scale jobs n = 100, 200; and the number of machines m = 5. For different job and machine combinations, 10 test data are generated for each group, and 60 test data are generated for all scales.

4.2. Parameters Tuning

The parameter values that affect the performance of MODABC and NSGA-III algorithms are determined by Taguchi design experiments. The algorithm parameter values under different problem scales are shown in Table 1. N represents the population size, MAXT represents the maximum number of iterations, and p_C and p_m represent the crossover rate and mutation rate (as MODABC is a meta-heuristic algorithm based on neighborhood search, there is no p_C and p_m).

Table 1. Parameter values for algorithms.

Parameter	MODABC			NSGA-III		
	Small	Medium	Large	Small	Medium	Large
N	100	100	100	100	100	100
MAXT	30	50	50	150	150	200
p_C	\	\	\	0.65	0.65	0.7
p_m	\	\	\	0.2	0.15	0.2

4.3. Computational Experiments and Discussion

In this study, each problem instance was ran 10 times, and the average Inverted Generational Distance IGD, non-dominated rate NR, and the number of Pareto solutions Nd obtained by different algorithms were calculated. The results are shown in Table 2.

Table 2. The performance of the algorithms. (n*m indicates the size of the problem).

Size	n*m	MODABC			NSGA-III		
		IGD	NR	Nd	IGD	NR	Nd
Small	n10m2	13.7824	0.9623	6.4	167.2477	0.0433	1.1
	n20m2	47.6821	0.9714	6.1	616.5514	0.0368	1.1
Medium	n40m3	117.3464	0.9733	6.9	2933.09	0.0314	1.1
	n80m3	416.3641	0.9718	4.2	10,980.8	0.0301	1.3
Large	n100m5	425.2905	0.9823	3.7	11,595.26	0.0214	1.6
	n200m5	1122.9323	0.9853	6.8	29,313.3	0.0203	2.1

It can be found from Table 2 that the IGD, NR, and Nd values of the MODABC algorithm are better than other algorithms, and the advantages increase with the increase in scale.

5. Conclusions

The MODABC algorithm is introduced to address the multi-objective optimization problem of unrelated parallel batch machine scheduling. It aims to minimize the objectives of Cmax, TEC, and ET concurrently. The experimental results show that the MODABC algorithm can improve the diversity and convergence of the solution. In the future, it is necessary to study the method proposed in this paper to solve the practical problems considering the constraints of machine failure and workpiece compatibility.

Author Contributions: Conceptualization, Y.C.; methodology, Y.C. and K.K.; software, K.K.; validation, Y.C. and K.K.; formal analysis, J.M.; investigation, S.H. and Y.C.; resources, Y.C.; data—curation, K.K.; writing—original draft preparation, K.K., writing—review and editing, Y.C. and J.M..; visualization, K.K.; supervision, J.M. and Y.C., project administration, Y.C.; funding acquisition: Y.C. All authors have read and agreed to the published version of the manuscript.

Funding: This research was funded by National Natural Science Foundation of China grant number (No. 51705370).

Institutional Review Board Statement: Not applicable.

Informed Consent Statement: Not applicable.

Data Availability Statement: The data that support the findings of this study are available from the corresponding author upon reasonable request.

Conflicts of Interest: The authors declare no conflict of interest.

References

1. Wang, H.; Der Chou, F. Solving the parallel batch-processing machines with different release times, job sizes, and capacity limits by metaheuristics. *Expert Syst. Appl.* **2010**, *37*, 1510–1521. [CrossRef]
2. Beldar, P.; Moghtader, M.; Giret, A.; Ansaripoor, A.H. Non-identical parallel machines batch processing problem with release dates, due dates and variable maintenance activity to minimize total tardiness. *Comput. Ind. Eng.* **2022**, *168*, 108135. [CrossRef]
3. Zhou, S.; Li, X.; Du, N.; Pang, Y.; Chen, H. A multi-objective differential evolution algorithm for parallel batch processing machine scheduling considering electricity consumption cost. *Comput. Oper. Res.* **2018**, *96*, 55–68. [CrossRef]
4. Li, K.; Zhang, H.; Chu, C.; Zhao, H.; Chen, J. A bi-objective evolutionary algorithm scheduled on uniform parallel batch processing machines. *Expert Syst. Appl.* **2022**, *204*, 117487. [CrossRef]

engineering proceedings

MDPI

Proceeding Paper

A Flexible Job Shop Scheduling Method Based on Multi-Fidelity Optimization †

Liuyan Zhong, Yarong Chen and Jabir Mumtaz *

School of Mechanical and Electrical Engineering, Wenzhou University, Wenzhou 325035, China; liuyanzhong2020@163.com (L.Z.); yarongchen@126.com (Y.C.)

* Correspondence: jabirmumtaz@live.com

† Presented at the Third International Conference on Advances in Mechanical Engineering 2023 (ICAME-23), Islamabad, Pakistan, 24 August 2023.

Abstract: Aiming at the impact of machine failure on scheduling schemes in actual production, this paper proposes a multi-fidelity optimization approach considering the preventive maintenance of the machine (MOAPMM). The genetic algorithm (GA) is used as a low-fidelity model to generate a number of feasible solutions; the feasible solutions are sorted, grouped, and selected as high-quality solutions according to the EOCBA method; the high-fidelity model considering machine preventive maintenance (PM) is constructed by using the FlexSim® software; and the high-quality solutions are simulated to obtain the optimal scheduling scheme. Experimental results show that our proposed method outperforms approaches that do not consider machine PM in terms of completion time for the flexible job shop scheduling problem with machine failure.

Keywords: multi-fidelity optimization; flexible job shop; workshop scheduling

Citation: Zhong, L.; Chen, Y.; Mumtaz, J. A Flexible Job Shop Scheduling Method Based on Multi-Fidelity Optimization. *Eng. Proc.* **2023**, *45*, 47. https://doi.org/10.3390/engproc2023045047

Academic Editors: Mohammad Javed Hyder, Muhammad Mahabat Khan, Muhammad Irfan and Manzar Masud

Published: 18 September 2023

1. Introduction

Job shop scheduling is a critical component of machine scheduling. Classical methods like mathematical programming and branch bounding [1] face limitations in computational efficiency as the problem size increases. Swarm intelligence algorithms offer a popular approach to solving job shop scheduling problems by obtaining near-optimal solutions within a limited time frame. For example, Ali et al. [2] designed a GA based on a virtual crossover operator for the dynamic event of the random arrival of jobs, showing an improved average completion time. Demir et al. [3] combined the ant colony algorithm with GA to minimize the lead time and delay time of each job. In actual production, disturbance events such as machine failures and emergency jobs often render the original optimal solution infeasible. With the development of computer technology, simulation-based scheduling methods have been widely used. Discrete event simulation accurately represents workshop elements by creating a model close to reality. However, it is obviously too time-consuming to simulate all the scheduling schemes, and Xu et al. [4] proposed a multi-fidelity model based on ordinal transformation and optimal sampling (MO^{2}TOS), which quickly screens high-quality solutions in low-fidelity models, samples from high-quality solutions for high-fidelity simulation, and selects the optimal solution based on the high-fidelity simulation results to reduce the computation time under the premise of guaranteeing the solution quality. Currently, there are fewer studies on high-fidelity optimization methods considering disturbance events. In this paper, we propose MOAPMM for solving flexible job shop scheduling considering machine failure.

2. Problem Description

The flexible job shop scheduling problem studied in this paper can be described as follows: there are n jobs $J = \{J_1, J_2, \ldots, J_i, \ldots, J_n\}$ and m machines $M = \{M_1,\ M_2, \ldots, M_k, \ldots, M_m\}$ in the workshop, each of job has n_i operations $J_i = \{O_{i1}, O_{i2}, \ldots O_{in_i}\}$; each operation has its corresponding set of optional machines and is processed at different times on different machines. Consider the production reality that machines require flexible PM, i.e., the cumulative running time of the machine cannot exceed the machine maintenance threshold u_m; otherwise, the machine fails, and the time of post-failure maintenance (FM) is much larger than the PM time. The goal of flexible job shop scheduling is to rationally arrange each job to be processed on the machine to minimize the makespan.

The production process follows the following constraints: (1) all jobs can be processed at the initial time; (2) each machine can only process one job at a time; (3) each job cannot be processed by more than one machine at the same time; (4) there is a sequential constraint on the processes for the same job, and there is no sequential constraint on the processes for the different jobs; (5) the priority of all jobs is the same.

3. Multi-Fidelity Optimization Approach Considering Preventive Maintenance of the Machine

Multi-fidelity optimization involves a low-fidelity model (a simplified mathematical representation) and a high-fidelity model (a discrete event simulation). In this paper, we propose MOAPMM, inspired by Xu et al.'s [4] MO^2TOS.

It consists of two modules: feasible solution generation and optimal solution selection. The feasible solution generation module employs a GA to quickly generate feasible solutions. The optimal solution selection module uses the Enhanced Optimal Computing Budget Allocation (EOCBA) method to select high-quality solutions; the high-fidelity simulation, considering the PM of the machine, prevents the increase in makespan caused by machine failures; and the optimal scheduling scheme is chosen from the high-quality solutions based on the simulation results. The flowchart of MOAPMM is shown in Figure 1.

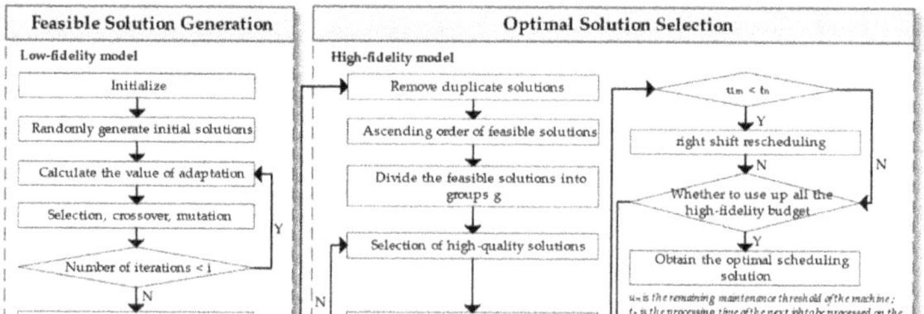

Figure 1. Flowchart of MOAPMM.

3.1. Feasible Solution Generation

The solution space of the flexible job shop scheduling problem is large, and the mathematical model cannot get the near-optimal solution of the problem in finite time. In this paper, GA is used as a low-fidelity model to generate feasible solutions directionally through the operations of selection, crossover, and mutation, and each generation of feasible solutions is saved to the set of feasible solutions.

3.2. Optimal Solution Selection

To avoid a time-consuming simulation of all feasible solutions, it is essential to select a suitable number of high-quality solutions for simulation. Optimal Computing Budget Allocation (OCBA) is a classic approach that divides feasible solutions equally into groups and randomly selects the same number of solutions from each group for high-fidelity simulation, and each selected solution consumes a high-fidelity budget. This operation is repeated until the high-fidelity budget is exhausted.

However, this method assigns an equal probability to all feasible solutions, which may not be conducive to the selection of better solutions. In this paper, we propose the EOCBA method for high-quality solution selection. This approach considers that while high-fidelity and low-fidelity models may diverge in target values, their results are often in agreement [5]. Therefore, in accordance with a sequentially decreasing ratio, a number of solutions are chosen from each group for high-fidelity simulation, following the ratio design principle: the ratio of high-quality solutions, denoted s_c selected from group 1 to g follows the pattern $g : g-1 : \ldots : 2 : 1$. This process is repeated until the high-fidelity budget is exhausted. If the remaining viable solutions, denoted s_r, in a group are less than the number of s_c solutions to be selected, the remaining unselected solutions $s_c - s_r$ will be included in the next group. Accordingly, the next group is required to select $s_c - s_r + s_c'$ feasible solutions for high-fidelity simulation, where s_c' indicates the originally planned number of high-quality solutions to be selected for the next group.

To align with actual workshop production, this paper considers the degradation of the machine's performance over time. If not maintained quickly, the machine becomes unusable. Therefore, in high-fidelity simulation, PM of the machine is taken into account. A flexible job shop is created using flexSim@ 20.0.10, with each machine having a maintenance threshold, denoted u_m. PM is performed before the machine's cumulative processing time exceeds u_m, resulting in a shorter maintenance time. Otherwise, the machine fails, resulting in longer maintenance times. When a job cannot be processed on a machine due to PM, the work is rescheduled using right shift rescheduling, as shown in Figure 2.

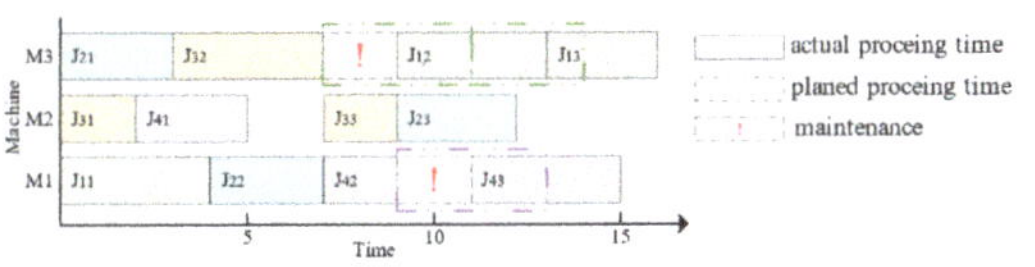

Figure 2. Schematic of right shift rescheduling.

Supposing u_m is 10 in Figure 2, the cumulative machining time t_{a_m} of M_3 is 7 after J_{32} finishes processing. However, since J_{12} requires a processing time of 4 on M_3, t_{a_m} exceeds 10, triggering the need for PM. With a maintenance time t_{pm} of 2 units, M_3's t_{a_m} resets to 0 after maintenance. Consequently, J_{12} must wait for the machine to undergo maintenance, causing a 2-unit right shift at its start processing time, and other jobs to be processed also need to be shifted to the right because the processing of jobs is subject to the conditions in Section 2.

4. Numerical Example and Analysis

4.1. Parameter Settings

In order to verify the effectiveness of the proposed method, this paper takes the two maintenance methods of PM and FM as comparison experiments in high-fidelity simulation and combines the above two groups of experiments with EOCBA and OCBA high-quality solution selection methods to form four groups of comparison experiments, and each group of experiments is carried out with different high-fidelity budgets, respectively. The budgets are strategically chosen to maximize the minimum budget required for the optimal objective value using pre-experimental data. Multiple budget sets are generated at 5-interval intervals to facilitate the process.

In this study, we conducted experiments using the literature [5] with n = 8, n_i = 4, and m = 8. Set u_m to 20, t_{pm} to 2, and FM time t_{fm} to 10, all time units are in hours. The feasible solutions are divided into five groups, and the high-quality solutions are selected for simulation from the five groups according to a ratio of 5:4:3:2:1. If the number of solutions to be selected is a non-integer calculated according to the ratio, the solutions will be selected in accordance with the rounding principle for selection.

4.2. Results Analysis

Results are presented in Table 1 based on the experiments with the given parameters.

Table 1. Individual program performance table.

Budge	PM						FM					
	OCBA			EOCBA			OCBA			EOCBA		
	Avg	Max	Min	Avg	Max	Min	Avg	Max	Min	Avg	Max	Min
5	44.0	59	34	37.6	48	31	56.2	83	42	50.4	72	39
10	39.3	46	31	36.5	49	26	50.5	61	39	47.8	65	34
15	41.2	67	29	37.0	52	26	54.8	91	37	48.3	68	34
20	40.6	58	30	35.7	50	26	53.4	82	38	46.9	66	34
25	39.4	56	26	36.4	50	26	52.5	72	34	47.6	64	34

PM consistently outperforms FM in terms of performance, as shown in Table 1. By proactively considering equipment availability and reducing downtime caused by failures, PM yields superior results. Notably, both optimal solution selection methods, EOCBA and OCBA, can obtain optimal scheduling solutions, but EOCBA uses a lower high-fidelity simulation budget than OCBA, which can obtain optimal scheduling solutions faster and reduce computing time.

5. Conclusions

This study proposes MOAPMM as a solution to the flexible shop scheduling problem, considering machine failures. The method selects high-quality solutions using the EOCBA method and conducts high-fidelity simulations considering machine PM to obtain the optimal scheduling scheme. Experimental results show that this method is superior to the FM simulation method. In addition, the EOCBA method reduces the high-fidelity simulation budget and computation time while ensuring the quality of the solutions obtained.

Author Contributions: Conceptualization, L.Z. and Y.C.; methodology, L.Z. and Y.C.; software, L.Z. and J.M.; validation, L.Z. and J.M.; formal analysis, L.Z. and Y.C.; writing—review and editing, Y.C. and L.Z.; visualization, L.Z.; supervision, J.M.; funding acquisition, Y.C. All authors have read and agreed to the published version of the manuscript.

Funding: This research was funded by the National Natural Science Foundation of China grant number [No. 51705370].

Institutional Review Board Statement: Not applicable.

Informed Consent Statement: Not applicable.

Data Availability Statement: Not applicable.

Conflicts of Interest: The authors declare no conflict of interest.

References

1. Luo, J.C.; Liu, Z.Q.; Xing, K.Y. Hybrid branch and bound algorithms for the two-stage assembly scheduling problem with separated setup times. *Int. J. Prod. Res.* **2019**, *57*, 1398–1412. [CrossRef]
2. Ali, B.K.; Telmoudi, A.J.; Gattoufi, S. Improved genetic algorithm approach based on new virtual crossover operators for dynamic job shop scheduling. *IEEE Access.* **2020**, *8*, 213318–213329. [CrossRef]
3. Demir, H.I.; Erden, C. Dynamic integrated process planning, scheduling and due-date assignment using ant colony optimization. *Comput. Ind. Eng.* **2020**, *149*, 106799. [CrossRef]
4. Xu, J.; Zhang, S.; Huang, E.; Chen, C.H.; Lee, L.H.; Celik, N. MO^{2}TOS: Multi-fidelity optimization with ordinal transformation and optimal sampling. *Asia. Pac. J. Oper. Res.* **2016**, *33*, 1650017. [CrossRef]
5. Wang, J.; Liu, Y.; Ren, S.; Wang, C.; Ma, S.Y. Edge computing-based real-time scheduling for digital twin flexible job shop with variable time window. *Robot. Comput.-Integr. Manuf.* **2023**, *79*, 102435. [CrossRef]

engineering proceedings

MDPI

Proceeding Paper

Optimizing Waste Collection and Transportation in Islamabad: Efficient Vehicle Routing for Sustainable Waste Management †

Tahiyyah Rashid [1,*], Saleem Ullah [1], Umair Habib [1], Nusrat Rehman [1], Usamah Rashid Qureshi [2] and Farzana Jafar [3]

1 Department of Mathematics, Air University, Sector E-9, Islamabad 44000, Pakistan; saleemullah@mail.au.edu.pk (S.U.); umair.habib@students.au.edu.pk (U.H.); nusrat.reshman@students.au.edu.pk (N.R.)

2 Department of Electrical and Information Engineering, Politecnico di Bari, 70126 Bari, Italy; usamahrashid.qureshi@poliba.it

3 Department of Physics, Air University, Sector E-9, Islamabad 44000, Pakistan; biameezab@gmail.com

* Correspondence: tahiyyah.rashid@gmail.com

† Presented at the Third International Conference on Advances in Mechanical Engineering 2023 (ICAME-23), Islamabad, Pakistan, 24 August 2023.

Abstract: The municipal solid waste management is key for preserving the environment, protecting public health, and maintaining public cleanliness. Apart from these benefits, developing countries are more attracted to such systems as they offer an economical solution for waste management. This article intends to present a cost-effective and optimized waste management solution for Islamabad, Pakistan. The key steps of this research include; 1. acquisition of data; 2. transformation of data into coordinate form; 3. modeling of optimized waste collection routes and transportation; 4. modification of the capacitated vehicle routing algorithm to yield the optimized vehicle routes. The simulations are performed in MATLAB 2017B. Trial results show the efficacy of the recommended method.

Keywords: capacitated vehicle routing problem (CVRP); Islamabad waste management; vehicle routing problem (VRP)

Citation: Rashid, T.; Ullah, S.; Habib, U.; Rehman, N.; Qureshi, U.R.; Jafar, F. Optimizing Waste Collection and Transportation in Islamabad: Efficient Vehicle Routing for Sustainable Waste Management. *Eng. Proc.* **2023**, *45*, 41. https://doi.org/10.3390/engproc2023045041

Academic Editors: Mohammad Javed Hyder, Muhammad Mahabat Khan, Muhammad Irfan and Manzar Masud

Published: 14 September 2023

1. Introduction

The collection, transportation, and disposal of solid waste (SW) produced by houses is a process known as municipal solid waste management. Waste collection and transportation are accomplished using a fleet of trucks. This conventional waste collection is based on speculation as to whether the filling levels of waste bins could vary from overflowing, partial filling, to completely emptying, which would result in unnecessary resource consumption. Improvements will reduce municipal spending as the cost of waste collection/transportation accounts for 60–80% of overall waste management system (WMS) expenses.

VRP research was conducted in 1959, and its goal was to model the distribution process and map out the most efficient pathways to deliver products to consumers [1]. In addition to the VRP's constraints, the CVRP adds the constraint of the vehicle's capacity.

A significant amount of garbage is produced due to the development in product diversity and consumption, endangering the ecosystem [2]. E-waste seriously harms the ecosystem when combine with the air and soil [3]. Burning/burying plastic garbage could endanger the land and air due to the hazardous chemical gases released [4]. Some 4 to 12 million metric tons of plastic trash are dumped into the oceans annually, leading to water contamination [5]. When medical waste is disposed of, it might cause major environmental issues [6].

It is estimated that the global daily production of medical waste due to the pandemic was over 1.5 million tons [7]. It is now necessary to collect this waste in a systematic manner [8], in order to lessen its harmful impact [9], along with other forms of e-trash [10]. One Indian city's annual e-waste contained valuable materials worth $65,000 [11], showing the economic advantages of employing demolition waste, which is between 50 and 70% of the world's total SW [12].

In this study the method of optimizing routes is used not the waste. The remainder of this article is outlined as follows: Section 2 includes the acquisition of a dataset and an overview of the research methodology; Section 3 contains the model design; the simulation results are presented in Section 4; finally, the article is concluded in Section 5.

The MATLAB 2017B algorithm CVRP is used for simulation.

2. Data Acquisition and Research Methodology

In this section a brief overview of steps to acquire data and the employed research method are provided. The following steps were taken to acquire the waste bin and vehicle data from the concerned authorities: 1. visit MCI; 2. enlist their challenges and collect data; 3. locate the bins on Google Maps according to the acquired dataset; 4. import data to the web plot digitizer, draw axes, locate bins, and find coordinates; 5. create CSV and TXT files so the data becomes ready for the mathematical computing and analysis tools.

The dataset is ready, the only problem left was to employ the research method for obtaining the optimized route. Amongst the available resources, CVRP is widely used as discussed in Section 1. The following steps were taken to obtain an optimized route: 1. import geographic data into MATLAB 2017B; 2. transform data, i.e., the number of vehicles, capacities of vehicles, distances between the bins, expected expenditure, etc. into the MATLAB 2017B format; 3. locate bins and input the amount of waste in each bin into MATLAB 2017B; 4. run the CVRP simulation in MATLAB 2017B; 5. find optimal results; 6. based on these results, we plan to submit the proposal to MCI.

3. Model Design

3.1. Description of the Problem

The aim of this research is to minimize the overall distance traveled and the total comprehensive cost, including vehicle costs. The vehicles are located at a depot, and its routes begin there. When a waste truck is full, or the assigned task is completed, it returns to the depot.

Assumptions: In addition to the above, we assume the following: 1. one vehicle collects each bin once; 2. there is a single depot; 3. the vehicles leave the depot and return when the job is completed; 4. the vehicles are of various capacities; 5. the location of the depot and each bin is known.

3.2. Model Construction

In this subsection, the complete mathematical model for the optimized waste collection problem is provided. The main objectives are to minimize the vehicle distance as well as the cost of waste collection. Therefore, the objective function are as follows:

$$\min TD = \sum_{m\in(B\cup D)} \sum_{n\in(B\cup D)} x^k_{mn} d_{mn} \tag{1}$$

$$\min TC = \sum_{k\in V} \sum_{n\in(B\cup D)} x^k_{0n} P_{fixed} + \sum_{m\in(B\cup D)} \sum_{n\in(B\cup D)} x^k_{mn} d_{mn} r_{mn} P_{fuel} \tag{2}$$

These objective functions are constrained by following conditions:

$$\sum_{k\in V} \sum_{m\in(B\cup D)} x^k_{mn} = 1; \quad \forall\, n \in (B\cup D) \tag{3}$$

$$\sum_{k\in V} \sum_{n\in(B\cup D)} x^k_{mn} = 1; \quad \forall\, m \in (B\cup D) \tag{4}$$

$$\sum_{m\in(B\cup D)} x^k_{mn} = \sum_{n\in(B\cup D)} x^k_{mn} = 1; \quad \forall\, m \in (B\cup D), k \in V \tag{5}$$

$$\sum_{m\in(B\cup D)}\sum_{n\in(B\cup D)} x_{mn}^{k} q_n \le C_p; \quad \forall\, k \in V \tag{6}$$

$$x_{mn}^{k} \in \{0,\, 1\}; \quad \forall\, m,\, n \in (B \cup D), k \in V \tag{7}$$

To begin, we have two objective functions, Equation (1) and Equation (2), their respective goals are the minimum total distance and minimum total cost. Constraint Equation (3) ensures that each waste bin is collected by a single vehicle. All routes will respect the maximum capacity by Equation (4) to Equation (6). Equation (7), defines the variable types, where B represents bins, D is depot, V is a set of vehicles, P_{fixed} is the expenses attached to each vehicle, d_{mn} is the distance between bins 'm' and 'n', r_{mn} is fuel consumption rate per unit kilometer, P_{fuel} is fuel consumption cost, x_{mn}^{k} checks whether a vehicle 'k' moves from waste bin 'm' to waste bin 'n'. q_n shows Waste collected at waste bin 'n'. C_p is a vehicle's maximum carrying capacity.

4. Results and Discussion

The major results of this research are presented in this section. The map of the waste collection points and depot at G-6/1 is presented in Figure 1a. Seventy bins are located throughout the territory. The red truck represents the MCI and the blue dots indicate the waste bins.

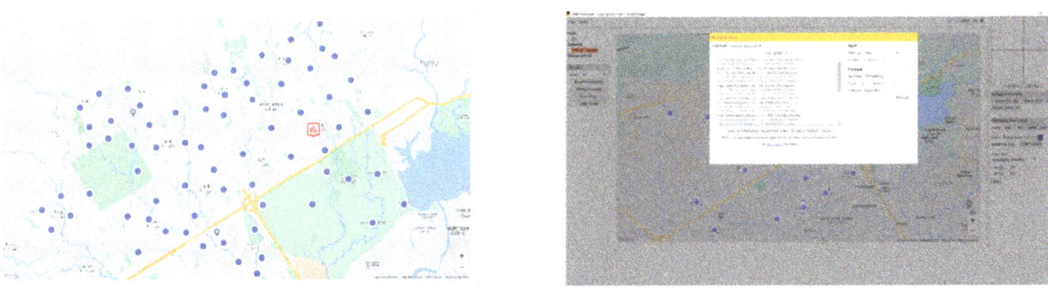

(a) (b)

Figure 1. (**a**) Locating the collection points on Google Maps; (**b**) importing to WebPlotDigitizer 4.1, drawing axes, locating collection points, and finding coordinates.

The CVRP algorithm is used for analysis to determine the exact method of solving.

In Figure 1b, the Google Maps points were transferred into the WebPlotDigitizer 4.1 to transform the location of bins into a Cartesian plane dataset. The depot was defined at $(0, 0)$, the $x - axis$ parallel to the Kashmir Highway and the $y - axis$ perpendicular to it. The geographic coordinates were retrieved and saved in both CSV and TXT file formats.

In Figure 2a, bins are displayed graphically within MATLAB 2017B, after the geographical coordinates were imported. The red circles are the location of the bins and the depot at $(0, 0)$.

In Figure 2b, the yellow square shows the depot and rest of the colors show the optimized routes. Figure 2c shows a monotonic reduction, illustrating a steady decline in cost.

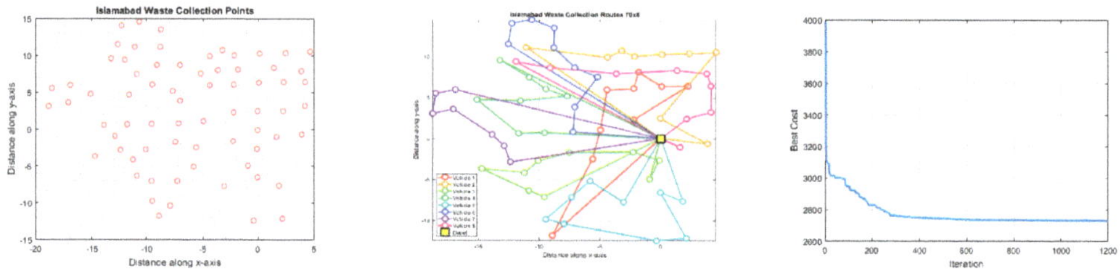

(a) (b) (c)

Figure 2. (**a**) Islamabad waste collection points in MATLAB 2017B; (**b**) Islamabad waste collection routes (70 × 8); (**c**) Islamabad waste collection optimized cost (70 × 8).

5. Conclusions

A case study in which the environmental problem of SWM was optimized by employing CVRP. The case involved SWM in the Islamabad Capital Territory (ICT) administered by MCI supervised by Capital Development Authority (CDA). The algorithm implemented for optimizing the routes shows a consistent and gradual decrease in cost with each iteration. It is crucial to emphasize that the algorithm ensures no further cost reduction is possible. After 1200 iterations, the algorithm identified the optimum paths, and no additional route optimization was achievable. This finding highlights the algorithm's effectiveness in enhancing SWM efficiency.

Author Contributions: Conceptualization, T.R., S.U. and U.R.Q.; methodology, T.R. and U.H.; software, T.R. and F.J.; validation, U.H. and N.R.; formal analysis, T.R. and U.H.; investigation, S.U.; resources, T.R. and U.R.Q.; writing—original draft preparation, T.R., N.R. and F.J.; writing—review and editing, U.H. and U.R.Q.; visualization, F.J.; supervision, S.U.; All authors have read and agreed to the published version of the manuscript.

Funding: This research received no external funding.

Institutional Review Board Statement: Not applicable.

Informed Consent Statement: Not applicable.

Data Availability Statement: Dataset is available upon request.

Acknowledgments: The authors are thankful to the Sanitation Directorate (MCI), CDA, Islamabad for their continuous support during the course of this research.

Conflicts of Interest: The authors declare no conflict of interest.

References

1. Dantzig, G.B.; Ramser, J.H. The truck dispatching problem. *Manag. Sci.* **1959**, *6*, 80–91. [CrossRef]
2. Misra, V.; Pandey, S. Hazardous waste, impact on health and environment for development of better waste management strategies in future in India. *Environ. Int.* **2005**, *31*, 417–431. [CrossRef]
3. Sivaramanan, S. E-Waste Management, Disposal and Its Impacts on the Environment. *Univers. J. Environ. Res. Technol.* **2013**, *3*, 531–537.
4. Khan, F.; Ahmed, W.; Najmi, A. Understanding consumers' behavior intentions towards dealing with the plastic waste: Perspective of a developing country. *Resour. Conserv. Recycl.* **2019**, *142*, 49–58. [CrossRef]
5. Brooks, A.L.; Wang, S.; Jambeck, J.R. The Chinese import ban and its impact on global plastic waste trade. *Sci. Adv.* **2018**, *4*, eaat0131. [CrossRef] [PubMed]
6. Manyele, S.V.; Tanzania, V. Effects of improper hospital-waste management on occupational health and safety. *Afr. Newsl. Occup. Health Saf.* **2004**, *14*, 30–33.
7. Benson, N.U.; Bassey, D.E.; Palanisami, T. COVID pollution: Impact of COVID-19 pandemic on global plastic waste footprint. *Heliyon* **2021**, *7*, e06343. [CrossRef] [PubMed]
8. Liap, V.F.L.; Michele, M.; Sanjaya, E.; Dadoui, Y.I.; Putri, V.N. Medical Waste Management During COVID-19 Pandemic. In Proceedings of the 3rd Tarumanagara International Conference on the Applications of Social Sciences and Humanities (TICASH 2021), online, 25 August 2021; pp. 1815–1822.

9. Zorpas, A.A.; Inglezakis, V.J. Automotive industry challenges in meeting EU 2015 environmental standard. *Technol. Soc.* **2012**, *34*, 55–83. [CrossRef]
10. Perkins, D.N.; Drisse, M.-N.B.; Nxele, T.; Sly, P.D. E-waste: A global hazard. *Ann. Glob. Health* **2014**, *80*, 286–295. [CrossRef] [PubMed]
11. Ravindra, K.; Mor, S. Distribution and health risk assessment of arsenic and selected heavy metals in Groundwater of Chandigarh, India. *Environ. Pollut.* **2019**, *250*, 820–830. [CrossRef] [PubMed]
12. Paschoalin Filho, J.A.; Camelo, D.G.; de Carvalho, D.; Guerner Dias, A.J.; Marcondes Versolatto, B.A. Use of construction and demolition solid wastes for basket gabion filling. *Waste Manag. Research.* **2020**, *38*, 1321–1330. [CrossRef] [PubMed]

engineering proceedings

MDPI

Proceeding Paper

The Development and Evaluation of a High-Frequency Toroidal Transformer for Solid-State Transformer Applications †

Abdul Shakoor *, Azhar Ul Haq and Taosif Iqbal

Department of Electrical Engineering, College of EME, National University of Science and Technology, Islamabad 4600, Pakistan; azhar.ulhaq@ceme.nust.edu.pk (A.U.H.); taosifiqbal@ceme.nust.edu.pk (T.I.)

* Correspondence: abdulshakoor@ceme.nust.edu.pk

† Presented at the Third International Conference on Advances in Mechanical Engineering 2023 (ICAME-23), Islamabad, Pakistan, 24 August 2023.

Abstract: The performance and efficiency of high-frequency transformers (HFTs) are significantly influenced by leakage inductance. To improve the efficiency of HFTs, it is crucial to consider the effects of leakage inductance during the design and analysis processes. This research study aims to investigate a high-frequency toroidal transformer by examining different magnetic materials (Ferrite, Amorphas, and Iron-Powdered Alloy), winding configurations (solid, twisted, and LITZ wire), and operating frequencies (10 kHz and 50 kHz). To validate the effectiveness of parametric optimization in enhancing the system efficiency, the designed toroidal HFT was constructed and tested in a 600/300 V 3 kW dual active bridge (DAB) converter. The leakage inductances were measured using a frequency sweeping LCR meter.

Keywords: dual active bridge; parasitic capacitance; leakage inductance; high-frequency transformer

Citation: Shakoor, A.; Haq, A.U.; Iqbal, T. The Development and Evaluation of a High-Frequency Toroidal Transformer for Solid-State Transformer Applications. *Eng. Proc.* **2023**, *45*, 11. https://doi.org/10.3390/engproc2023045011

Academic Editors: Mohammad Javed Hyder, Muhammad Mahabat Khan, Muhammad Irfan and Manzar Masud

Published: 8 September 2023

1. Introduction

Electromagnetics play a crucial role in various applications, particularly in transformer-based power electronics equipment. Parameters such as parasitic capacitance and leakage inductance significantly affect the efficiency and performance of these devices. Power electronics often utilize galvanic isolation-based topologies due to their compactness, active/reactive power flow management, and protection features. However, if a transformer has improperly linked primary and secondary windings, it experiences increased power losses and current and voltage spikes, leading to a lower efficiency. Therefore, it is important to consider the leakage inductance and parasitic components when developing efficient high-frequency transformers. In fact, incorporating leakage inductance and parasitic capacitance into a transformer can help in building LC tank circuits, resulting in a 15% decrease in the volume of soft-switching converters [1]. When working with higher frequencies, passive components can be made smaller at the cost of higher parasitic effects, which may experience current spikes [2]. To address these issues, various core materials can be employed. Ferrites are durable ceramics based on metal oxides that exhibit fewer core losses at a low cost [3]. However, ferrites have limitations such as a low tensile strength and significant permeability roll-off. Another category of core material are powdered magnetic cores, which are composed of alloyed powdered cores, soft magnetic powder composites, and powdered iron. Alloy powdered cores such as Moly-permalloy (MPP), High-flux, Kool Mμ, and XFlux offer stability at higher temperatures, a reduced leakage flux, low magnetostriction, and lower prices [4]. Powdered iron cores are the least expensive, but can be sensitive to temperature increases. Amorphous alloy cores are more robust and corrosion-resistant, making them suitable for medium-frequency transformers with temperature-dependent saturation flux density concerns [5]. In transformers, the leakage inductance energy is divided into two parts, one stored in the winding and the other in the insulation zone. Analytical techniques often ignore leakage inductance at high frequencies due to the magnetic field unevenness caused by eddy and proximity effects [6].

In this research article, various core materials (ferrite, iron-powdered alloy, and amorphas,) wire structures (round, twisted, and LITZ wire), and winding arrangements (single or multi-layer) are analyzed for the optimization of the leakage inductance behavior of different transformers.

2. Design and Analysis Procedure

The dual active bridge converter is composed of two full-bridge inverters that are linked through a high-frequency transformer. V_p and V_s are the primary and secondary square-wave voltage amplitudes, f_{sw} is the switching frequency, and D is the duty cycle. Square-wave voltages generated by full bridges are used to calculate the leakage inductance L_{leak} , which is most appropriate variable for maximizing the power transfer abilities of DAB.

$$P_{ps} = \frac{nV_pV_sD(1-D)}{2\pi f_{sw}L_{\text{leak}}} \quad (1)$$

The transformer is designed using the geometrical constant of the core, which can be seen from Table 1. The leakage inductance of a transformer is adjusted by experimenting with the value of the mutual inductance. The number of turns and geometrical arrangement of the core of a transformer are the few key parameters for determining the leakage inductances. The below section provides a collection of the essential formulae used to select an appropriate core material. The effective area and effective length of the toroidal core are calculated using Equations (2) and (3), while the energy stored in the core and volume from Equations (4) and (5).

$$A_{\text{core}} = h\left(\frac{OD-ID}{2}\right) \quad (2)$$

$$l_{\text{core}} = \pi\left(\frac{OD+ID}{2}\right) \quad (3)$$

$$E_{\text{store}} = \frac{A_{\text{core}}l_{\text{core}}B_{\text{max}}^2}{2\mu} \quad (4)$$

$$V_{\text{core}} = \frac{2E_{\text{store}}\mu}{B_{\text{max}}^2} \quad (5)$$

Table 1. HFT design specifications.

Test Cases	Case 1	Case 2	Case 3
Core Material	Ferrite	Iron-Powdered Alloy (MPP)	Amorphas (Microlite)
Optimal Flux Density (Tesla)	0.25	0.9	0.7
Dimension (OD, ID, H)	85, 53, 20 mm	85, 53, 20 mm	85, 53, 20 mm
Number of Turns (Primary, Secondary)	100, 50	100, 50	100, 50
Wire Gauge (Primary, Secondary)	21, 21	21, 21	21, 21
Numbers of Strands (Primary, Secondary)	5, 3	5, 3	5, 3
Voltage (Primary, Secondary)	600, 300	600, 300	600, 300
Frequency (kHz)	10	10	10

The magnetic core volume is proportional to the permeability μ of the core material and inversely proportional to the square of B_{max}, as described in Equation (5). Thus, magnetic materials with a greater saturation flux density B_{sat} are appropriate for small-volume transformers. The B_{max} and μ values were taken from the material data sheet. Three HFTs are simulated in ANSYS using three cores of different materials wound with three wire arrangements such as round, twisted, and LITZ wire. Excitation is assigned by sectioning the copper coil and specifying turns. The leakage inductance and parasitic capacitance are calculated using the eddy current and magnetostatic solvers. The HFT used in dual active bridge converters is shown in Figure 1a, and the leakage inductance measurement setup in Figure 1b.

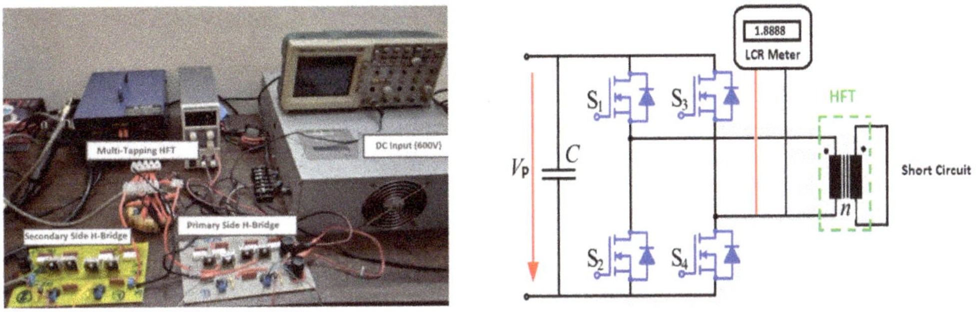

(a) **(b)**

Figure 1. (**a**) Hardware setup of dual active bridge and (**b**) leakage inductance measurement.

Leakage Inductance Calculations: The leakage inductance levels in a transformer are influenced by the material and core structure. In this study, ANSYS Maxwell was used to calculate the leakage and magnetizing inductance for different core materials. To facilitate an easy comparison of the leakage and parasitic qualities, the magnetizing inductance of the transformer designs was deliberately kept similar during the simulations. Based on the simulation analysis, the Ferrite core wound with a twisted wire construction exhibited the best results as tabulated in Table 2. Consequently, a hardware prototype was built using a ferrite core wound with isolated twisted wire to validate the results. Table 3 demonstrates that the hardware results for the leakage inductance and parasitic capacitance closely match the simulation results.

Table 2. HFT Simulations of different core materials and wire structures at 10 kHz frequency (Np = 100, Ns = 50).

Windings Configuration	Core Material	Magnetizing Inductance (mH)		Leakage Inductance (µH)	
		Primary Winding	Secondary Winding	Primary Winding	Secondary Winding
Round Wire	Ferrite	47.3	11.5	89.00	24.20
	Microlite	34.7	8.32	191.2	127.7
	MPP	33.1	8.91	191.1	126.4
Twisted Wire (isolated strands)	Ferrite	46.1	11.1	69.00	14.00
	Microlite	35.3	8.31	189.2	119.7
	MPP	34.7	9.12	191.1	120.4
LITZ wire	Ferrite	47.9	10.9	61.00	12.10
	Microlite	34.9	8.12	181.2	117.1
	MPP	34.8	9.01	181.9	117.9

Table 3. Comparison of simulation and experimental results for twisted wire (isolated strands) wound HFT.

Parameters	Unit	10 kHz		50 kHz	
		Simulation	Experimental	Simulation	Experimental
Inductance (Primary)	mH	46.1	46.5	40.5	42.9
Inductance (Secondary)	mH	11.1	11.3	9.70	9.95
Leakage Inductance (Primary)	μH	69.0	70.2	1.0	1.18
Leakage Inductance (Secondary)	μH	14.1	14.8	4.00	4.73

3. Results and Discussion

The simulation of a high-frequency transformer was conducted using ferrite, microlite, and MPP cores, wound with solid round wire, twisted wire (isolated strands), and LITZ wire. Power electronics magnetics (PEmag TM) in ANSYS calculated the leakage inductance through a finite element analysis (FEA). To validate the results, a hardware prototype was fabricated using a ferrite core wound with a twisted pair (isolated strands).

Table 3 presents a comparison between the simulation and experimental results, noting that the leakage inductance values apply to both the primary and secondary windings. The prototype transformer was measured for three scenarios using a BK Precision 880 LCR meter. The minimal difference between the simulation and experimental results was attributed to the careful implementation of the hardware prototype. The high-frequency transformer design demonstrated positive characteristics for leakage inductance at 10–50 kHz switching frequencies. This eliminated the need for a large series inductor to optimize the power transfer for a DAB converter.

4. Conclusions

Achieving a balance between leakage inductances is a challenging task that heavily relies on design parameters. Adjusting the core size, material, or number of turns alone is not enough to effectively reduce leakage inductance. A Finite Element Analysis (FEA) provides more accurate results compared to traditional analytical methods when computing these values. Simulations on various test cases helped us to determine the optimal parameters. A high-frequency transformer prototype was then implemented for validation. At an operating frequency of 50 kHz, optimal values for leakage inductance were achieved in a ferrite core wound with twisted wire.

Author Contributions: Conceptualization, A.S. and T.I.; methodology, A.S. and T.I.; software, A.S.; validation, A.S.; formal analysis, A.S. and T.I.; investigation, A.S.; data curation, A.S.; writing—original draft preparation, A.S.; writing—review and editing, A.S.; supervision, T.I. and A.U.H.; project administration, A.U.H. All authors have read and agreed to the published version of the manuscript.

Funding: This research received no external funding.

Institutional Review Board Statement: Not applicable.

Informed Consent Statement: Not applicable.

Data Availability Statement: Not applicable.

Conflicts of Interest: The authors declare no conflict of interest.

References

1. Amini, M.; Almassalkhi, M. Trading off robustness and performance in receding horizon control with uncertain energy resources. In Proceedings of the 2018 Power Systems Computation Conference (PSCC), Dublin, Ireland, 11–15 June 2018; pp. 1–7.
2. Saadatmand, S.; Shamsi, P. Adaptive critic design-based reinforcement learning approach in controlling virtual inertia-based grid-connected inverters. *Int. J. Electr. Power Energy Syst.* **2021**, *127*, 106657. [CrossRef]
3. Rhaman, M.; Matin, M.; Hossain, M. Ferromagnetic, electric, and ferroelectric properties of samarium and cobalt co-doped bismuth ferrite nanoparticles. *J. Phys. Chem. Solids* **2020**, *147*, 109607. [CrossRef]

4. Imaoka, J.; Okamoto, K.; Shoyama, M. Modeling, magnetic design, simulation methods, and experimental evaluation of various powder cores used in power converters considering their dc superimposition characteristics. *IEEE Trans. Power Electron.* **2019**, *34*, 9033–9051. [CrossRef]
5. Ahmed, S.; Chang, K.-C.; Nguyen, T.-T.; Chu, K.-C.; Chang, F.-H.; Wang, H.-C. *Interleaved DC/DC Boost Converter for Hybrid Electric and Electrical Vehicles*; Springer: Cham, Switzerland, 2021; pp. 708–716.
6. Nan, X.; Sullivan, C.R. An equivalent complex permeability model for Litz-wire windings. *IEEE Trans. Ind. Appl.* **2009**, *45*, 854–860. [CrossRef]

engineering proceedings

MDPI

Proceeding Paper

Output Feedback Control of Two-Time-Scale Permanent-Magnet DC Motor Using High-Gain Observers †

Abid Raza [1,*], Naveed Mazhar [1], Fahad Mumtaz Malik [1], Rameez Khan [1], Arslan Khan [1] and Hameed Ullah [2]

1 College of Electrical and Mechanical Engineering, National University of Sciences & Technology, Islamabad 44000, Pakistan
2 PRISMA LAB, University of Naples Federico II, 80138 Naples, Italy
* Correspondence: abid.raza@ceme.nust.edu.pk
† Presented at the Third International Conference on Advances in Mechanical Engineering 2023 (ICAME-23), Islamabad, Pakistan, 24 August 2023.

Abstract: In this paper, we propose a methodology for implementing an output feedback control (OFC) strategy for a two-time-scale nonlinear system. A permanent-magnet DC motor model is utilized. We estimated the states using a high-gain observer for a two-time-scale nonlinear system. The results demonstrate the robustness and efficacy of the suggested OFC technique. Furthermore, the output feedback control approach exhibits robustness to parametric uncertainties, thereby making it feasible for practical implementation in a two-time-scale nonlinear system.

Keywords: output feedback control; high-gain observer; permanent-magnet motor; singularly perturbed system; two-time-scale system

Citation: Raza, A.; Mazhar, N.; Malik, F.M.; Khan, R.; Khan, A.; Ullah, H. Output Feedback Control of Two-Time-Scale Permanent-Magnet DC Motor Using High-Gain Observers. *Eng. Proc.* **2023**, *45*, 20. https://doi.org/10.3390/engproc2023045020

Academic Editors: Mohammad Javed Hyder, Muhammad Mahabat Khan, Muhammad Irfan and Manzar Masud

Published: 11 September 2023

1. Introduction

Small parameters such as capacitances and time constants are the cause of the development of singular perturbation theory. These parameters are usually ignored in dynamic models to simplify the models; however, this over-simplification comes at a cost. Singular perturbation theory ensures that this over-simplified model remains useful for analysis and control design [1]. Multiple-time-scale systems are common in science and engineering as many scientific and engineering processes exhibit multiple-time-scale behavior [2]. In a two-time-scale system, some states evolve slowly with time, while other states evolve at a faster rate with time. The state model of a system depends on a small perturbation parameter ε, which is either the system's parameter or injected artificially to produce a two-time-scale system. Setting $\varepsilon = 0$ brings an abrupt change in the dynamic behavior of a system and produces a reduced-order system [3].

The measurement of all the variables of interest in a dynamic system is not feasible due to cost-related considerations or technical limitations. These constraints necessitate the use of output feedback control (OFC) in place of full-state feedback control (SFC). The most widely used solution to the problem, when states are required but not measured, is to use an observer for state estimation. The Luenberger observer and Kalman filter are commonly used for the state estimation of linear systems. For nonlinear systems, linear observers are applied after linearizing the system about the equilibrium point. However, this technique is local and only works when the states are evolved in the vicinity of the point of linearization. For global results, nonlinear observers are used for nonlinear systems.

The theory of multiple-time-scale behavior has been developed and analyzed widely; however, the nonlinear state estimation of multiple-time-scale systems is still an open area of research. There are limited results on nonlinear observer design for nonlinear singularly perturbed systems. However, these nonlinear state estimation techniques are not appropriate in terms of observer gains when used for multiple-time-scale systems.

With the advancements in nonlinear control theory, special attention has been given to nonlinear observers. High-gain observers (HGOs), being robust to parametric and modeling uncertainties, have gained more attention in nonlinear control systems [4]. Through the utility of the separation principle, controls and observers can be designed separately, and the results under SFC are similar to those of OFC given that the so-called high-gain parameter approaches zero. In recent years, extensive research has been conducted on the design, analysis and implementation of high-gain observers for nonlinear systems; however, there is a lack of research and literature available on the topic of robust observer design, for both slow and fast states simultaneously, using HGO.

In this article, we addressed the problem of state estimation via HGO for a class of standard two-time-scale systems. We propose an HGO-based OFC technique for a two-time-scale model of a permanent-magnet direct current (PMDC) motor. We show that the proposed HGO follows the separation principle, i.e., as the value of the high-gain parameter is reduced sufficiently, the response of OFC approaches that of SFC.

2. Output Feedback Control Design

Permanent-magnet motors are widely used in a large number of electrical and mechanical applications. Figure 1 shows the equivalent circuit of a PMDC motor. The dynamic model of a PMDC motor at no load is represented by the following state equations [3]:

$$J\frac{d\omega}{dt} = k_t i, \tag{1}$$

$$L\frac{di}{dt} = -k_b\omega - Ri + E, \tag{2}$$

where ω represents the angular speed, i denotes the armature circuit current, L represents the armature inductance, J represents the rotor moment, R is the armature circuit resistance, k_b is the back emf constant and k_t is the torque constant. The input of the system is input voltage E, whereas the angular speed is the output of the system, defined as ω. Singular perturbation theory states that the perturbation parameter ε should be dimensionless. Thus, defining the dimensionless parameters as $\omega_r = \omega/\ \Theta$ and $i_r = iR/k_b\Theta$ and $E_r = E/k_b\Theta$, systems (1) and (2) take the following form:

$$\mathcal{T}_m\frac{d\omega_r}{dt} = i_r, \tag{3}$$

$$\mathcal{T}_e\frac{di_r}{dt} = -\omega_r - i_r + E_r, \tag{4}$$

where $\mathcal{T}_m = JR/k_tk_b$ and $\mathcal{T}_e = L/R$ are mechanical and electrical time constants, respectively. Since $\mathcal{T}_m \gg \mathcal{T}_e$, defining a dimensionless time variable $t_r = t/\mathcal{T}_m$, the systems (3) and (4) takes the following form:

$$\frac{d\omega_r}{dt_r} = i_r, \tag{5}$$

$$\varepsilon\frac{di_r}{dt_r} = -\omega_r - i_r + E_r, \tag{6}$$

where the dimensionless perturbation parameter ε is defined as:

$$\varepsilon = \frac{\tau_e}{\tau_m} = \frac{Lk_tk_b}{JR^2}, \tag{7}$$

Setting $\varepsilon = 0$ in (6),

$$i_r = -\omega_r + E_r = h(x), \tag{8}$$

and with a change in coordinates $y = i_r - h(x) = i_r + \omega_r - E_r$, and taking the input voltage $E_r = -k_1\omega_r - k_2 i_r$, the boundary-layer model

$$\frac{dy}{d\tau} = -(1 + k_2)y, \tag{9}$$

is globally exponentially stable for $k_1 = 1$ and $k_2 > -1$, and the HGOs for the slow and fast variables are taken as:

$$\frac{d\widehat{\omega}_r}{dt_r} = \hat{i}_r + \frac{\alpha_1}{\varepsilon}(\omega_r - \widehat{\omega}_r), \tag{10}$$

$$\varepsilon\frac{d\hat{y}}{dt_r} = -(1 + k_2)\hat{y} + \frac{\beta_1}{\bar{\varepsilon}}(\omega_r - \widehat{\omega}_r), \tag{11}$$

with the globally bounded control law

$$E_r = -k_1\widehat{\omega}_r - k_2\hat{i}_r, \tag{12}$$

whereas the control input can be globally bounded by saturating it outside the region of interest and $\hat{i}_r = \hat{y} - \widehat{\omega}_r - E_r$.

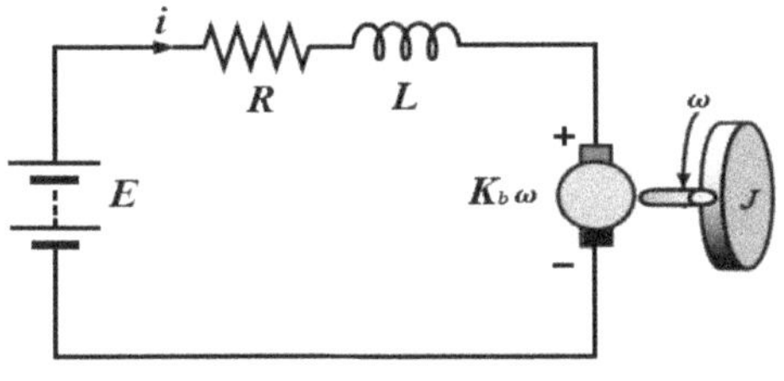

Figure 1. Equivalent Circuit of permanent-magnet DC Motor.

3. Simulation Results

This section presents the simulation results of the OFC designed for the PMDC motor model. To investigate the efficacy of the designed HGOs on the PMDC model, it is simulated in MATLAB. The parameters utilized are: $J = 1.8 \times 10^{-4}$ kgm^2, $k_t = 0.057$ Nm/A, $k_b = 0.057$ Vs/rad, $L = 0.9$ mH, $R = 10\ \Omega$, $\omega_r(0) = 0.1$, $i_r(0) = 0.5$ and $\hat{\omega}_r(0) = \hat{i}_r(0) = 0$. The controller parameters $k_1 = k_2 = 1$ and the observer parameters are taken as $\alpha_1 = \beta_1 = 2$, whereas the high-gain parameters are chosen such that $\bar{\varepsilon} = \varepsilon = 0.1$. The control input is constrained within the range of 0 to 10. Within a 20% range, the parameters deviate from their nominal values.

Figure 2a shows a comparison of the angular speed under SFC and OFC. The slow state ω shows a slow transient in the response. The response of the fast state is shown in Figure 2b. The current exhibits a fast transient initially due to the two-time-scale system. The response under SFC approaches the response under OFC for angular velocity and current. Thus, the state estimation using the high-gain observer shows the robustness of the observer. Figure 2c shows the estimation error of the angular velocity and current. The input voltage response under OFC is presented in Figure 2d.

Figure 2 shows that the estimation error of the fast state converges more rapidly compared to the slow state. Hence, as the estimation error becomes zero quickly, the slow state estimation is guaranteed to converge in finite time. The input voltage depicts a peak due to the fast convergence of the fast state. However, this could be reduced by restricting the input signal inside the region of interest and making the input globally bounded.

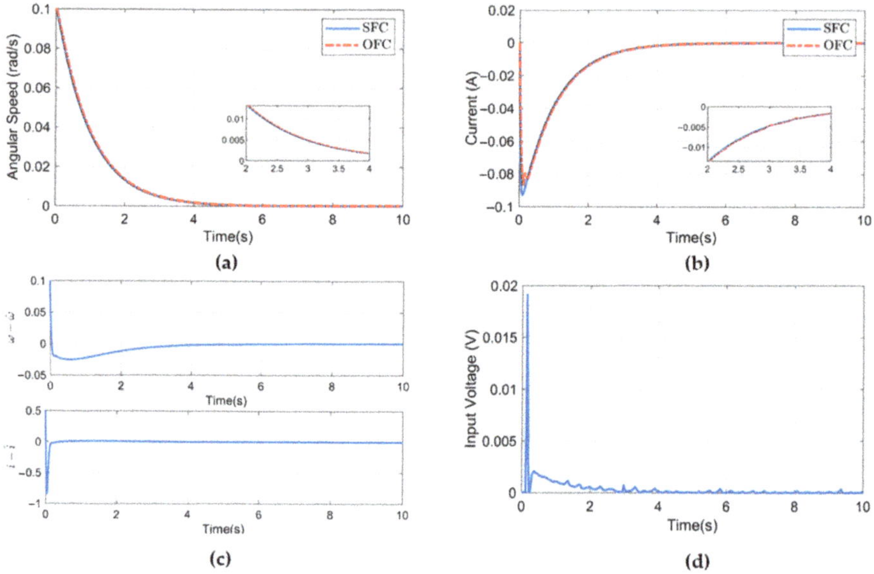

Figure 2. (**a**) Angular speed of the PMDC under SFC and OFC. (**b**) Armature current under SFC and OFC. (**c**) Estimation error of the estimated states with measured states. (**d**) Input voltage under OFC.

4. Conclusions

In this article, an OFC is designed for a two-time-scale nonlinear system. The proposed OFC is demonstrated on a PMDC motor example to provide empirical evidence regarding the effectiveness and robustness of the observer. The results demonstrate that the response under OFC not only matches that under state feedback control, but also shows immunity to parametric uncertainties, which makes it practically implementable for two-time-scale nonlinear systems.

Author Contributions: Conceptualization, A.R. and F.M.M.; methodology, N.M. and F.M.M.; software, A.R. and H.U.; writing—original draft preparation, A.R.; writing—review and editing, H.U.; validation, R.K.; formal analysis, N.M. and A.K.; investigation, H.U. and A.K.; data curation, R.K. All authors have read and agreed to the published version of the manuscript.

Funding: This research received no external funding.

Institutional Review Board Statement: Not applicable.

Informed Consent Statement: Not applicable

Data Availability Statement: Not applicable.

Conflicts of Interest: The authors declare no conflict of interest.

References

1. Kokotović, P.; Khalil, H.K.; O'reilly, J. *Singular Perturbation Methods in Control: Analysis and Design*; SIAM: New York, NY, USA, 1999.
2. Raza, A.; Malik, F.M.; Mazhar, N.; Khan, R. Two-time-scale robust output feedback control for aircraft longitudinal dynamics via sliding mode control and high-gain observer. *Alex. Eng. J.* **2022**, *61*, 4573–4583. [CrossRef]
3. Khalil, H.K. *Nonlinear Systems*, 3rd ed.; Prentice Hall: Englewood Cliffs, NJ, USA, 2002.
4. Khalil, H.K. *High-Gain Observers in Nonlinear Feedback Control*; SIAM: Philadelphia, PA, USA, 2017.

engineering proceedings

MDPI

Proceeding Paper

Obstructions in BIM Implementation for Developing Countries—A Mini-Review †

Usman Aftab [1], Farrokh Jaleel [1], Rafiq Mansoor [1], Muhammad Haroon [1,2,*] and Mughees Aslam [3]

1 Department of Mechanical Engineering, International Islamic University, Islamabad 44000, Pakistan; usman.phdem18@iiu.edu.pk (U.A.); farrokh.jaleel@iiu.edu.pk (F.J.); rafiq.mansoor@iiu.edu.pk (R.M.)
2 Department of Mechanical Engineering, Capital University of Science and Technology (CUST), Islamabad 44000, Pakistan
3 Department of Construction Engineering and Management, National University of Sciences and Technology, Risalpur 24080, Pakistan; maslam@mce.nust.edu.pk
* Correspondence: muhammad.haroon@cust.edu.pk
† Presented at the Third International Conference on Advances in Mechanical Engineering 2023 (ICAME-23), Islamabad, Pakistan, 24 August 2023.

Abstract: Technologically advanced countries are accruing benefits from the adoption of Building Information Modelling (BIM) in the Architecture, Engineering, and Construction (AEC) industry after decades of struggling for productivity enhancement using innovation and automation. Most of the developing countries have not been able to embrace technology in the AEC Sector and, consequently, are unable to ameliorate prevalent performance-related issues in construction projects. This review article identifies hurdles to BIM implementation in selected developing countries via an examination of the latest studies. The most significant challenges observed in this study are Lack of Training for Professionals, Lack of Awareness, Huge Capital Cost, Resistance to Change, and Complexity of BIM Software. This study is an update on previous studies conducted with the aim of assisting the implementation of BIM in developing countries.

Keywords: building information modeling; AEC industry; BIM implementation

Citation: Aftab, U.; Jaleel, F.; Mansoor, R.; Haroon, M.; Aslam, M. Obstructions in BIM Implementation for Developing Countries—A Mini-Review. *Eng. Proc.* **2023**, *45*, 26. https://doi.org/10.3390/engproc2023045026

Academic Editors: Mohammad Javed Hyder, Muhammad Mahabat Khan, Muhammad Irfan and Manzar Masud

Published: 11 September 2023

1. Introduction

Building Information Modeling (BIM) is believed to be one of the most significant evolutions in the Architecture, Engineering, and Construction (AEC) industry. It is an innovative way for virtual designing, project management, and performance management [1]. In developed countries like Finland, Singapore, Korea, the USA, the UK, Australia, and Germany, BIM adoption has accrued many benefits to the stakeholders, which include better cost estimation, better overall design understanding, reduced construction cost, improved construction planning and monitoring and project quality enhancement [2]. However, most of the developing countries are lagging in BIM adoption due to numerous cultural, financial, and organizational constraints. Sustainable adoption of BIM in developing countries is essentially needed for economic progression, technological advancement, and performance improvement [3]. Governments of technologically advanced countries have played an important role in overcoming implementation challenges [4].

This review paper will endeavor to highlight the most significant challenges to BIM implementation being encountered in developing countries, which need more attention from relevant stakeholders. In this regard, literature from the last five years is surveyed with the latest information regarding BIM implementation in 15 developing countries, namely Ethiopia, Hongkong, Nigeria, GCC, Pakistan, Yemen, India, Indonesia, Seychelles, Poland, Cambodia, Jordan, Malaysia, Turkey, and Iran.

2. Main Findings and Discussion

Important findings of this review study are as follows:

1. The literature confirms that the BIM adoption rate is low in developing countries because of the distinct challenges faced by each one of them.
2. The literature review suggests that most of the developing countries (excluding Turkey, Malaysia, Hongkong, and Poland) are at the "infancy stage" of BIM adoption (pre-implementation stage) and are still struggling with basic issues of BIM awareness and capacity building.
3. Most of the studies highlight that the lead role of the government in the implementation of BIM can be instrumental in overcoming the challenges.
4. This study identifies four categories of challenges to implementing BIM in developing countries from the literature, i.e., (1) legal and contractual, (2) process-based, (3) cultural and organizational, and (4) government-related. The main categories are further divided into 27 sub-categories. Figure 1 illustrates the frequency of identified challenges reported in the literature.
5. The most significant challenges to the implementation of BIM in the selected studies are described as follows as per the frequency of reporting. Details are enclosed in Table 1 below.
    (1) Lack of Training and BIM Expertise. The most repeated barrier in the selected studies is the lack of training opportunities for the professionals. Developing countries do not possess an adequate number of BIM experts for implementing BIM.
    (2) Lack of Awareness. Most of the stakeholders in developing countries are still uninformed about the significance of BIM and its associated advantages.
    (3) Huge Capital Cost. The initial investment in training of Human Resources and installation of technical infrastructure is substantially high for the AEC firms in developing countries.
    (4) Resistance to Change/unwillingness/Bias. Resistance to change falls in the realm of cultural domain. Decision makers in the AEC Industry in most of the developing countries show their reluctance to implement BIM.
    (5) The complexity of BIM software. BIM implementation requires the integration of multiple computer software with independent technical requirements and inputs, making it complex, especially for beginners.

Table 1. Most Significant Challenges to BIM Implementation in Developing Countries.

Barriers/Risks/Challenges	**Ethiopia**	**Hongkong**	**Nigeria**	**GCC**	**Pak**	**Yemen**	**India**	**Indonesia**	**Seychelles**	**Poland**	**Cambodia**	**Jordan**	**Malaysia**	**Turkey**	**Iran**	**Frequency**
References	[5]	[2]	[6]	[7]	[8]	[9]	[10]	[11]	[12]	[13]	[14]	[15]	[16]	[17]	[18]	
Lack of Awareness	▲		▲	▲		▲	▲		▲	▲		▲		▲	▲	10
Lack of Training/BIM Expertise	▲		▲		▲	▲	▲		▲	▲	▲	▲	▲	▲	▲	12
Huge capital cost			▲		▲	▲	▲	▲			▲	▲	▲			8
Resistance to Change/unwillingness/Bias	▲	▲		▲				▲			▲			▲	▲	7
Complexity of BIM software	▲	▲					▲	▲						▲	▲	6

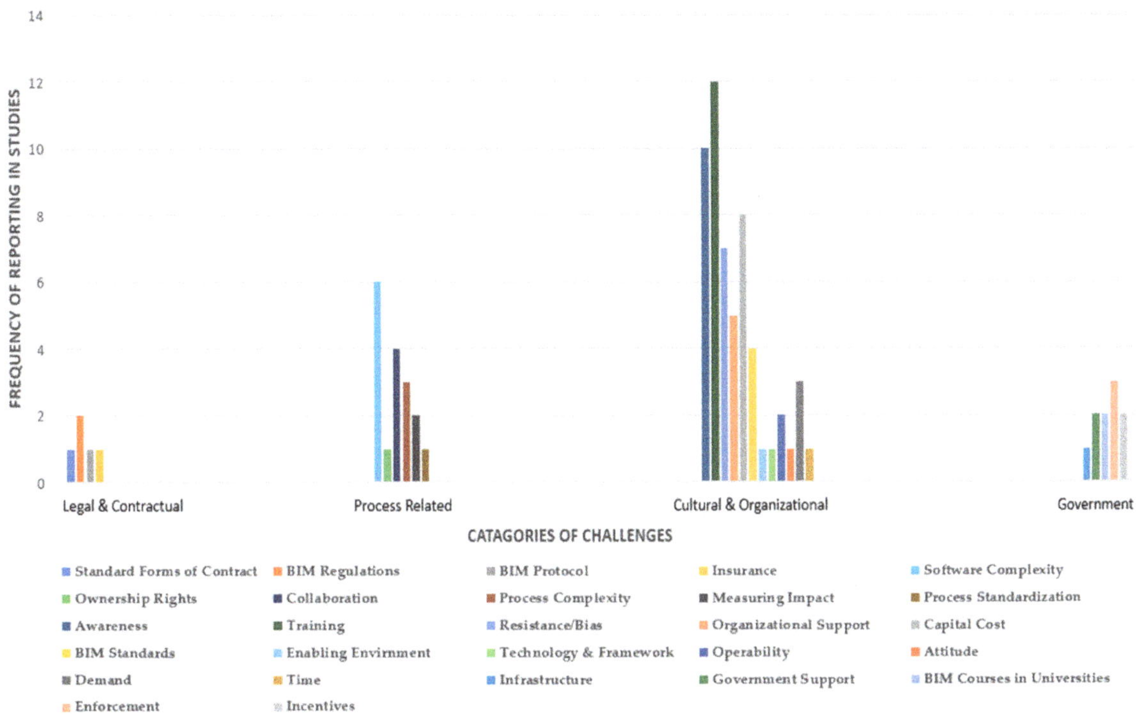

Figure 1. Frequency of Challenges to BIM Implementation in Developing Countries.

3. Conclusions and Recommendations

Out of fifteen selected developing countries, only Turkey, Malaysia, Poland, and Hong Kong have been able to achieve preliminary milestones for BIM implementation, whereas the remaining countries are still struggling with teething problems due to technological, cultural, and financial challenges. Following are the recommendations for facilitating BIM implementation in the developing countries:

A. Developing countries need to take guidance from the experiences in the developed countries regarding digitalization/automation required for BIM implementation [19].
B. BIM needs to be incorporated into the curriculum of relevant engineering and technical education areas for general awareness and training of manpower [6].
C. Developing countries are required to formulate a roadmap for BIM implementation as per their distinct environment [5].
D. New adopter countries are required to make BIM adoption up to a particular level mandatory for AEC firms [20].
E. Cost Benefit Analysis of BIM implementation must be studied by all stakeholders to financially and technically evaluate BIM adoption [21].

Author Contributions: Conceptualization, U.A. and F.J.; methodology, R.M. and M.A.; formal analysis, M.H. and M.A.; investigation, M.H. and U.A.; data curation, R.M. and U.A.; writing—original draft preparation, U.A.; writing—review and editing, F.J., R.M. and M.H.; supervision, F.J. All authors have read and agreed to the published version of the manuscript.

Funding: This research received no external funding.

Institutional Review Board Statement: Not applicable.

Informed Consent Statement: Not applicable.

Data Availability Statement: Not applicable.

Conflicts of Interest: The authors declare no conflict of interest.

References

1. Azhar, S. Building Information Modeling (BIM): Trends, Benefits, Risks, and Challenges for the AEC Industry. *Leadersh. Manag. Eng.* **2011**, *11*, 241–252. [CrossRef]
2. Chan, D.W.; Olawumi, T.O.; Ho, A.M. Perceived benefits of and barriers to Building Information Modelling (BIM) implementation in construction: The case of Hong Kong. *J. Build. Eng.* **2019**, *25*, 100764.
3. Saka, A.B.; Chan, D.W.M.; Siu, F.M.F. Drivers of sustainable adoption of building information modelling (BIM) in the Nigerian construction small and medium-sized enterprises (SMEs). *Sustainability* **2020**, *12*, 3710. [CrossRef]
4. Lee, G.; Borrmann, A. BIM policy and management. *Constr. Manag. Econ.* **2020**, *38*, 413–419. [CrossRef]
5. Belay, S.; Goedert, J.; Woldesenbet, A.; Rokooei, S. Enhancing BIM implementation in the Ethiopian public construction sector: An empirical study. *Cogent Eng.* **2021**, *8*, 1886476. [CrossRef]
6. Babatunde, S.O.; Udeaja, C.; Adekunle, A.O. Barriers to BIM implementation and ways forward to improve its adoption in the Nigerian AEC firms. *Int. J. Build. Pathol. Adapt.* **2021**, *39*, 48–71. [CrossRef]
7. Umar, T. Challenges of BIM implementation in GCC construction industry. *Eng. Constr. Arch. Manag.* **2022**, *29*, 1139–1168. [CrossRef]
8. Akdag, S.G.; Maqsood, U. A roadmap for BIM adoption and implementation in developing countries: The Pakistan case. *Archnet-IJAR* **2020**, *14*, 112–132. [CrossRef]
9. Gamil, Y.; Rahman, I.A.R. Awareness and challenges of building information modelling (BIM) implementation in the Yemen construction industry. *J. Eng. Des. Technol.* **2019**, *17*, 1077–1084. [CrossRef]
10. Ahuja, R.; Sawhney, A.; Jain, M.; Arif, M.; Rakshit, S. Factors influencing BIM adoption in emerging markets-the case of India. *Int. J. Constr. Manag.* **2020**, *20*, 65–76. [CrossRef]
11. Telaga, A.S. A review of BIM (Building Information Modeling) implementation in Indonesia construction industry. In *IOP Conference Series: Materials Science and Engineering*; Institute of Physics Publishing: Bristol, UK, 2018.
12. Adam, V.; Manu, P.; Mahamadu, A.-M.; Dziekonski, K.; Kissi, E.; Emuze, F.; Lee, S. Building information modelling (BIM) readiness of construction professionals: The context of the Seychelles construction industry. *J. Eng. Des. Technol.* **2021**, *20*, 823–840. [CrossRef]
13. Leśniak, A.; Górka, M.; Skrzypczak, I. Barriers to bim implementation in architecture, construction, and engineering projects—The Polish study. *Energies* **2021**, *14*, 2090. [CrossRef]
14. Durdyev, S.; Mbachu, J.; Thurnell, D.; Zhao, L.; Hosseini, M.R. BIM adoption in the cambodian construction industry: Key drivers and barriers. *ISPRS Int. J. Geo-Inf.* **2021**, *10*, 215. [CrossRef]
15. Hyarat, E.; Hyarat, T.; Al Kuisi, M. Barriers to the Implementation of Building Information Modeling among Jordanian AEC Companies. *Buildings* **2022**, *12*, 150. [CrossRef]
16. Manzoor, B.; Othman, I.; Gardezi, S.S.S.; Altan, H.; Abdalla, S.B. BIM-Based Research Framework for Sustainable Building Projects: A Strategy for Mitigating BIM Implementation Barriers. *Appl. Sci.* **2021**, *11*, 5397. [CrossRef]
17. Tan, S.; Ayalp, G.G. Root factors limiting BIM implementation in developing countries: Sampling the Turkish AEC industry. *Open House Int.* **2022**, *47*, 732–762. [CrossRef]
18. Khanzadi, M.; Sheikhkhoshkar, M.; Banihashemi, S. BIM applications toward key performance indicators of construction projects in Iran. *Int. J. Constr. Manag.* **2020**, *20*, 305–320. [CrossRef]
19. Borrmann, A.; König, M.; Koch, C.; Beetz, J. Building Information Modeling: Why? What? How? In *Building Information Modeling*; Springer International Publishing: Cham, Switzerland, 2018; pp. 1–24.
20. Hamma-Adama, M.; Kouider, T. Comparative Analysis of BIM Adoption Efforts by Developed Countries as Precedent for New Adopter Countries. *Curr. J. Appl. Sci. Technol.* **2019**, *36*, 1–5. [CrossRef]
21. Al-Ashmori, Y.Y.; Othman, I.; Rahmawati, Y.; Amran, Y.H.M.; Sabah, S.H.A.; Rafindadi, A.D.; Mikić, M. BIM benefits and its influence on the BIM implementation in Malaysia. *Ain Shams Eng. J.* **2020**, *11*, 1013–1019. [CrossRef]

engineering proceedings

MDPI

Proceeding Paper

Real-Time Vehicle Lateral Dynamics Estimation Using State Observer and Adaptive Filter †

Malik Kamal Mazhar *, Muhammad Jawad Khan, Karam Dad Kallu and Yasar Ayaz

School of Mechanical & Manufacturing Engineering (SMME), National University of Sciences & Technology, Islamabad 44000, Pakistan; jawad.khan@smme.nust.edu.pk (M.J.K.); karamdad.kallu@smme.nust.edu.pk (K.D.K.); yasar@smme.nust.edu.pk (Y.A.)

* Correspondence: kamalmazhar.pg@smme.edu.pk

† Presented at the Third International Conference on Advances in Mechanical Engineering 2023 (ICAME-23), Islamabad, Pakistan, 24 August 2023.

Abstract: Accurate state estimation of a vehicle is essential for ensuring the effective operation of stability control systems, particularly in dynamic road conditions. The side-slip angle serves as a crucial parameter for vehicle handling and safety control. However, the commercially available sensors for measuring side-slip angle are often expensive, prompting the utilization of estimation methods that rely on vehicle dynamics and the available sensor measurements. This paper introduces a novel observer for side-slip angles that employs a bicycle model and directly incorporates the lateral accelerometer signal through roll angle estimation. Roll angle estimates are obtained using novel complementary filters (NCF). Complementary filter tuning parameters are adjusted automatically using the recursive least square estimation technique. The estimation performance of the mentioned algorithms is verified using standard maneuvers through CarSim®.

Keywords: attitude estimation; non-linear complementary filter; sensor fusion; lateral acceleration compensation; roll angle; side-slip angle

Citation: Kamal Mazhar, M.; Khan, M.J.; Kallu, K.D.; Ayaz, Y. Real-Time Vehicle Lateral Dynamics Estimation Using State Observer and Adaptive Filter. *Eng. Proc.* **2023**, *45*, 27. https://doi.org/10.3390/engproc2023045027

Academic Editors: Mohammad Javed Hyder, Muhammad Mahabat Khan, Muhammad Irfan and Manzar Masud

Published: 12 September 2023

1. Introduction

Accurate estimation of sideslip angles is crucial for effective control of the vehicle's steering stability [1,2]. The measurement of the side-slip angle can be accomplished through direct or indirect means, such as utilizing GPS or optical sensors. Nevertheless, the measurements obtained through these methods exhibit high sensitivity to changes in the surrounding environment and weather conditions. As a consequence, the reliability of such measurements is considerably diminished. Moreover, this reliance on indirect measurement techniques introduces additional costs to the overall system.

In order to address this challenge, several research endeavors have been undertaken to explore feasible strategies for indirect estimation of the side-slip angle [3–7].

Fukada [4] approached entails limitations regarding vehicle dynamics due to the non-linear region characteristics of tires in maneuvers like J-turns. Farrelly and P. Wellstead developed an observer utilizing the kinematic model, but they did not account for the influence of accelerometer errors and road disturbances on the roll and pitch angles [8]. H. Lee proposed a simple bicycle-model-based lateral velocity observer but it neglected uncertainties related to accelerometers [9].

The paper comprises the following components. In Section 2, mathematical modeling is discussed. In Section 3, a roll angle estimation method is introduced, which dynamically determines the gain parameters, eliminating the necessity for manual tuning of the filter and a novel side-slip angle observer is proposed using an altered bicycle model. The simulation results under various conditions and steer inputs are discussed in Section 4. Section 5 of the paper provides a summary that serves as the concluding section.

2. Mathematical Modeling

The bicycle model shown in Figure 1 is commonly used to describe the lateral motion of a vehicle [10]. It is a simplified linearized model that assumes a constant longitudinal velocity v_x and symmetrical lateral tire cornering stiffness of C_f and C_r. l_f and l_r are the distances from the wheel center to the front and rear. m is the mass. I_z is the yaw moment of inertia. $u = \delta_f$ is the steer angle input.a_y is the lateral acceleration and r is the yaw rate.

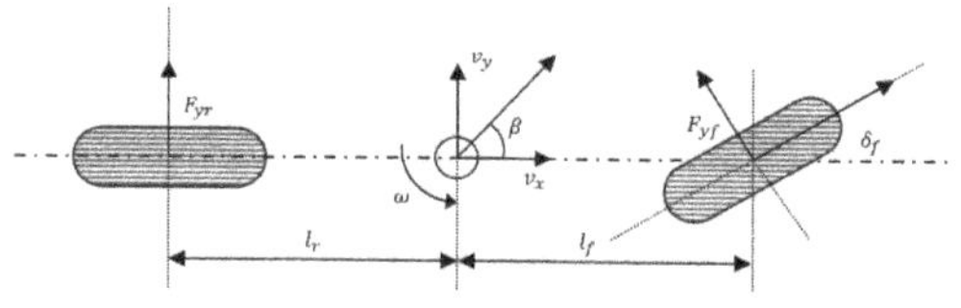

Figure 1. Bicycle model for vehicles.

The system can be represented by the state space matrix formulation as follows:

$$A = \begin{bmatrix} -\frac{2(C_f + C_r)}{mv_x} & \frac{2(C_r l_r - C_f l_f)}{m{v_x}^2} - 1 \\ \frac{2(C_r l_r - C_f l_f)}{I_z} & -\frac{2(C_f l_f^2 + C_r {l_r}^2)}{I_z v_x} \end{bmatrix} \quad (1)$$

$$B = \begin{bmatrix} \frac{2C_f}{mv_x} \\ \frac{2C_f l_f}{I_z} \end{bmatrix} \quad (2)$$

$$C = \begin{bmatrix} 0 & 1 \\ -\frac{2(C_f + C_r)}{m} & -\frac{2(C_f l_f - C_r l_r)}{mv_x} \end{bmatrix} \quad (3)$$

$$D = \begin{bmatrix} 0 \\ \frac{2C_f}{m} \end{bmatrix}$$

$$u = \delta_f andy = \begin{bmatrix} r \\ a_y \end{bmatrix} \quad (4)$$

3. Roll Angle and Side-Slip Angle Estimation

3.1. Adaptive Complementary Filter for Roll Angle of Ground Vehicle

The process of obtaining the roll angle estimate involves blending two initial estimates, each applicable within different operational ranges. During this blending procedure, the weights are chosen to consistently prioritize the more accurate estimate. Figure 2 depicts the block diagram of a second-order complementary filter for real-time implementation. It is important to ensure that the combined output of both filters results in a unity gain. As a result, the low-pass filter can be represented by the following equation of $G_2(s)$:

$$G_2(s) = \frac{s^2}{s^2 + as + b} \quad (5)$$

Similarly, the equation for high pass filter is written as

$$1 - G_2(s) = \frac{as + b}{s^2 + as + b} \quad (6)$$

Variables a and b are deciding factors for this estimate, which rely on operating conditions for different maneuvers.

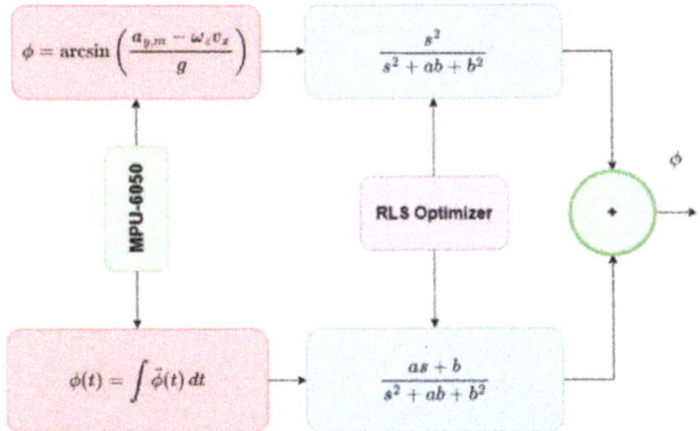

Figure 2. Adaptive complementary filter estimation block diagram.

3.2. *Side-Slip Angle Estimation*

The state estimate equation can be written as

$$\hat{x} = A\hat{x} + Bu + L(y - C\hat{x}) \quad (7)$$

where A, B and C matrices are state space model parameters. The estimation error, e, can be expressed as Equation (8).

$$e = x - \hat{x},\ \dot{e} = (A - LC)e = \hat{A}e \quad (8)$$

The observer gain matrix L, which can be determined using the pole placement method, plays a crucial role in achieving this convergence.

4. Experiment and Results

The validation process involved conducting multiple standardized tests, including step steer and double lane change maneuvers, and comparing the results obtained from our analytical model with those generated by CarSim®.

Figure 3a shows the estimation of roll angles using NCF for the step input. RMSE for roll angles for the step input is 0.0030 deg. Similarly, Figure 3b shows the results of the DLC maneuver and the RMSE error for the roll angle is observed as 0.0022 deg. Once the roll angle is estimated, we used the lateral acceleration equation to obtain the side-slip angle estimate using the Luenberger observer. The step steer input was fed to estimate the side-slip angle in Figure 4a and the RMSE for step input estimate is 6.7881×10^{-5}. We further investigated the algorithm efficacy by using the DLC input as seen in Figure 4b, where an RMSE error of 5.2537×10^{-5} deg can be observed.

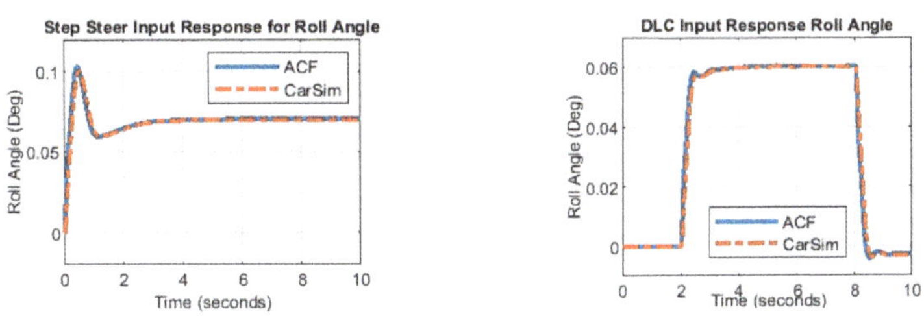

(**a**) (**b**)

Figure 3. Roll angle estimate for step steer (**a**) and DLC (**b**).

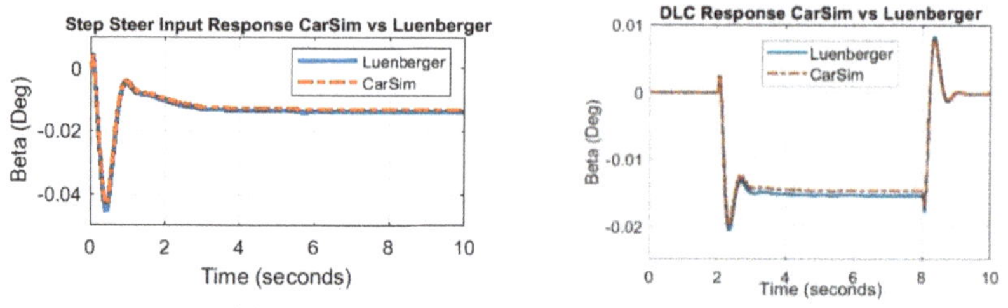

(**a**) (**b**)

Figure 4. Side-slip angle estimate for step steer (**a**) and DLC (**b**).

5. Conclusions

This research paper presents an innovative method for directly utilizing the lateral accelerometer signal to estimate the side-slip angle. In order to mitigate the influence of gravity-induced bias on the lateral accelerometer signal, we propose a vehicle roll angle estimator. This method offers a valuable contribution to the field and demonstrates promising potential for application. Observer performance is verified through CarSim® using different tests conducted under diverse road conditions and steering inputs. Incorporation of a lateral acceleration compensator improved the estimation results for the side-slip angle.

Author Contributions: Conceptualization, M.K.M. and M.J.K.; supervision, Y.A.; review and editing, M.K.M. and K.D.K. All authors have read and agreed to the published version of the manuscript.

Funding: This research received no external funding.

Institutional Review Board Statement: Not applicable.

Informed Consent Statement: Not applicable.

Data Availability Statement: Not applicable.

Conflicts of Interest: The authors affirm that the study was conducted without any commercial or financial affiliations that could be interpreted as a potential conflict of interest.

References

1. Zhang, W.; Wang, Z.; Zou, C.; Drugge, L.; Nybacka, M. Advanced Vehicle State Monitoring: Evaluating Moving Horizon Estimators and Unscented Kalman Filter. *IEEE Trans. Veh. Technol.* **2019**, *68*, 5430–5442. [CrossRef]
2. Katriniok, A.; Abel, D. Adaptive EKF-Based Vehicle State Estimation with Online Assessment of Local Observability. *IEEE Trans. Control. Syst. Technol.* **2016**, *24*, 1368–1381. [CrossRef]
3. Ungoren, A.Y.; Peng, H.; Tseng, H.E. A study of lateral speed estimation methods. *Int. J. Veh. Auto.* **2004**, *2*, 126–144. [CrossRef]
4. Fukada, Y. Slip-angle estimation for vehicle stability control. *Veh. Syst. Dyn.* **1999**, *32*, 375–388. [CrossRef]
5. Nam, K.; Oh, S.; Fujimoto, H.; Hori, Y. Estimation of sideslip and roll angles of electric vehicles using lateral tire force sensors through RLS and Kalman filter approaches. *IEEE Trans. Ind. Electron.* **2013**, *60*, 988–1000. [CrossRef]
6. Zhang, B.; Du, H.; Lam, J.; Zhang, N.; Li, W. A novel observer design for simultaneous estimation of vehicle steering angle and sideslip angle. *IEEE Trans. Ind. Electron.* **2016**, *63*, 4357–4366. [CrossRef]
7. Zhang, C.; Chen, Q.; Qiu, J. Robust $\mathcal{H}\infty$ filtering for vehicle sideslip angle estimation with sampled-data measurements. *Trans. Inst. Meas. Control* **2017**, *39*, 1059–1070. [CrossRef]
8. Farrelly, J.; Wellstead, P. Estimation of vehicle lateral velocity. In Proceedings of the 1996 IEEE International Conference on Control Applications IEEE International Conference on Control Applications held together with IEEE International Symposium on Intelligent Contro, Dearborn, MI, USA, 15–18 November 1996; pp. 552–557.
9. Lee, H. Reliability indexed sensor fusion and its application to vehicle velocity estimation. *J. Dyn. Syst. Meas. Control* **2006**, *128*, 236–243. [CrossRef]
10. Rajamani, R. *Vehicle Dynamics and Control*; Springer: Berlin/Heidelberg, Germany, 2012.

engineering proceedings

MDPI

Proceeding Paper

Shape Memory-Based Smart Extra Ocular Muscles for Ophthalmological Studies †

Zainab Ali * and Jahan Zeb Gul

Department of Mechatronics and Biomedical Engineering, Air University Main Campus PAF Complex, E-9, Islamabad 44230, Pakistan; jahanzeb@mail.au.edu.pk
* Correspondence: zainab.ali20@hotmail.com
† Presented at the Third International Conference on Advances in Mechanical Engineering 2023 (ICAME-23), Islamabad, Pakistan, 24 August 2023.

Abstract: Vision relies on precise eye movements controlled by six extraocular muscles (EOMs). Training kits and functional eye models are crucial for ophthalmological studies. The study investigates the potential of shape memory material (SMM) in ophthalmology, focusing on eye model mechanics and four thermally actuated smart EOMs. SMM was chosen over other actuators because of its high-power density (36 W/kg), high actuation forces (100 gf), and high displacements (200%). Finite element method (FEM) simulation on a 3D eye model was performed to analyze the thermo-mechanical properties of smart muscle, revealing the shape memory effect and super elasticity which result in the four eye movements of elevation, depression, adduction, and abduction. FEM simulation, compared to the eye model prototype, yielded a 16 mm deformation in smart muscle with a thermal actuation of 313 K (40 °C). FEM analysis, while combining smart muscle with the eye model, resulted in a 6.8517 mm deformation in the simulation. Prototype deformation was acquired as 7.262 mm. The successful actuation of the eye using thermally actuated smart muscles proves its potential for ophthalmology kits.

Keywords: extraocular muscles; ophthalmological models; shape memory material; thermally actuated smart muscles; finite element method; extraocular movements

Citation: Ali, Z.; Gul, J.Z. Shape Memory-Based Smart Extra Ocular Muscles for Ophthalmological Studies. *Eng. Proc.* **2023**, *45*, 54. https://doi.org/10.3390/engproc2023045054

Academic Editors: Mohammad Javed Hyder, Muhammad Mahabat Khan, Muhammad Irfan and Manzar Masud

Published: 21 September 2023

1. Introduction

The study focuses on the importance of understanding EOM mechanics for scientific and ophthalmological purposes. Existing models have limitations in replicating the complex functionality of human eye movements [1]. Training kits and models are necessary for teaching EOM assessment [2]. Previous biomimetic devices have attempted to mimic realistic eye movements using different approaches such as robotics [3], and rubber-based models [4]. SMMs have gained popularity in soft robotics and biomedical fields due to their unique characteristics, allowing them to recover their shape after deformation [5]. This study explores the potential of using SMMs in ophthalmology, considering their high-power density, actuation forces, and displacements. FEM simulations on a 3D eye model are conducted to analyze the shape memory effect and super elasticity of smart muscles. The simulations demonstrate that SMMs can enable realistic eye movements, including elevation, depression, adduction, and abduction. The results are evaluated by comparing the FEM simulation with an eye model prototype.

2. Methodology

A digital three-dimensional (3D) eye model was created using Solid Works computer-aided design (CAD) software, accurately representing the sclera, cornea, lens, and optic nerve [6]. The rectus muscles were replaced with SMMs from the BMX 150 series by Toki Corporation [7]. The oblique muscles were not included, as the scope of this study does not include the rotation motions of the eye. The SMM helices were attached to the eye sclera as muscle replacements, with the other end secured on top of the optic nerve. In this study, shape memory-based helix structures mimic four EOM, including the superior, inferior, medial, and lateral recti. The dimensions of the eye model and muscle helix are shown in Table 1.

Table 1. Dimensions of helical spring mimicking extraocular muscle and eyeball along with the support shaft mimicking the optic nerve.

Characteristics	Unit	Specification
Coil diameter (D)	mm	0.62
Wire diameter (d)	mm	0.15
Coil diameter to wire diameter ratio D/d	-	4.1
Number of turns	mm	6
Diameter (Eyeball)	mm	24
Diameter (shaft)	mm	3
Radius (eyeball)	mm	12
Radius (shaft)	mm	1.5
Length	mm	10
Area	mm^2	1802.461
Perimeter	mm	9.428

2.1. Simulation and Fabrication

A simplified eye model was created for efficient simulation in Ansys Workbench 22 R1. The model utilized SMM for the helical spring and rubber for the eye material, as shown in Figure 1a,b. Material properties, such as elastic modulus and temperature scaling parameters, were defined. After successful meshing, the model underwent thermomechanical analysis with applied forces and varying temperatures. After a successful simulation, and the acquired results, the next step was to 3D-print a CAD model. The Stereo Lithography (STL) file of the model was evaluated prior to the printing. The meshing size and printing parameters were observed. Finite difference method (FDM)-based 3D printing was performed on a reality Ender 3D printer using a material polylactic acid (PLA).

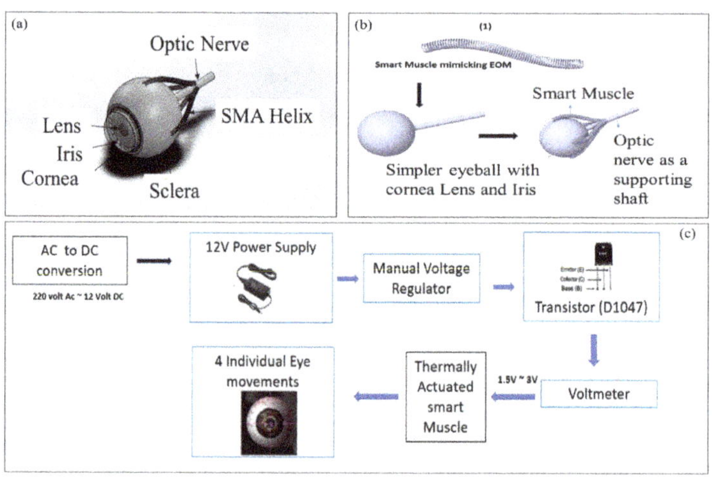

Figure 1. Complete schematic: (**a**) finished 3D model with labeled anatomy; (**b**) complete schematic of the model used for simulation; (**c**) Block diagram of an open loop-based control system used for prototype testing.

2.2. Prototype Testing

The model underwent testing and physical analysis using an open-loop control system. A fixed voltage of 0–12 V was acquired as an output whereas the muscle was activated on 2.8 V and 1.5 A. The muscle was stretched to twice its original length using hand force, and then contracted back by applying a stimulus in terms of joule heating via voltage from a control system. The smart muscles' shape memory effect enabled them to deform and move the eye. Four muscles were given individual stimuli to observe desired movements, including elevation, depression, adduction, and abduction. Figure 1d shows the block diagram of a control system.

3. Results and Discussion

3.1. Simulation Results

The simulation results are shown in Figure 1a–c. It was observed that the maximum displacement in the case of SMM was 43 mm, greater than the original length of 28 mm, i.e., 28 − 43 = 15 mm. The results have been validated with the constitutive model based on Hencky strain [8]. Meanwhile, the shape memory helix attached to an eyeball gave a maximum displacement of 6.8157 mm in all four cases but with different axes, two movements in the horizontal axis (+x, −x) and two in the vertical (+y, −y). Figure 2 shows the simulation results.

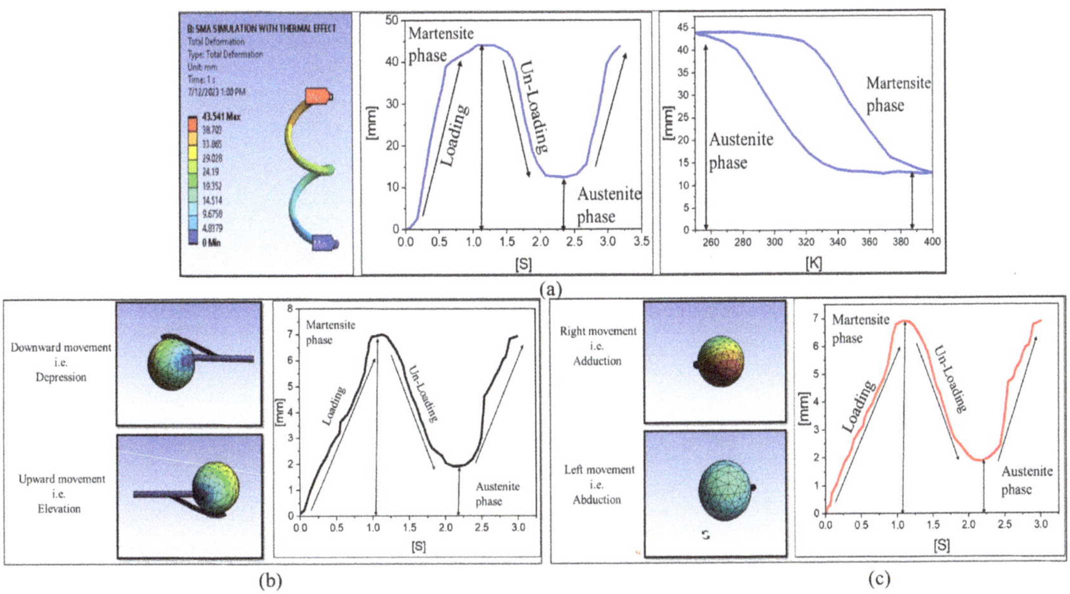

Figure 2. Simulation results: (**a**) smart muscle showing deformation vs. time, and deformation vs. temperature; (**b**) smart muscle attached with eyeball showing shape memory effect in terms of deformation vs. steps for two vertical movements; (**c**) smart muscle attached with eyeball showing shape memory effect in terms of deformation vs. steps for two horizontal movements.

3.2. Prototype Results

The eyeballs with four shape memory helixes were subjected to the control system, as shown in Figure 1d, to provide a stimulus in terms of joule heating so that SMM could contract and relax just like a muscle and achieve the necessary movements. The primary action against each muscle was observed using tracker software and a modeling tool. The results acquired in the form of primary eye movements and their deviation in length over time are shown in Figure 3 and Table 2.

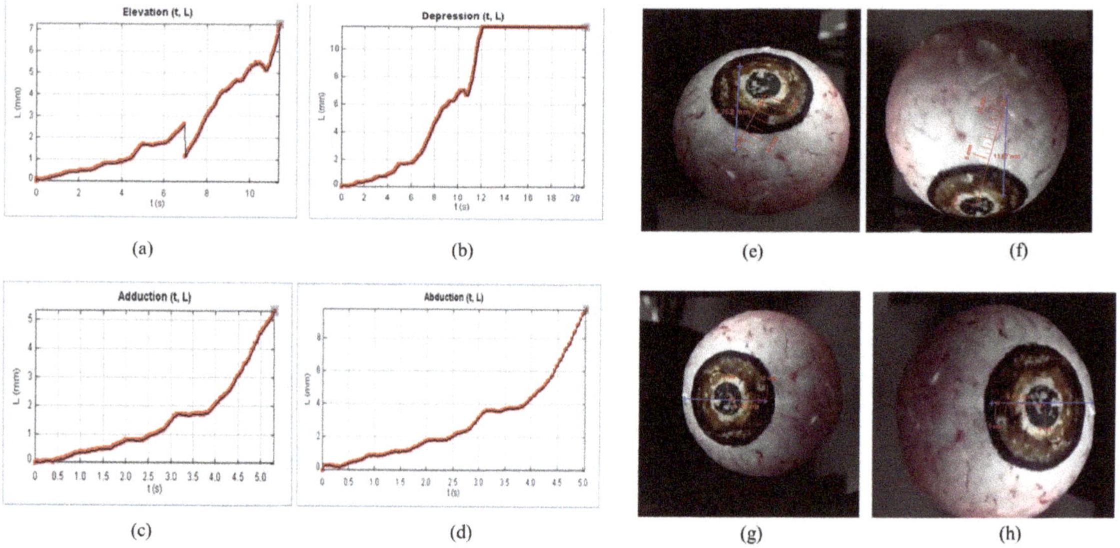

Figure 3. Tracker graphs acquired against all primary eye movements: (**a**) elevation caused by superior rectus (**b**) depression caused by inferior rectus (**c**) adduction caused by medial rectus (**d**) abduction caused by lateral rectus Prototype eye model movements: (**e**) upward movement, i.e., elevation; (**f**) downward movement, i.e., depression; (**g**) leftward movement, i.e., abduction; (**h**) rightward movement, i.e., adduction.

Table 2. Primary eye movements and their deviation in length over time.

Primary Movements	Time [s]	Length [mm]	Speed [mm/s]
Elevation	11.380	7.262	0.638
Depression	12.180	11.67	0.957
Adduction	5.033	9.793	1.941
Abduction	5.300	5.357	1.010

4. Conclusions

The following information was obtained:

i. Tracking graphs were acquired (distance covered over time using video analysis and modeling tool).
ii. The speeds of four extraocular smart muscles (SR = 0.638 mm/s, IR = 0.957 mm/s, MR = 1.94 mm/s, LR = 1.010 mm/s).
iii. Deformation of smart muscle (simulation = 6.8517 mm, prototype = 7.262 mm).
iv. Motion of an eyeball (elevation/upward, depression/downward, adduction/right, abduction/left).
v. Material phase characterizations were observed (martensite phase to austenite phase).

Author Contributions: Conceptualization, Z.A. and J.Z.G.; methodology, Z.A. and J.Z.G.; software, Z.A; validation, Z.A. and J.Z.G.; formal analysis, Z.A. and J.Z.G.; investigation, Z.A. and J.Z.G.; data curation, Z.A.; writing—original draft preparation, Z.A.; writing—review and editing, Z.A.; supervision, J.Z.G.; project administration, J.Z.G. All authors have read and agreed to the published version of the manuscript.

Funding: This research received no external funding.

Institutional Review Board Statement: Not applicable.

Informed Consent Statement: Not applicable.

Data Availability Statement: The data will be available on request.

Conflicts of Interest: The authors declare no conflict of interest.

References

1. Regal, S.; Troughton, J.; Djenizian, T.; Ramuz, M. Biomimetic models of the human eye, and their applications. *Nanotechnology* **2021**, *32*, 302001. [CrossRef] [PubMed]
2. Bernd, P.; Jakway, J. A simplified approach to teaching medical students ocular movements and the rationale in testing the oculomotor, trochlear, and abducent nerves. *Anat. Sci. Educ.* **2008**, *1*, 126–129. [CrossRef] [PubMed]
3. Gu, J.J.; Meng, M.; Cook, A.M.; Liu, P.X. Design, Sensing and Control of a Robotic Prosthetic Eye for Natural Eye Movement. *Appl. Bionics Biomech.* **2006**, *3*, 134949. [CrossRef]
4. Wood, A.J.; Dayal, M.R. Using a Model to Understand the Symptoms of Ophthalmoplegia. *J. Undergrad. Neurosci. Educ.* **2018**, *16*, R33–R38. [PubMed]
5. Copaci, D.-S.; Blanco, D.; Martin-Clemente, A.; Moreno, L. Flexible Shape Memory Alloy Actuators for Soft Robotics: Modelling and Control. *Int. J. Adv. Robot. Syst.* **2020**, *17*, 1729881419886747. [CrossRef]
6. Bekerman, I.; Gottlieb, P.; Vaiman, M. Variations in Eyeball Diameters of the Healthy Adults. *J. Ophthalmol.* **2014**, *2014*, 503645. [CrossRef] [PubMed]
7. Biometal Toki Corporation Co., Ltd. Available online: https://www.toki.co.jp/biometal/english/contents.php (accessed on 10 December 2022).
8. Arghavani, J.; Auricchio, F.; Auricchio, F.; Naghdabadi, R. A Finite Strain Kinematic Hardening Constitutive Model Based on Hencky Strain: General Framework, Solution Algorithm and Application to Shape Memory Alloys. *Int. J. Plast.* **2011**, *27*, 940–961. [CrossRef]

engineering proceedings

MDPI

Proceeding Paper

Role of the Pakistan Cement Industry towards the Achievement of Net Zero Goal by Mid-Century: A Review from a Waste Heat Recovery Perspective †

Muhammad Haroon [1,2,*] and Abubakr Ayub [2]

1 Department of Mechanical Engineering, Capital University of Science and Technology (CUST), Islamabad 44000, Pakistan
2 Department of Mechanical Engineering, International Islamic University, Islamabad 44000, Pakistan; abubakrayub@gmail.com
* Correspondence: muhammad.haroon@cust.edu.pk

† Presented at the Third International Conference on Advances in Mechanical Engineering 2023 (ICAME-23), Islamabad, Pakistan, 24 August 2023.

Abstract: Owing to high energy-intensive operations, cement production is responsible for global 6–8% of CO_2 emissions and, thus, can be a major contributor in the net zero mission. Pakistan's cement industry can produce 69 million tons of cement per year and has an overall share of 5.3% in economy. One ton of cement production releases approximately one ton of carbon dioxide. Therefore, it is necessary to decarbonize this industry. Two strategies can be employed (waste heat utilization and CO_2 capturing) for the decarbonization of the cement industry. This comprehensive review article is focused on the waste heat recovery potential and the technologies employed to utilize this potential for the cement industry of Pakistan.

Keywords: net zero goal; decarbonization; waste heat recovery; steam Rankine cycle; cement industry; CO_2 emissions; carbon neutralization; solar power generation; multigeneration systems

Citation: Haroon, M.; Ayub, A. Role of the Pakistan Cement Industry towards the Achievement of Net Zero Goal by Mid-Century: A Review from a Waste Heat Recovery Perspective. *Eng. Proc.* **2023**, *45*, 45. https://doi.org/10.3390/engproc2023045045

Academic Editors: Mohammad Javed Hyder, Muhammad Mahabat Khan, Muhammad Irfan and Manzar Masud

Published: 15 September 2023

1. Introduction

Over the past three decades, a steep incremental trend in carbon dioxide emissions has been observed, which is the main driving factor for global climate change. The two major contributors in CO_2 emissions are fossil fuels and industry. To protect humanity from the dangerous effects of climate change, all stakeholders should create strategies to minimize the CO_2 emissions. The main strategy for CO_2 emissions minimization is the net zero goal. All stakeholders (researchers, scientists, engineers, technologists, and companies) around the globe are increasingly trying to achieve the net zero goal. It simply means the removal of CO_2 as we produce by 2050. Achieving net zero means obtaining equity between greenhouse gases' production and removal from the atmosphere. It can only be operationalized through political, social, and improved techno-economic systems. Net zero has two components, namely, waste heat recovery and CO_2 capturing. This review study will only focus on waste heat utilization from cement production plants in Pakistan.

The net zero strategy is applicable for energy and nonenergy sectors. Despite substantial improvement in the energy sector, the nonenergy industry produces 31% of all process, with combustion CO_2 emissions and essential materials (e.g., steel, iron, cement and lime, nonferrous metals, chemicals, paper, and pulp) being responsible for approximately 22% of global CO_2 emissions over the last few decades [1]. Steel, iron, cement, chemicals and some other sectors have high needs to process heat and chemical reaction emissions, thus are named hard-to-abate industrial sectors. That is the reason these industrial sectors are the big candidates for CO_2 emissions and, thus, waste heat utilization.

Waste gases from cement production plants are usually released at high temperatures and mass flow rates and thus have significant potential for utilization. The quantity of heat released from the cement production plants to the atmosphere can be a considerable amount of the consumed energy for production process. Power production from waste gas utilization can provide a significant amount of the energy requirement for cement plants and can play a significantly important role in the net zero goal achievement.

Conventional technologies employed globally for power production, by utilizing waste heat from the industrial sector and other sources, are steam-driven Rankine cycles (single flash and dual pressure cycles), organic fluid-driven Rankine cycles [2], and Kalina cycles. Steam-driven Rankine cycles are mostly employed for the utilization of waste heat from cement industries situated in South Asia and Asia because the temperature of the exit gases from the air quenching cooler (AQC) and suspension preheater (SP) is high. Furthermore, the working fluid employed in these cycles is water, which is safe and environmentally friendly. Supercritical CO_2 (sCO_2)-driven Brayton power cycles are the emerging technology for power generation from waste heat. These cycles are beneficial in terms of high thermal efficiency, compact turbo-machinery, eco-friendly attributes, and cost-effectiveness [3,4].

The basic purpose of this review article is (1) to keep an eye on the contribution of the Pakistan cement sector in the net zero goal achievement, (2) to identify the shortcomings in this global cause, and (3) to propose a methodology which can fulfill the shortcomings in the decarbonization of the Pakistan cement industry.

2. Materials and Methods

The Pakistan cement industry's energy consumption is approximately 11% of the total industrial energy consumption in Pakistan. Depending on the technology and the age of cement production units, the average annual electricity consumption lies between 90 and 130 kWh. Out of 25 existing cement production plants (with 50 production lines), 11 waste heat recovery (WHR) systems are installed, possessing a worth of more than 110 MW capacity. The remaining waste heat utilization potential from the cement industry is 100–200 MW [5].

2.1. Recent Developments in the Cement WHR Sector of Pakistan

Researchers are continuously proposing solutions for the betterment of the Pakistan cement WHR sector. Ali et al. [6] used back propagation neural networks to predict the power of the waste heat utilization system of a Attock Cement Pakistan Limited and compared the results with the actual WHR cycle that was thermodynamically analyzed. Their results show 19.75% thermal efficiency and 10.06 MW power generation. Most significantly, the comparison of the actual WHR cycle and the predictive model concluded that data science is a strong alternative for thermodynamic modeling to evade hefty calculations.

2.2. Bestway, the Pioneer Cement Industry of Pakistan in Carbon Neutralization

Bestway is the largest cement production group in Pakistan with five production sites and eight production lines. The production capacity of the Bestway group is approximately 15.3 MT per 300 days. All the sites of Bestway are equipped with solar power generation and waste heat recovery (WHR) plants, which is the biggest step towards the net zero goal. The Bestway group fulfill their 45–50% energy needs through WHR and solar power generation. The Bestway group recently inaugurated its 7200 tpd plant in Mianwali, which fulfills its 50% energy requirement from solar renewable power generation and is hence called the greenfield cement production plant. Details of the Bestway group are enclosed in Table 1 below.

Table 1. Contribution of Bestway cement group for net zero goal.

Sr. No.	Bestway Cement Production Sites	Production Lines	Installed Solar Power Capacity	Installed WHR Capacity (MW)	Important Information Related to Solar Installations
1	Mianwali	1	20 MW	9	**Technology:** Roen captive solar energy.
2	Hattar	2	6.4 MW	6 (line 1), 9 (line 2)	**Annual power generation:** 97,992 MWH.
3	Chakwal	2	15.2 MW	15	**Reduction in levelized cost of energy:** 40%.
4	Farooqia	2	14.4 MW	7.5	**Annual CO_2 emissions reductions:** 60,265 tonnes.
5	Kallar Khar	1	14.8 MW	9.8	

2.3. *Proposed Methodology for Pakistan Cement Industry Decarbonization*

No one can deny the importance of the stand-alone waste heat utilization system; however, for more efficient and intelligent waste heat utilization, a multigeneration/sectorial integration system can be a good solution. These systems can produce multiple outputs at the same time with the same heat source. In this study, the same type of multigeneration system is proposed for the decarbonization of the Pakistan cement industry, i.e., to maximize the power output from WHR system and CO_2 capturing at the same time without disturbing cement production (Figure 1). Table 2 summarizes the installed waste heat recovery technologies for cement plants in Pakistan.

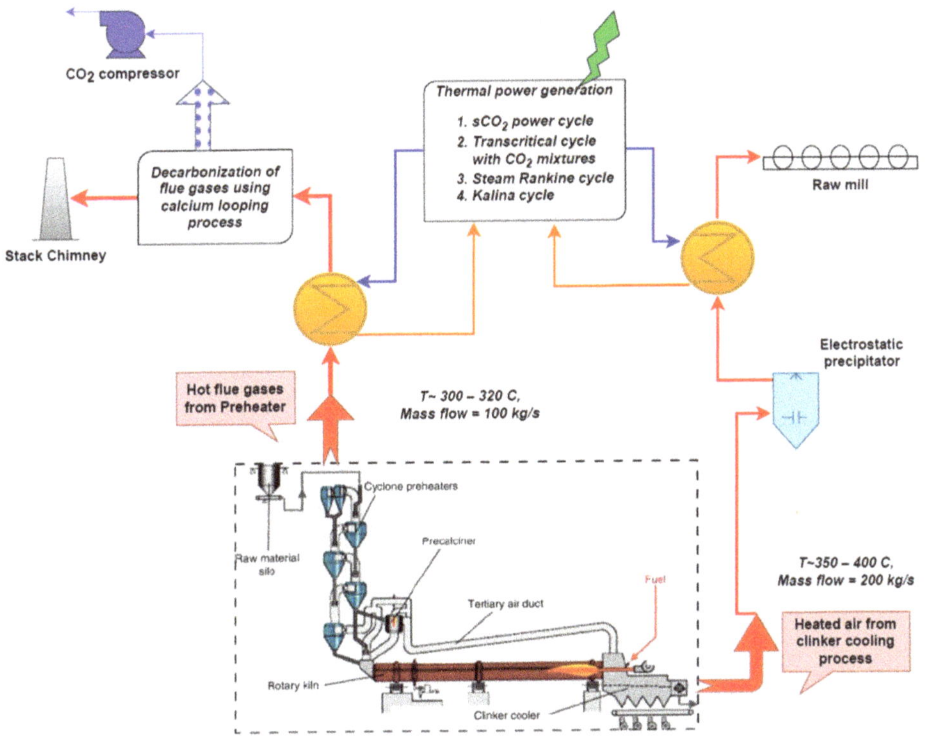

Figure 1. Proposed methodology for Pakistan cement industry decarbonization.

Table 2. Summary of waste heat recovery technologies for cement industries in Pakistan [5].

Cement Industry	Kiln Type/No of Lines/ Capacity (tpd)	Total Installation Cost (USD)	WHR Capacity (MW)	Net Zero Goal in Terms of CO_2 Savings (t/y)	Power Generation (MWh/y)
D.G Cement (D.G. khan Plant)	4/5 stage preheater/2/6700	15 million	10.4	40,332	70,088
(Khairpur plant)	Rotax 2/1/6700	11.8 million	8.5	28,542	61,301
Cherat Cement	Rotary/1/3200	9.3 million	7	25,761	41,730
Attock cement (Hub Plant)	4 × 64 Rotary/2/5200	18.6 million	12	37,908	58,320
Fecto Cement	Dry/1/2600	7.2 million	6	19,584	38,400
Karachi Plant	Rotary/2/6600	9.1 million	10	33,820	58,291
Lucky Cement	Dry/3/9000	12.54 million	15	42,992	87,437

3. Conclusions and Future Recommendations

The purpose of highlighting the waste heat potential of the Pakistan cement industry is to grab the attention of the research community, policy makers and investors in order to create strategies to decarbonize the industry. This can only be operationalized by also considering CO_2 capturing techniques. No CO_2 capturing facility is available for any cement production plant in Pakistan. So, conducting initial research and creating feasibility plans is recommended for the implementation of CO_2 capturing technology retrofit with waste heat recovery technologies for the Pakistan cement industry.

Author Contributions: M.H. and A.A. contributed equally to the conceptualization, methodology, data curation, and writing. All authors have read and agreed to the published version of the manuscript.

Funding: This research received no external funding.

Institutional Review Board Statement: Not applicable.

Informed Consent Statement: Not applicable.

Data Availability Statement: Not applicable.

Conflicts of Interest: The authors declare no conflict of interest.

References

1. Bataille, C.G.F. Physical and policy pathways to net-zero emissions industry. *Wiley Interdiscip. Rev. Clim. Chang.* **2020**, *11*. [CrossRef]
2. Haroon, M.; Ayub, A.; Sheikh, N.A.; Ahmed, M.; Shalaby, A.-B. Harnessing Ocean Thermal Energy from Offshore Locations in Pakistan Using an Organic Rankine Cycle. *Eng. Proc.* **2022**, *23*, 24. [CrossRef]
3. Haroon, M.; Ayub, A.; Sheikh, N.A.; Imran, M. Exergetic performance and comparative assessment of bottoming power cycles operating with carbon dioxide-based binary mixture as working fluid. *Int. J. Energy Res.* **2020**, *44*, 7957–7973. [CrossRef]
4. Haroon, M.; Sheikh, N.A.; Ayub, A.; Tariq, R.; Sher, F.; Baheta, A.T.; Imran, M. Exergetic, Economic and Exergo-Environmental Analysis of Bottoming Power Cycles Operating with CO_2-Based Binary Mixture. *Energies* **2020**, *13*, 5080. [CrossRef]
5. Zeb, K.; Ali, S.M.; Khan, B.; Mehmood, C.A.; Tareen, N.; Din, W.; Farid, U.; Haider, A. A survey on waste heat recovery: Electric power generation and potential prospects within Pakistan. *Renew. Sustain. Energy Rev.* **2017**, *75*, 1142–1155. [CrossRef]
6. Ali, A.; Kamal, K.; Ratlamwala, T.A.H.; Sheikh, M.F.; Arsalan, M. Power prediction of waste heat recovery system for a cement plant using back propagation neural network and its thermodynamic modeling. *Int. J. Energy Res.* **2021**, *45*, 9162–9178. [CrossRef]

engineering proceedings

MDPI

Proceeding Paper

Design of a Chassis Dynamometer Facility for the European Type-Approval of Passenger Cars Manufactured in Pakistan †

Kashif Usman [1,2], Muhammad Kashif [1], Muhammad Haroon [2,3,*], Muhammad Ahmed [3], Haseeb Ahmad [4] and Al-Bara Shalaby [2]

1 Department of Mechanical Engineering, University of Central Punjab, Lahore 54782, Pakistan; kashif.phdme51@iiu.edu.pk (K.U.); muhammad.kashif@ucp.edu.pk (M.K.)
2 Department of Mechanical Engineering, International Islamic University, Islamabad 44000, Pakistan; albara.phdme55@iiu.edu.pk
3 Department of Mechanical Engineering, Capital University of Science and Technology (CUST), Islamabad 44000, Pakistan; ahmedislam123@gmail.com
4 Department of Mechanical Engineering, Superior University, Lahore 54782, Pakistan; haseeb.ahmed@superior.edu.pk
* Correspondence: muhammad.haroon@cust.edu.pk
† Presented at the Third International Conference on Advances in Mechanical Engineering 2023 (ICAME-23), Islamabad, Pakistan, 24 August 2023.

Abstract: The objective of this research is to develop a suitable chassis dynamometer to test new vehicles for compliance with emission standards for type-approval in Pakistan. There is a lack of facilities and protocols for testing new passenger vehicles to measure vehicle exhaust gas emissions. This study is performed to fill this gap and might help local authorities in the implementation of exhaust emission test procedures. Assessment of exhaust emissions can be performed by placing the vehicle over the chassis dynamometer. An air-cooled eddy current dynamometer is selected for this purpose. The road load equation is used to simulate the real-life performance of a vehicle while driving at different speeds.

Keywords: chassis dynamometer; emission regulations; European type-approval

Citation: Usman, K.; Kashif, M.; Haroon, M.; Ahmed, M.; Ahmad, H.; Shalaby, A.-B. Design of a Chassis Dynamometer Facility for the European Type-Approval of Passenger Cars Manufactured in Pakistan. *Eng. Proc.* **2023**, *45*, 16. https://doi.org/10.3390/engproc2023045016

Academic Editors: Mohammad Javed Hyder, Muhammad Mahabat Khan, Muhammad Irfan and Manzar Masud

Published: 11 September 2023

1. Introduction

Pakistan currently has emission standards of Tiers Pak-II that are equivalent to standards of EURO 2. The measuring method for Tier Pak-II is the New European Driving Cycles (NEDC). Therefore, it is the latest technology recently adopted by automobile manufacturers in Pakistan. Now, it is mandatory for all automobile manufacturers in Pakistan to follow the emission standards. However, the problem is that there is no independent approval authority within Pakistan that can offer testing services to vehicle manufacturers to make sure that the specific emission standards are met. Table 1 (first two columns) shows the specifications shared by one of the leading manufacturers in Pakistan, in which the emission standards followed by the manufacturer can be seen clearly. However, on the other hand, any CO_2 emission data in compliance with EURO-II cannot be seen.

Researchers are performing experimental and numerical investigations on a chassis dynamometer design. Lourenco et al. [1] proposed a model for a vehicle and twin roller chassis dynamometer to improve mobility systems. Zhang et al. [2] enclosed the latest research (on vehicle chassis dynamometer development) in a review article and proposed the AC chassis dynamometer as a mainstream trend. Different measurement aspects, road simulation, and control systems are also discussed.

This research is targeted to design a chassis dynamometer for type-approval of new passenger vehicles manufactured in Pakistan underneath environmental legislations in force. This chassis dynamometer could be used as a test platform for direct performance testing of new passenger vehicles in the laboratory. Another feature of this study is to obtain insight into the real-world emissions behavior of road vehicles in varying operating situations. Exhaust emissions are analyzed by a vehicle driving on the chassis dynamometer inside the modern emission testing laboratory (SGS Pakistan Pvt. Ltd.) at Karachi to verify the fitness of the vehicle in terms of pollutant emissions. Initially, roller design calculations are performed using numerical design tools. Then, the structure of the chassis dynamometer is designed for the power absorption from the chassis dynamometer at the time of the emission test.

Table 1. Technical specifications of passenger cars manufactured in Pakistan (first two columns) and other important parameters of various vehicles manufactured in Pakistan (last five columns).

Description	Specifications	Manufacturer	Model	Max Power (hp)	Weight (Kg)	Tire Reference
Model	Toyota Corolla Altis	Suzuki	Mehran	40	800	145/70R12
Power	138 hp	Suzuki	Swift	90	1050	185/70R15
RPM @ Max Power	6400	Honda	Civic Oriel	140	1273	215/55R16
Max Torque	173 nm	Honda	BRV	120	1230	195/60R15
RPM @ Max Torque	4000	Toyota	Altis	138	1275	205/55R16
Engine displacement	1798 cc	Toyota	GLI	84	1275	196/65R15
Fuel type	Gasoline					
Number of cylinders	Four					
Top Speed	240 km/h					
Tire	205/55R16					
Kerb Weight	1275 kg					
Emission Standards	Euro-II					
CO_2 Emissions (g/km)	-------					

2. Materials and Methods

2.1. Design of Chassis Dynamometer

The vehicle behavior under different road conditions is described by the road load equation (RLE). This RLE (Equation (1)) is a fundamental requirement of a chassis dynamometer. In order to simulate the real-life performance of a vehicle, RLE is used to calculate the change in torque with the change in vehicle speed. Its primary importance is that it provides a linkage between performance on the road and performance in the test cell.

$$F_{roadload} = a + bV + cV^2 + M\frac{dV}{dt} + Mg\sin\theta \quad (1)$$

where $F_{roadload}$ is the resistance to progress (N), V is the vehicle speed (km/h), a is the value equivalent to rolling resistance (N), b is the value equivalent to frictional resistance N/(km/h), c is the value equivalent to the coefficient of air resistance N/(km/h)^2, M is the mass of the vehicle (kg), and θ is the road slope in radians.

Power required at any vehicle speed is shown in Equation (2) [3]:

$$\text{Power} = F_{roadload} * V \quad (2)$$

First, there is the need to conduct an elementary survey of vehicle specifications manufactured in Pakistan. In this context, Table 1 (last five columns) presents power and some other performance parameters for many light-duty cars manufactured in Pakistan.

2.2. Components Selection

Detailed information regarding the selection of the dynamometer, transmission shafts, and roller design is included in Table 2. Figure 1a shows the position of front and rear rollers with maximum and minimum angles of tires, and Figure 1b shows the isometric view of rollers and shaft with designed dimensions.

Table 2. Important final designed specifications of chassis dynamometer.

Sr. No	Parameters	Optimum Designed Values
1.	Maximum power	250 hp
2.	Power absorber dynamometer	Mustang air-cooled eddy current dynamometer (model: MDK-70);
3.	Rollers	Precision-machined and dynamically balanced rollers; Four rollers of diameter 217 mm; rollers center-to-center distance is 434 mm; Inner and outer track widths of 762 and 2692 mm, respectively;
4.	Inertia	Base mechanical inertia of 2200 lbs and rollers rotational inertia of 5.92 kg.m^2;
5.	Transmission shafts	Two shafts of 44 mm diameter and 2692 mm length;
6.	Driving cycle	New European driving cycle (NEDC) with 460 lb ft maximum torque and 3500 maximum roller rpm at the speed of 140 km/h;
7.	Control system	Dyn Pro2 from Dyne systems;
8.	Structure	Durable steel structure frame and restrained system;
9.	Air requirement	A blower fan with maximum airflow rate of 8000 m^3/h;
10.	Electricity requirement	230 V AC (30 Amp) for the eddy current dynamometer and 15 Amp (115 V AC) for the computer system.

(**a**) (**b**)

Figure 1. (**a**) Position of front and rear rollers with maximum and minimum angles of tires. (**b**) Isometric view of rollers and shaft with designed dimensions.

A wheel's rpm (N) is calculated by using Equation (3) [3], where R is the tire radius.

$$N = \frac{V}{2\pi R} \tag{3}$$

As in this research, the design of the chassis dynamometer is for a European type-approval test in which the maximum vehicle speed is 120 km/h, but to incorporate some tolerances, a 140 km/h maximum is considered. To calculate roller rpm, Equation (4) is used [3]:

$$N1\ D1 = N2\ D2 \tag{4}$$

where $N1$ and $N2$ are the rpms of the tire and roller, and $D1$ and $D2$ are the diameters of the tire and roller. For rollers (having 889 mm face length), a mild steel schedule 40 pipe is employed.

3. Results and Discussion

The final designed specifications of the chassis dynamometer are enclosed in Table 2. This test platform will be able to measure exhaust emissions when integrated with emission testing equipment. It will be the first step in testing the performance and ensuring the emission limits of new passenger cars manufactured in Pakistan.

As per NEDC, the complete testing requires around 20 min if performed without any interruption. NEDC consists of two segments. In the first segment, the vehicle under testing is driven through an urban driving cycle (ECE), which is completed four times. Every ECE comprises 15 phases. In the second segment, one extra-urban driving cycle

(EUDC) is completed, which comprises 13 phases. Table 3 encloses the sample testing data for gaseous pollutants and the EUDC gear shift breakdown summary.

Table 3. Sample testing data for gaseous pollutants and EUDC gear shift breakdown summary.

Gaseous Pollutants Measurement for FWD Transmission			Gear Shift Breakdown Summary of EUDC		
	ECE	EUDC	Operations	Duration (s)	Percent
Cycle distance (km)	4.06	6.65	Idle phases	20	5%
Sampling time (s)	780	400	Idle phases for gear shifts	20	5%
CVS venturi (m^3/min)	12.2	12.2	Gear shifts	6	1.5%
CVS volume at 20 °C (m^3)	137.3	70.4	1st gear	5	1.3%
PM filter volume (L)	1040.29	532.59	2nd gear	9	2.2%
CVS ambient baggage (ppm)			3rd gear	8	2%
CO	1.45	1.66	4th gear	99	24.8%
HC	4.1	4.56	5th gear	233	58.2%
NOx	0.141	0.121	Complete cycle	400	100%
CO_2	455.44	448.06			
CVS sampling baggage (ppm)					
CO	28.51	23.94			
HC	10.01	11.68			
NOx	15.18	22.14			
CO_2	4279.3	6850.1			

4. Conclusions

This study is focused on the design of a chassis dynamometer facility for European type-approval of new passenger vehicles manufactured in Pakistan. Exhaust emissions examination can be performed by putting the vehicle over this chassis dynamometer. For this purpose, an air-cooled eddy current dynamometer from Mustang having a maximum load absorption capability of 250 hp is selected, and RLE is used to simulate the real-time performance of a vehicle while driving at varying speeds.

Author Contributions: Conceptualization; K.U. and M.K.; methodology; M.H.; software; A.-B.S.; validation; H.A., M.A. and M.H.; formal analysis; M.A.; investigation; K.U.; writing; K.U. and M.H.; supervision.; M.K. All authors have read and agreed to the published version of the manuscript.

Funding: This research received no external funding.

Institutional Review Board Statement: Not applicable.

Informed Consent Statement: Not applicable.

Data Availability Statement: Not applicable.

Conflicts of Interest: The authors declare no conflict of interest.

References

1. de Menezes Lourenço, M.A.; Eckert, J.J.; Silva, F.L.; Santiciolli, F.M.; Silva, L.C.A. Vehicle and Twin-Roller Chassis Dynamometer Model Considering Slip Tire Interactions. *Mech. Based Des. Struct. Mach.* **2022**, *51*, 6166–6183. [CrossRef]
2. Zhang, X.; Zhou, Z. Research on Development of Vehicle Chassis Dynamometer. *J. Phys. Conf. Ser.* **2020**, *1626*, 012150. [CrossRef]
3. Martyr, A.J.; Plint, M.A. *Engine Testing: Theory and Practice*; Elsevier: Amsterdam, The Netherlands, 2011.

engineering proceedings

MDPI

Proceeding Paper

A First Step towards the Performance Enhancement of IC Engines Using a Desmodromic Valve System †

Shummaila Rasheed, Muhammad Haroon *, Muhammad Irfan, Zayyan Ibrar, Usama Bin Ramzan and Danish Hanif

Department of Mechanical Engineering, Capital University of Science and Technology (CUST), Islamabad 44000, Pakistan; shummaila@cust.edu.pk (S.R.); m.irfan@cust.edu.pk (M.I.); zayyansatti@gmail.com (Z.I.); usamabinramzan@gmail.com (U.B.R.); danishhanif056@gmail.com (D.H.)

* Correspondence: muhammad.haroon@cust.edu.pk

† Presented at the Third International Conference on Advances in Mechanical Engineering 2023 (ICAME-23), Islamabad, Pakistan, 24 August 2023.

Abstract: The traditional valve train system consists of a spring mechanism with a specific level of stiffness. When compressing the spring, a significant amount of power is required. However, by replacing the conventional valve spring mechanism with a desmodromic valve system, the stiffness is reduced, leading to improved engine performance. This study analyzed the impact of this replacement on the engine's performance and overall weight using finite element modeling. A comparative study was also conducted, which revealed that the desmodromic valve system generates a torque of more than 120 Nm, surpassing the 105 Nm produced by the traditional valve system. Furthermore, the weight of the desmo engine only increased by 2.75 kg.

Keywords: desmodromic valve system; valve train mechanism; finite element analysis; IC engines; engine torque

Citation: Rasheed, S.; Haroon, M.; Irfan, M.; Ibrar, Z.; Ramzan, U.B.; Hanif, D. A First Step towards the Performance Enhancement of IC Engines Using a Desmodromic Valve System. *Eng. Proc.* **2023**, *45*, 46. https://doi.org/10.3390/engproc2023045046

Academic Editors: Mohammad Javed Hyder, Muhammad Mahabat Khan and Manzar Masud

Published: 18 September 2023

1. Introduction

The internal combustion engine relies on precise valve timing for optimal performance [1]. Traditional cam-driven valve systems use springs to return the valves to their closed position after being opened by the camshaft lobes. However, these valve springs have certain limitations, especially at high engine speeds, which can lead to valve float and reduced performance [2]. In conventional cam-driven systems, the valve springs act as elastic components, providing the force necessary to close the valves. A higher spring stiffness results in a stronger force for a given valve displacement, while a lower stiffness leads to a weaker force [3]. At high engine speeds, the camshaft rotates rapidly, causing the valve to follow the cam profile closely during its opening phase. However, during the closing phase, the valve spring's inertia and mechanical limitations can lead to valve bounce or float. Valve float occurs when the valve fails to close quickly enough, reducing the engine's ability to maintain proper compression and combustion, ultimately limiting its performance [4]. A desmodromic valve system overcomes the limitations of conventional valve springs by eliminating the reliance on springs for valve closure. Instead, the desmodromic system uses two separate cams, one for valve opening and another for valve closing, ensuring precise control over the valve's movement without relying on spring force. With a desmodromic system, the spring stiffness is not a limiting factor in achieving high engine speeds and performance. Since the valve closure is mechanically actuated using a separate cam, there is no reliance on spring forces to close the valves, eliminating the possibility of valve float [2]. This precise valve control allows for optimized valve timings, improved volumetric efficiency, and better combustion characteristics, contributing to enhanced power output, fuel efficiency, and reduced emissions. Ensuring the appropriate stiffness of the desmodromic actuation mechanism is essential to achieving accurate valve timing and reliable operation under various engine conditions [5].

In this work, a desmodromic valve system is designed for 4-stroke internal combustion engines. Unlike the standard valve system that utilizes a spring-loaded mechanism, the spring is replaced with two cam profiles for each valve in the desmodromic valve system. An end cam is incorporated into this valve system to prevent cam hop when used alongside a conventional valve system.

2. Methodology

2.1. FE Model and Meshing

For the analysis, 3D CAD models of both mechanisms had to be created. To accomplish this, MATLAB software was used, which takes inputs such as base circle diameter, lift, angles of rise and fall, and dwell angles, and then generates the cam profiles according to the chosen function. The resulting profile is then imported into SolidWorks to generate the 3D shape of the camshaft. For other parts, CREO software was employed to create their models by taking dimensions from the physically existing engine (HONDA D13B2) presented in Figure 1. The meshing of the part in ANSYS was carried out using a linear mechanical meshing approach with an average element size of approximately 2 mm. The resulting mesh comprised 10,207 nodes and 22,729 elements, ensuring a sufficient resolution to capture the intricate features and complexities of the parts' geometry. To validate the mesh quality, a series of convergence tests were conducted, ensuring that the simulation results remained consistent and independent of the mesh refinement.

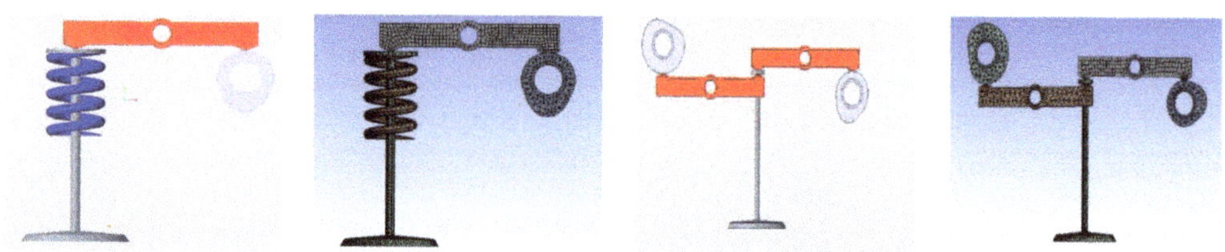

(a) (b) (c) (d)

Figure 1. (**a**,**b**) Geometry and meshing of traditional valve system. (**c**,**d**) Geometry and meshing of desmodromic valve system.

2.2. Boundary Conditions

In the FE analysis of the desmodromic valve train, frictionless contact regions are considered. Figure 2a,b illustrates the applied boundary conditions for both mechanisms.

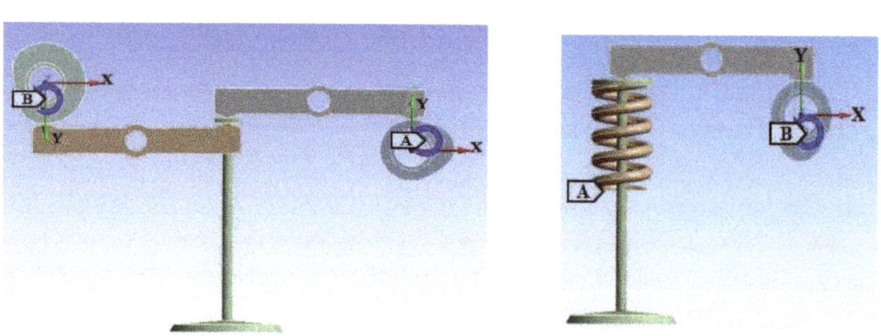

(a) (b)

Figure 2. (**a**) Boundary conditions for desmodromic valve system and (**b**) for traditional valve system.

The connections between the ground and camshaft, as well as the rocker arms, allow for rotational movement, while the connection between the valve and ground enables linear motion. A rotational velocity of 1 rad/s is used to simulate the camshaft's rotation. Similarly, the traditional valve train mechanism also assumes frictionless contact regions and employs revolute joints between the ground, camshaft, and rocker arm. The spring is supported at the bottom and connected to the valve at the upper end.

2.3. Regular and Desmodromic CAM Profiles

The regular cam shape (Figure 3a,c), which is capable of opening and closing the valve using a single cam, was generated using MATLAB code. During the rise phase of a standard cam mechanism, the lift angle is 65 degrees, and the duration of lift is 1 unit. Similarly, during the fall phase, the lift angle is 65 degrees, and the duration of lift is 229 units. In a desmodromic cam mechanism (Figure 3b,d), the ascent angle is 65 degrees, and the dwell period during the rise phase is 229 units. Similarly, the descent angle is 65 degrees, and the dwell period during the fall phase is 1 unit.

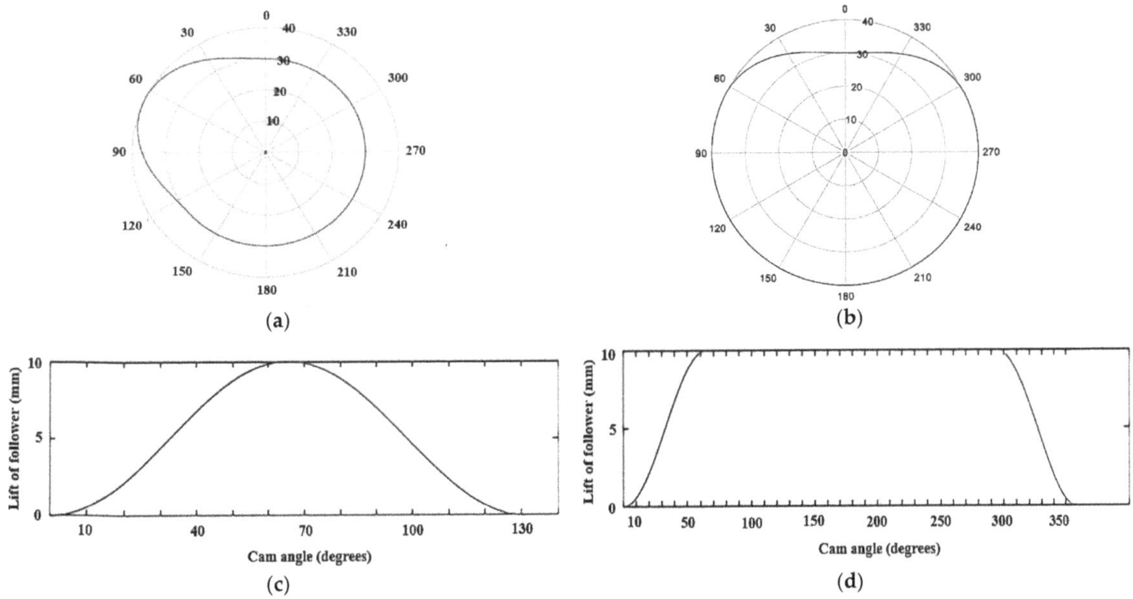

Figure 3. (**a**) Regular CAM profile; (**b**) desmodromic CAM profile; (**c**) lift vs. CAM angle for regular CAM and for (**d**) desmodromic valve mechanism.

3. Results and Discussion

In the case of the conventional valve mechanism, there are two primary locations shown in Figure 4 where stresses occurred: the lower end of the spring and the base circle of the cam profile. Upon analysis, the maximum stress observed at these points was found to be 3.3566×10^{-6} MPa, while the minimum stress recorded was 2.5399×10^{-19} MPa. Within the desmodromic valve system, stress was distributed across various components such as the rocker arms and cam profiles. However, the valve stem bore the highest stress levels. With a maximum stress of 113.56 MPa and a minimum stress of 3.9438×10^{-11} MPa, it is imperative for the valve stem to exhibit sufficient strength to withstand these stresses and avoid any deformation or failure. When comparing the engine torque, the traditional valve system produced an engine output torque of 105 Nm, while the desmodromic valve system generated 120 Nm, which was calculated by considering the forces acting on the system and the moment arms associated with those forces. The mass of engine was increased by only 2.75 kg using the desmodromic valve mechanism instead of the traditional spring-based mechanism.

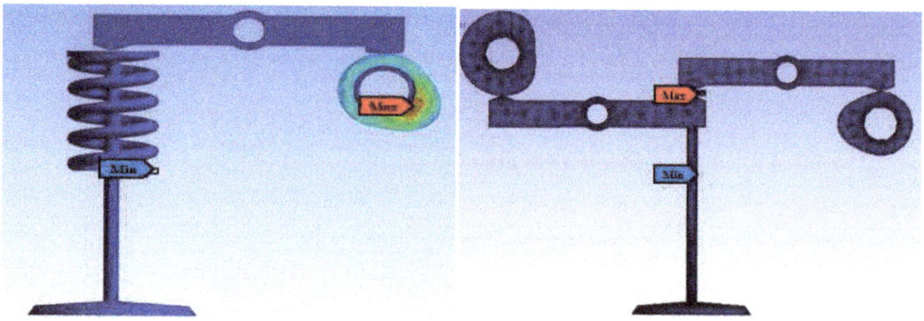

(**a**) (**b**)

Figure 4. (**a**) Equivalent stresses for traditional valve system and for (**b**) desmodromic valve system.

4. Conclusions

It is evident that the desmodromic valve system offers advantages over the spring valve system. The stress on the rocker arm was reduced to -0.127785 MPa from 8.774×10^{-4} MPa (conventional mechanism), and the engine torque was increased to 120 Nm. Lower stress levels were typically experienced by the desmodromic spring-less valve system in comparison to a regular valve system. This is due to the elimination of valve float at high engine speeds, preventing excessive stress on the valve components. Precise and consistent valve opening and closing are ensured by the direct mechanical linkage in the desmodromic system, thereby minimizing the chances of valve impacts or collisions.

Author Contributions: Conceptualization; S.R., M.H. and Z.I.; methodology; S.R., M.I. and U.B.R.; formal analysis; M.H. and S.R.; investigation; M.H., D.H. and M.I.; data curation; Z.I., U.B.R. and D.H.; writing—original draft preparation; S.R., D.H. and U.B.R.; writing—review and editing; M.H. and M.I.; supervision; M.H. and S.R.; All authors have read and agreed to the published version of the manuscript.

Funding: This research received no external funding.

Institutional Review Board Statement: Not applicable.

Informed Consent Statement: Not applicable.

Data Availability Statement: Not applicable.

Conflicts of Interest: The authors declare no conflict of interest.

References

1. RDahham, Y.; Wei, H.; Pan, J. Improving Thermal Efficiency of Internal Combustion Engines: Recent Progress and Remaining Challenges. *Energies* **2022**, *15*, 6222. [CrossRef]
2. Rahul, W.; Sourabh, S.; Swapnil, S.; Nikhil, J. Study of Design & Manufacturing of Cam Operated Spring Less Valve System for IC Engine. *IJSRDV* **2020**, *7*, 385–388.
3. Zewen, G.; Xiaonan, H.; Elspeth, K.; Jianqiao, Y. Non-linear finite element model for dynamic analysis of high-speed valve train and coil collisions. *Int. J. Mech. Sci.* **2020**, *173*, 105476.
4. Gu, Z. Static and Dynamic Analysis of Nonlinear Valve Springs Based on Finite Element Analysis and Machine Learning Algorithm. Ph.D. Thesis, Lancaster University, Lancaster, UK, 2022; p. 201.
5. Hu, B.; Li, Y.; Yin, L. Theoretical and Experimental Analysis of Dynamic Characteristics for a Valve Train System. *Sensors* **2021**, *21*, 6328. [CrossRef]

engineering proceedings

Proceeding Paper

Integration of Cyber-Security Locks with SCADA Software for Smart Surveillance Management †

Duaa Ayesha *, Tayyaba Iqbal and Ghulam Asghar

Mechanical Engineering Department, Capital University of Science and Technology (CUST), Islamabad 44000, Pakistan; tayyabaiqbal7724@gmail.com (T.I.); ghulam.asghar@cust.edu.pk (G.A.)
* Correspondence: duaaayesha68@gmail.com
† Presented at the Third International Conference on Advances in Mechanical Engineering 2023 (ICAME-23), Islamabad, Pakistan, 24 August 2023.

Abstract: Due to the augmented engrossment of technology in security, everything moves towards a cohesive system. Smart cyber-lock technology is one of the eminent concepts with the potential to solve emerging safety and security issues. This work aims to establish an efficient system of safety and security through the integration of smart cyber-locks with a software package. The smart monitoring and evaluation of security-intensive sites are accomplished through cyber-link software, and the audit reports of security processes are generated on a quarterly basis. The comparative analyses of the case study data represent that the illegal security breaches decline quite noticeably after the successful implementation and synchronization of cyber-security locks with SCADA software.

Keywords: cyber-lock; cyber-security; cyber-link; surveillance; smart technology; monitoring and evaluation

Citation: Ayesha, D.; Iqbal, T.; Asghar, G. Integration of Cyber-Security Locks with SCADA Software for Smart Surveillance Management. *Eng. Proc.* **2023**, *45*, 17. https://doi.org/10.3390/engproc2023045017

Academic Editors: Mohammad Javed Hyder, Muhammad Mahabat Khan, Muhammad Irfan and Manzar Masud

Published: 11 September 2023

1. Introduction

Cyber-locks are known as key-centric access control systems considered to increase security, liability, and authorized control of a system. The inimitable design of electronic-lock cylinders and programmable smart-keys of cyber-locks resolve the security difficulties that no other scheme can [1]. Cyber-audit is a software that is used in project security management and allows for the assignment of keys by setting expirations and staff monitoring, audit trails, access schedules, and custom reports. It manages the schedules with administrative and customized access to the individuals who hold a key in the concerned department. The generated output is used to create custom audit reports with email notifications for specific events. The questions of when, who, and why are evaluated through data analysis using an authorized identification process [2].

The Internet of Things (IoT)-enabled door lock system is used to manage automation and has various advanced features. This smart-door locking system has the ability to open and close the door through an authentication process [3]. Another study [4] addressed the intelligent door lock system built on raspberry Pi (RPI). The system was synchronized with Android and Java (enterprise edition) EE to scan the concerned persons' fingerprint data as the input interface, and RPI cameras were used for the process of monitoring in the door lock. Hadis et al. [5] demonstrated that the dependence related to the access is reduced with the control over the server in a cyber-lock system. Attributes are set that have the role, date, location, and fine-granted control for the access permissions.

The application and usefulness of cyber-locks have been addressed by numerous studies; however, there is a scarcity in research focused on the synchronization of smart-locks with software for efficient vigilance. Therefore, this study would attempt to integrate cyber-locks and SCADA (supervisory control and data acquisition) software for the proficient monitoring and enhancement of the security of an asset.

2. Methodology

This research scheme is divided into two parts: (i) monitoring and (ii) evaluation. Cyber-link and SCADA software are managed in parallel for the synchronization of results. For the live monitoring of sites where cyber-locks are installed, the SCADA system is controlled and managed from the control room of the parent organization. Default parameters are set in SCADA through which non-stop (24/7) data monitoring is accomplished. The SCADA is operated through the telecommunication signals, so if any signal issue occurs, its modem stops the communication. Due to this issue, the SCADA software requires a reset, and an ON-SCAN alarm is produced for enduring the communication process. On completion of the reset, the SCADA is again managed from the control room and further developments are performed. Cyber-link software is checked and analyzed in parallel after every scheduled activity of the site.

The interface of cyber-link software is divided into three main portions for the monitoring process, as depicted in Figure 1. All data related to cyber-locks are communicated over the software that is checked and analyzed through three main sections. The first is about the display of a cyber-lock activity report, in which complete details of each activity over the site are noted with the proper date and time marked. The second section is related to the history of key/lock attempts performed over the site. The third section is associated with the access chart, as it shows the key assignments to the concerned lock and person. Only the concerned person with their unique key could open the lock; otherwise, an alarm is produced due to unauthorized attempts, in case any other/invalid key is inserted in the lock.

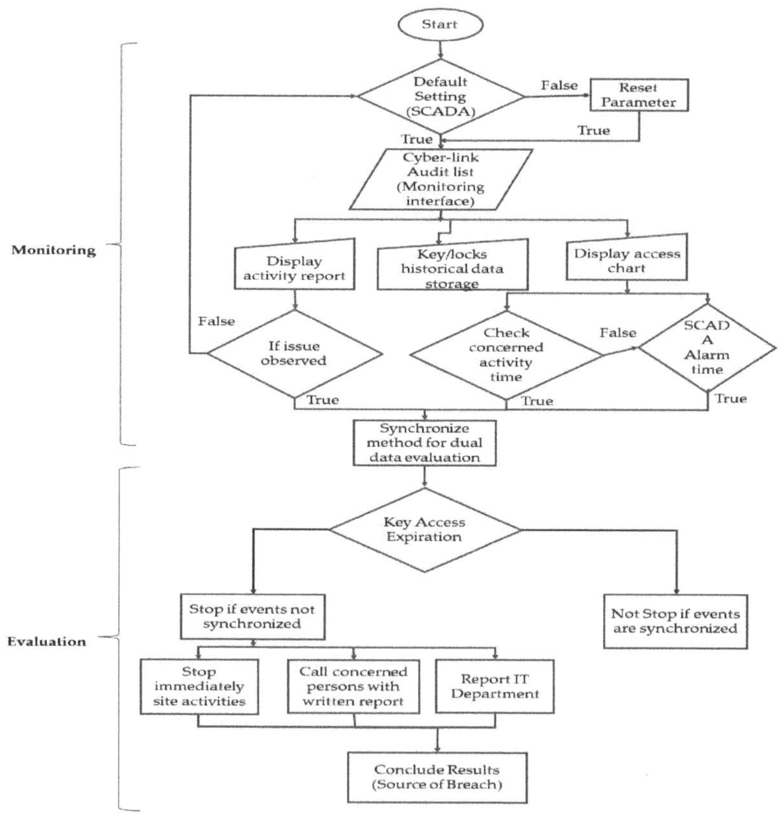

Figure 1. Flow chart of monitoring and evaluation process.

In case of an issue observed in an activity report, prompt action is taken by checking the access chart. Through an access chart, the two situations are analyzed in parallel by checking the concerted activities' times, both in the cyber-link and SCADA (alarm time). Both of them are synchronized; hence, the data evaluation process is accomplished. Through data analysis, a security breach is checked out, and as a result of this, key access is finalized. If any type of security breach is observed in the data, the following three steps are performed: (i) The organization reports to the IT department to check other parameters and security breaches for the clearance of other activities. (ii) The concerned supervisor is immediately called to stop the site activities without any delay. (iii) Written evidence is mandatory from the concerned person of the site and consumer; otherwise, the issue is not resolved. By examining all the results, if an issue is identified and then the responsibility is fixed regarding the breach source, the key for the concerned zone is stop If no issue is noticed and/or the issue is resolved at the same time, then the site activities are sustained without any interruption.

3. Results and Discussion

The quarterly audit data of five sites were used to analyze the results of breaches before and after the cyber-locks' synchronization with SCADA software. Additionally, three types of security breaches were observed: user premises access, cyber-attacks, and human error, as shown in Figure 2.

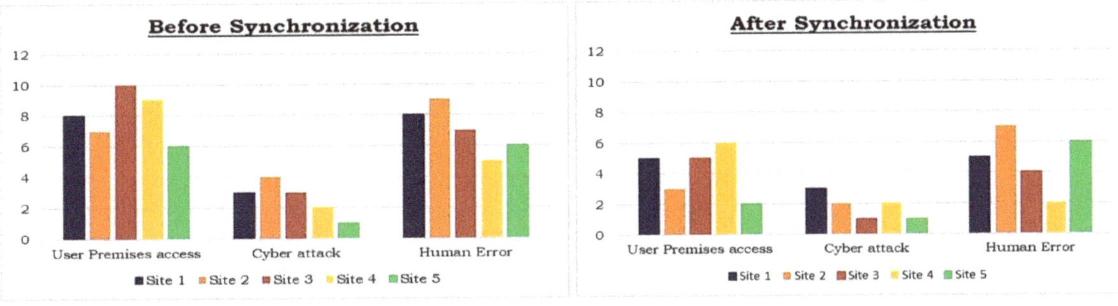

Figure 2. Results of security breaches before and after synchronization of cyber-link and SCADA.

User premises access attempts were large in number before the synchronization of cyber-locks and SCADA, as site supervisors approached the system easily and made physical changes on the site without any fear of being caught. However, after the installation and synchronization of cyber-locks and SCADA, user premises access without notice became very difficult, and more than one access in a day required written permission. Cyber-attacks happened on modem data and other site systems; hackers tried to gain access to site data. Cyber-attacks were also managed successfully without compromising on security because consumers and supervisors were both alerted after the synchronization of the system. Human errors occurred mostly due to the reset issue of the modem and other systems installed over the site. Team members neglected some issues during the troubleshooting and routine work on the site. Due to the negligence of supervisors and other team members, human errors were faced before the synchronization of the system. It was observed that human errors were also reduced to some extent after the modification of the system but not completely. The comparison of all security breach data in Figure 2 represents that the user premises access was reduced quite significantly, from 40 to 21, whereas the cyber-attacks and human errors were also decreased, from 13 to 9 and 35 to 24, respectively.

4. Conclusions

The comparative results of this study encourage security-intensive organizations to adopt cyber-security locks to enhance the safety and security of their assets. Through the implementation and synchronization of cyber-link and SCADA, security breaches were reduced to a significant extent. User premises access breaches were controlled properly, due to which site issues were decreased and efficient surveillance was maintained. Illegal approaches to the site by the employees of an organization or outsiders became difficult and incidents of security breaches were also minimized.

Author Contributions: Conceptualization, methodology: D.A.; investigation, resources: T.I.; writing—original draft preparation: D.A.; writing—review and editing: G.A.; visualization: T.I.; supervision: G.A. All authors have read and agreed to the published version of the manuscript.

Funding: This research received no external funding.

Institutional Review Board Statement: Not applicable.

Informed Consent Statement: Not applicable.

Data Availability Statement: Not applicable.

Conflicts of Interest: The authors declare no conflict of interest.

References

1. Bjartmar Hylta, S.; Söderberg, P. Smart Locks for Smart Customers?: A Study of the Diffusion of Smart Locks in an Urban Area. Master Thesis, School of Industrial Engineering and Management (ITM), KTH, Stockholm, Sweden, 2017.
2. CyberAudit-Web Management Software. Available online: https://cyberlock.com/product-lines/cyberaudit-web/ (accessed on 10 July 2023).
3. Adiono, T.; Fuada, S.; Anindya, S.F.; Purwanda, I.G.; Fathany, M.Y. IoT-enabled door lock system. *Int. J. Adv. Comput. Sci. Appl.* **2019**, *10*, 445–449. [CrossRef]
4. Zhang, X.; Song, M.; Xu, Y.; Dai, Z.; Zhang, W. Intelligent door lock system based on raspberry Pi. In Proceedings of the 2021 2nd International Conference on Artificial Intelligence and Information Systems, Chongqing, China, 28–30 May 2021; pp. 1–7.
5. Hadis, M.S.; Palantei, E.; Ilham, A.A.; Hendra, A. Design of smart lock system for doors with special features using bluetooth technology. In Proceedings of the 2018 International Conference on Information and Communications Technology, Yogyakarta, Indonesia, 6–7 March 2018; pp. 396–400.

MDPI
St. Alban-Anlage 66
4052 Basel
Switzerland
www.mdpi.com

Engineering Proceedings Editorial Office
E-mail: engproc@mdpi.com
www.mdpi.com/journal/engproc

www.ingramcontent.com/pod-product-compliance
Lightning Source LLC
Chambersburg PA
CBHW041946260926
51850CB00018B/975